Manual of Water Supply Practices

M41

Ductile-Iron Pipe and Fittings

AF772858

Fourth Edition

American Water Works Association

Manual of Water Supply Practices—M41, Fourth Edition

Ductile-Iron Pipe and Fittings

Sr. Editorial Manager, Books: Suzanne Snyder
Manager – Manuals: Molly Beach
Manager – Publishing Operations: Gillian Wink
Specialist – Copyright and Permissions: Peggy Tyler
Director of Publishing: John Fedor
Technical Editor: Anne Lanctôt
Cover Image: Sylverarts Vectors/Shutterstock.com
Production: Innodata

Library of Congress Cataloging-in-Publication Data

Names: American Water Works Association author
Title: M41--Ductile iron pipe and fittings / by American Water Works Association.
Other titles: Ductile iron pipe and fittings
Description: Fourth edition. | Denver, CO : American Water Works Association, [2025] | Revised edition of: Ductile iron pipe and fittings. | Includes bibliographical references and index.
Identifiers: LCCN 2025006262 (print) | LCCN 2025006263 (ebook) | ISBN 9781647172015 print | ISBN 9781613007266 electronic
Subjects: LCSH: Ductile iron pipe | Water-pipes | Cast-iron pipe | Pipe fittings
Classification: LCC TD491 .D84 2025 (print) | LCC TD491 (ebook) | DDC 628.1/5--dc23/eng/20250326
LC record available at https://lccn.loc.gov/2025006262
LC ebook record available at https://lccn.loc.gov/2025006263

ISBN 978-1-64717-201-5 ISBN, electronic 978-1-61300-726-6
http://doi.org/10.12999/AWWA.M41ed4

American Water Works Association

American Water Works Association
6666 West Quincy Avenue
Denver, CO 80235-3098
awwa.org

Contents

This page intentionally blank.

Figures

This page intentionally blank.

Tables

Preface

This is the fourth edition of AWWA M41, *Ductile-Iron Pipe and Fittings*. This manual provides the user with both technical and general information to aid in the design, specification, procurement, installation, and understanding of ductile-iron pipe and fittings. It is a discussion of recommended practice, not an AWWA standard calling for compliance with certain specifications. It is intended for use by utilities and municipalities of all sizes, whether as a reference book or textbook for those not fully familiar with ductile-iron pipe and fitting products. Design engineers and consultants may use this manual in preparing plans and specifications for new ductile-iron pipe projects.

The manual covers ductile-iron pipe and fitting products and certain appurtenances and their application to practical installations, whether of a standard or special nature. For adequate knowledge of these products, the entire manual should be studied. Readers will also find the manual a useful source of information when assistance is needed with specific or unusual conditions. The manual contains a list of applicable national standards, which may be purchased from the respective standards organizations (e.g., AWWA, ASTM). Standards referenced in this manual refer to the latest editions.

Credit is extended to the Ductile Iron Pipe Research Association (DIPRA) for granting permission to reprint numerous graphics and tables found throughout this manual.

This page intentionally blank.

Acknowledgments

This manual was prepared by the AWWA Standards Committee A21 on Ductile-Iron Pipe and Fittings. The membership of the committee at the time this manual was revised was as follows:

L. Gregg Horn, *Chair*

General Interest Members

M. Dunn, Crist Engineers, Inc., Little Rock, Ark.

J. Hebenstreit, Underwriters Laboratories, Inc., Northbrook, Ill.

M.B. Horsley, Horsley Engineering LLC, Overland Park, Kans.

J.A. Hrdlicka (*liaison, nonvoting*), Standards Engineer Liaison, AWWA, Denver, Colo.

D.H. Kroon, Aegion, Houston, Tex.

D.D. Lary (*alternate*), Wright-Pierce, Topsham, Maine

C.R. Laucamp, AECOM, Chicago, Ill.

P.J. Mourt, HDR, Plymouth Meeting, Pa.

J.R. Plattsmier, Consultant, New Port Richey, Fla.

S. Pool (alternate), HDR Engineering, Denver, Colo.

T.M. Stinson, Kleinfelder SEA Company, Holbrook, Mass.

Producer Members

L.R. Dunn (*alternate*), US Pipe & Foundry Company, Birmingham, Ala.

J.H. Eddings, McWane, Inc., Phillipsburg, N.J.

J.A. Fancher (*alternate*), American Cast Iron Pipe Company, Birmingham, Ala.

M.D. Gaston, American Cast Iron Pipe Company, Birmingham, Ala.

L.G. Horn, Ductile Iron Pipe Research Association, Birmingham, Ala.

D. Hughes, US Pipe & Foundry Company, Birmingham, Ala.

J.C. Jones, Georg Fischer, Dallas, Tex.

H. Kennedy, Jr., Harold Kennedy & Associates, Fuquay Varina, N.C.

J.S. Palombo, Everett J. Prescott, Inc., Gardiner, Maine

A. Rich (*alternate*), G.A. Rich and Sons, Deer Creek, Ill.

D. Swalley, Electrosteel, Birmingham, Ala.

User Members

M.J. Britch, Tualatin Valley Water District, Beaverton, Ore.

J. Fox, Tampa Bay Water, Clearwater, Fla.

D.D. Montgomery, Independence Water Department, Independence, Mo.

C.J. Patla, Connecticut Water Company, Inc., Clinton, Conn.

G. Ramon (*liaison, nonvoting*), Standards Council Liaison, Little Rock Water Reclamation Authority, Little Rock, Ark.

J.M. Ross, Denver Water, Denver, Colo.

Chapter 1

General Information and History

This manual provides the user with both technical and general information to aid in the design, specification, procurement, installation, and understanding of ductile-iron pipe.

The manual describes ductile-iron pipe and fitting products, appurtenances, and their application to practical installations, whether of a standard or special nature.

1.1 HISTORY

1.1.1 General

The history of pipe parallels that of civilization. The earliest communities were established near water sources that later became inadequate as populations increased. Hydraulic engineering was born of the necessity to transport water closer to the point of use. From ancient Babylon's clay pipes to the underground tunnels of ancient Greece to the aqueducts of the Roman Empire to the cast-iron mains of Versailles, France, to today's modern pipe, the search for an affordable piping material with reliable strength and durability has evolved. Cast-iron pipe rapidly became the standard material for water distribution mains. Today, in the United States and Canada alone, there are more than 584 utilities that have had cast-iron distribution mains with continuous service records of more than 100 years, and 32 utilities have had them for 150 years or more.

The production of iron pipe most likely developed from or coincided with the manufacture of cannons as early as the year 1313. There is an official record of cast-iron pipe being manufactured in the Siegerland area of Germany in 1455 for installation at the Dillenburg Castle.

In 1664, King Louis XIV of France ordered the construction of a cast-iron pipe main to extend 15 mi (24 km) from a pumping station at La Seine a Marly to Versailles to supply water for the palace fountains and town. This cast-iron pipe provided continuous service for more than 330 years. When the line was originally installed, the production of iron required the use of expensive charcoal for the reduction of the iron ore. By 1738, success had been achieved in producing lower-cost iron by using coke instead of charcoal. Following this success, the more progressive cities began to install cast-iron mains.

The use of cast-iron pipe was introduced in the United States in the early 1800s. Since that time, various other piping materials have been offered for water distribution. However, waterworks engineers continued to use cast iron until ductile iron, a stronger and equally durable piping material, became available.

1.1.2 Development of Joints

1.1.2.1 Flanged. Originally, cast-iron pipe was made with flanged joints, using lead gaskets. Improved joints of this type are still used for many aboveground plant installations and other specialized applications.

1.1.2.2 Bell and spigot. The bell and spigot joint was developed in 1785 and was used extensively until the 1950s. This joint was assembled by caulking yarn or braided hemp into the base of the annular bell cavity and then pouring molten lead into the remaining space inside the bell. Upon solidification, the lead was compacted by caulking, thus effecting a watertight seal. Materials other than lead have also been used to confine yarn or hemp in the base of the bell cavity.

1.1.2.3 Mechanical joint. The mechanical joint was developed for gas industry use in the late 1920s but has since been used extensively in the water industry. This joint has standardized dimensions and uses the basic principle of the stuffing box and gland with a rubber gasket being compressed by the gland.

1.1.2.4 Roll-on joint. The roll-on joint was developed in 1937 and was used for approximately 20 years before its manufacture was discontinued. Assembly of this joint involved a compressed rubber gasket rolled under a restriction ring followed by caulked square braided jute. The remainder of the joint was packed with a bituminous compound.

1.1.2.5 Push-on joint. The push-on joint was developed in 1956 and represented an important advancement in the water distribution field. This joint consists of a single rubber gasket placed in a groove inside the socket at the bell end of the pipe. After lubricating the joint in accordance with the manufacturer's instructions, the plain end of the pipe is pushed past the gasket, compressing it and forming a seal that is tight versus high internal pressures, vacuum, and external water head. Assembly of the push-on joint is simple and fast. Large bell holes are not required for this joint, and it can be assembled under wet-trench conditions or even under water.

1.1.2.6 Special joints. Several special joints are available. These joints include ball and socket for subaqueous crossings, seismic joints for earthquake applications, plain-end coupled, threaded and coupled, and other variations of restrained joints.

A more detailed description of the joints used with ductile-iron pipe is found in Chapter 5 of this manual.

1.1.3 Development of Ductile-Iron Pipe

The advent of ductile-iron pipe in 1948 was one of the most significant developments in the pressure pipe industry. Quickly recognized as a pipe material with all the beneficial qualities of gray cast-iron pipe plus additional strength and ductility, it was first used for special and severe conditions of high pressure, water hammer, and excessive external loads. For more than 70 years, it has proved to be virtually trouble and maintenance free as an underground pressure pipe material, and today it is used in the transportation of raw and potable water, sewage, slurries, and process chemicals. The strength and rugged durability of ductile iron result in a high level of operational reliability. For these reasons, it rapidly replaced gray cast iron in the marketplace and is the iron piping material now specified by waterworks professionals.

1.1.4 Manufacture of Ductile-Iron Pipe

The centrifugal casting methods used in manufacturing ductile-iron pipe have been in the process of commercial development and refinement since 1925. The steady improvements that led to the present state of the art have been covered by hundreds of patents and technical papers and represent the ingenuity of many dedicated engineers, metallurgists, and foundry personnel. A detailed description of the metallurgical and physical characteristics of ductile-iron pipe and fittings produced to ANSI/AWWA standards is in Chapter 3 of this manual.

1.2 APPLICATIONS AND APPURTENANCES

Ductile-iron pipe and related products convey water from its source to its point of use through the myriad stages of storage, transmission (Figure 1-1), filtration, pumping, chemical treatment, and distribution. After the water is used, ductile-iron pipe plays a similar part in collecting, conveying, pumping, treating, reusing, and discharging wastewater in the never-ending water cycle. Ductile-iron pipe has long been noted for its long-term economy, performance, reliability, and proven record in serving the public's health, safety, and daily water needs.

1.2.1 Applications

Ductile-iron pipe that meets the requirements of various pertinent AWWA standards is used for numerous applications, some of which are the following:

- Raw water supply lines
- Transmission mains
- Distribution mains
- Fire mains
- Sewer force mains
- Subaqueous crossings
- Air service
- Pump station piping
- Intake lines
- Brine handling
- Acid handling
- Sulfur handling
- Gravity sewers
- Treatment plant piping
- Outfall mains
- Reclaimed water lines
- Penstocks
- Various chemical processes piping
- Lift station piping

Source: Ductile Iron Pipe Research Association

Figure 1-1 A 48-in. ductile-iron pipe water transmission main

- Geothermal and hydronic piping
- Culverts
- Aerial crossings

Various papers and reports have been published concerning unusual installations, such as applications subject to buoyancy, low pressures, or partial vacuum, etc., as well as routine applications of ductile-iron pipe. Data on many of these applications are available from engineering handbooks, journals, manufacturers' catalogs, and the Ductile Iron Pipe Research Association (DIPRA).

1.2.2 Appurtenances

Ductile-iron piping systems have great versatility when combined with the diversity of fittings, joints, valves, hydrants, coatings, linings, gaskets, and specials available today. More than 15,000 variations of fitting patterns are currently available. Almost all of these fittings are manufactured under various ANSI/AWWA standards that ensure quality, performance, and standardization. ANSI/AWWA standards also provide requirements for a variety of joints, gaskets, linings, and corrosion control systems for virtually any water system application.

Valves to control flow, direct flow, shut off, vent, prevent backflow, and provide fire protection are available in joints and types compatible with ductile-iron pipe systems. These valves and hydrants are manufactured according to ANSI/AWWA standards, as well as insurance and independent test laboratory requirements in some cases.

Many special applications can be provided in ductile-iron pipe systems, such as restrained joints, flexible joint subaqueous pipe, base-mounted fittings, bonded joints, self-locking gaskets, epoxy or other special linings, zinc coating, and polyethylene encasement.

1.3 DUCTILE-IRON PIPE RESEARCH ASSOCIATION (DIPRA)

DIPRA, formerly the Cast Iron Pipe Research Association (CIPRA), is a nonprofit organization whose members are manufacturers of ductile-iron pressure pipe. Since 1915, the association has provided engineering and research information on cast- and ductile-iron products by

- participating in national and international standards development by ANSI, AWWA, ASTM International, American Society of Mechanical Engineers, NACE International, National Fire Protection Association, and the International Organization for Standardization (ISO);
- presenting engineering seminars on design, pipeline construction, corrosion control, and special applications and problems;
- providing field services, such as soil investigations, flow tests, and consultation with engineers on piping problems, including pipe and system design; and
- publishing research and engineering information in technical papers, brochures, and manuals.

This page intentionally blank.

Chapter 2

Applicable Standards

AWWA standards for cast-iron pipe date back to 1890. Today, AWWA publishes 11 standards related to ductile-iron pipe, ductile- and gray-iron fittings, and related items, including joint materials, design, and installation. These standards are individually revised, updated, and published on a five-year basis. Other organizations that publish standards related to ductile-iron pipe and fittings include ASTM and the International Organization for Standardization (ISO). Note: The standards listed in this chapter will not be listed in the reference sections of individual chapters.

2.1 RELATED STANDARDS

The following describes the current editions of standards related to ductile-iron pipe and fittings.

2.1.1 ANSI/AWWA C104/A21.4, Cement–Mortar Lining for Ductile-Iron Pipe and Fittings

First used in 1922 in Charleston, S.C., cement–mortar lining has proven to be a very effective means of preventing tuberculation and maintaining a high flow coefficient in both gray- and ductile-iron pipe and fittings. As a result, ductile-iron pipe as well as gray- and ductile-iron fittings used for potable water applications are normally furnished with a cement–mortar lining unless otherwise specified by the purchaser.

2.1.2 ANSI/AWWA C105/A21.5, Polyethylene Encasement for Ductile-Iron Pipe Systems

Loose polyethylene encasement was first used experimentally in the United States in 1951 to protect cast-iron pipe in corrosive environments. The first field installation of polyethylene encasement of cast-iron pipe in an operating water system was completed in 1958.

Since that time, thousands of installations have been made in severely corrosive soils throughout the United States in pipe sizes ranging from 3 in. to 64 in. (76 mm to 1,600 mm) in diameter. This history of usage, coupled with extensive and ongoing research, has clearly demonstrated the efficacy of polyethylene encasement as a corrosion protection means for both gray- and ductile-iron pipe. Polyethylene encasement is also used as a soil corrosion preventive in a number of other countries, and an international standard for polyethylene sleeving (ISO-8180) was adopted after the procedure was developed in the United States. ASTM A674 covers polyethylene encasement for ductile-iron pipe, which closely parallels the ANSI/ AWWA standard.

The ANSI/AWWA standard describes material requirements, effects of sunlight, tube size or sheet width, and installation methods. Appendix A to the standard describes a method for evaluating potentially corrosive environments to gray- and ductile-iron pipe to determine when polyethylene encasement should be used.

2.1.3 ANSI/AWWA C110/A21.10, Ductile-Iron and Gray-Iron Fittings

This standard describes requirements and dimensions of ductile-iron and gray-iron fittings in mechanical, flanged, and push-on joint configurations. The standard further describes markings, certification requirements, inspection, coatings and linings, and acceptance tests. Detailed tables include data on thickness requirements, joint dimensions, gasket seats, and laying lengths.

2.1.4 ANSI/AWWA C111/A21.11, Rubber-Gasket Joints for Ductile-Iron Pressure Pipe and Fittings

This standard describes rubber-gasket joints for gray- and ductile-iron pipe and fittings with mechanical joints, push-on joints, or modified mechanical or push-on joints. The standard includes detailed requirements for each joint type, such as dimensions, tolerances, lubrication, gaskets, marking, and packing. Also described in the standard are requirements for glands, gaskets, and bolts, along with tables designating tolerances, dimensions, physical specifications, thread length and depth, bolt sizes, and torque ranges. Notes on installing mechanical joints, including recommended bolt torques to be applied, are in the appendix of the standard.

2.1.5 ANSI/AWWA C115/A21.15, Flanged Ductile-Iron Pipe With Ductile-Iron or Gray-Iron Threaded Flanges

This standard pertains to flanged ductile-iron pipe with either gray- or ductile-iron threaded flanges. Tables are provided that describe required thickness, working pressure, class of pipe, and weights and dimensions. The standard describes inspection and certification by the manufacturer, acceptance on delivery, flanges to be used, fabrication procedures, coatings, linings, and markings. A table of flange gasket dimensions is also included in the appendix.

2.1.6 ANSI/AWWA C116/A21.16, Protective Fusion-Bonded Coatings for the Interior and Exterior Surfaces of Ductile-Iron and Gray-Iron Fittings

This standard describes protective fusion-bonded epoxy coatings for the interior and exterior surfaces of ductile- and gray-iron fittings used for water, wastewater, and reclaimed water systems. The standard describes the material, application, and performance requirements for these coatings. It does not describe coatings agreed on between the purchaser and the manufacturer for special service conditions, such as salt water, sewers, acid, high temperature, and so forth.

2.1.7 ANSI/AWWA C150/A21.50, Thickness Design of Ductile-Iron Pipe

The thickness design of ductile-iron pipe is presented in this standard, along with an example problem that illustrates proper application of the design method. Tables and charts are also provided that describe earth loads, truck loads, trench loads, design values for standard laying conditions, casting tolerances, diameter–thickness ratios, and other pertinent design parameters.

2.1.8 ANSI/AWWA C151/A21.51, Ductile-Iron Pipe, Centrifugally Cast

This standard describes the manufacture of ductile-iron pipe for water, wastewater, and reclaimed water systems with push-on joints and mechanical joints. It includes general requirements; inspection and certification by the manufacturer; inspection, delivery, and acceptance by the purchaser; tolerances or permitted variations; coatings and linings; testing requirements; marking and weighing of pipe; foundry records required; defective specimens and retests; and pipe rejection criteria. Tables are provided for standard thickness requirements, rated working pressure, maximum depth of cover, and standard dimensions and weights of push-on joint and mechanical joint ductile-iron pipe. Appendix A of this standard consists of two tables that show pipe thicknesses required for various tap sizes.

2.1.9 ANSI/AWWA C153/A21.53, Ductile-Iron Compact Fittings

Compact fittings are specifically designed to use the attendant strength of ductile iron. Compact fittings are alternatives to those covered by ANSI/AWWA C110/A21.10.

This standard describes compact ductile-iron fittings ranging in size from 3 in. through 48 in. (70 mm through 1200 mm) with mechanical joints, push-on joints, or such other joints (except flanged joints) as specified and 54 in. through 64 in. (1,400 mm through 1,600 mm) with push-on or flanged joints. The information provided closely parallels that given in ANSI/AWWA C110/A21.10.

2.1.10 ANSI/AWWA C600, Installation of Ductile-Iron Mains and Their Appurtenances

This standard pertains to the installation of ductile-iron water mains and their appurtenances and describes such topics as inspection, handling, storage, pipe installation, joint assembly, backfilling, hydrant installation, thrust restraint, flushing, pressure and leakage tests, highway and railroad crossings, and tapping. Many tables and diagrams are provided for clarification.

2.1.11 ANSI/AWWA C606, Grooved and Shouldered Joints

This standard, describing grooved and shouldered joints, includes sections on general requirements, materials, general design, inspection, testing, and rejection. Tables and illustrations provide grooving dimensions, coupling and joint configurations, and other pertinent information.

2.2 OTHER REFERENCE STANDARDS AND SPECIAL PRODUCTS

In addition to the ANSI/AWWA standards described previously, the following standards are commonly used by the industry:

- ASTM A377—Standard Index of Specifications for Ductile Iron Pressure Pipe.
- ASTM A674—Standard Practice for Polyethylene Encasement for Ductile Iron Piping for Water or Other Liquids.
- ASTM A716—Standard Specification for Ductile Iron Culvert Pipe.
- ASTM A746—Standard Specification for Ductile Iron Gravity Sewer Pipe.

Because of innovations in research, design, and manufacturing methods and in response to changing needs of the marketplace, manufacturers of ductile-iron pipe and related products also offer other products that are not currently included in ANSI/AWWA standards. These products include restrained joints, bell-less joints, special linings and coatings, special fittings, and welded-on bosses and outlets. Many of these products meet applicable parts of current standards, and some may be included in future revisions. Information on some of these products is included in this manual. For more details, consult the manufacturers.

Chapter 3

Manufacturing and Testing

3.1 MANUFACTURING

The process of manufacturing ductile-iron pipe and fittings begins in a pipe mill with the controlled melting of steel scrap, iron scrap, plant-returned scrap, and in some operations, pig iron. Because of the high volumes of iron required for efficient production, melting in most pipe and fittings foundries is performed in cupolas or various types of electric furnaces. The resulting molten iron must be desulfurized so that later treatment will produce the nodular or spheroidal graphite form that is characteristic of the ductile-iron microstructure. Desulfurization can be accomplished in several ways. In pipe-making, it is usually done by mixing calcium carbide or lime and fluorspar into the melt. The mixing mechanism can be a ceramic stirring beam (Ruhrstahl-Heraeus process) or the introduction of inert gas into the bottom of a ladle through a porous plug.

3.1.1 Metallurgical Terms

Several metallurgical terms have been used in this chapter with which the reader may not be familiar. These terms are defined as follows:

Austenitizing: A treatment that heats iron-base alloys to a temperature at which the carbon will enter solution, thus dissolving iron carbides. (The high-temperature phase of iron is called austenite.)

Carbide (iron): A hard compound of iron and carbon (Fe_3C), also called cementite.

Degenerated graphite: A graphite shape that is not spheroidal but also not lamellar.

Ductile iron: A cast ferrous material in which a major part of the carbon content occurs as free graphite in a substantially nodular or spheroidal form.

Ferrite: Mostly pure iron.

Ferritizing: Heat treatment to convert iron-base alloys to ferrite.

Flake graphite: The lamellar shape of graphite in nontreated irons.

Graphite: A crystalline form of carbon having a hexagonal structure. It is one of the two solid phases freezing out of liquid iron during eutectic solidification. (AFS 2019)

Nodular graphite: Rounded, spheroidal form of graphite created by treatment of low-sulfur-base iron with nodularizing agents. (AFS 2019)

Pearlite: A constituent of ductile iron, gray iron, and steel, the microstructure of which is composed of alternate layers of ferrite and iron carbide.

3.1.2 Magnesium Treatment

To produce ductile iron's spheroidal graphite microstructure, the desulfurized iron must be further treated with magnesium. In pipe-making, the magnesium is usually introduced either in pure form or by adding a magnesium-bearing ferrosilicon alloy in lump form. In the latter form, sometimes as much as half of the required magnesium will be replaced by rare earth metals that tend to mitigate the effects of deleterious "tramp" elements and/or reduce the flare and fume associated with the treatment.

The proper quantity of magnesium to be used is sensitive and critical. Insufficient magnesium will result in degenerated graphite nodules, while excess magnesium, in addition to being uneconomical, can lead to excess carbide formation and dross inclusions. Therefore, because the amount of magnesium necessary is proportional to the sulfur content, frequent and accurate sulfur analyses are necessary to maintain adequate control of the magnesium additions and the quality of the finished product.

Properly treated iron will solidify with the graphite in the form of nodules or spheroids. However, because magnesium has a low vaporization temperature and low solubility in iron, the treated metal must be cast relatively soon after magnesium addition for retention of its ductile characteristics. The distribution and shape of the graphite nodules are set when the metal solidifies in the mold and will remain essentially unchanged during subsequent processing.

3.1.3 Casting

The matrix (primary internal structure) of pipe centrifugally cast in water-cooled metal molds will contain various proportions of structures known as ferrite (which has relatively high ductility), pearlite (which has reduced ductility and toughness), and iron carbide (which is hard). Pipe centrifugally cast in other types of molds might contain proportions of ferrite and pearlite, but because of slower cooling, it can be produced without containing carbide. The structures obtained depend on the variations in iron chemistry, pouring temperature, solidification rate, and inoculation at the casting machines.

The essential elements of pipe-casting machines include the pipe mold and a means of rotating the mold at speeds sufficient to hold the molten iron against the mold wall by centrifugal force; delivering the iron to the inside of the mold, such as by a trough; applying inoculating material to the mold and/or iron; cooling the external surface of the mold; and extracting the solidified pipe (Figure 3-1).

At the beginning of a cast, a sand core that forms the inside contour of the pipe bell is inserted into the bell end of the mold and locked in place by one of several available mechanisms. For most casting machines, when the bell core is in place, the machine starts the rotation and delivers the mold over the length of the stationary trough until the spout end of the trough is near the bell core. Other casting machines differ in that the rotating mold is stationary and the trough is from a machine ladle designed and maintained to move through the mold. At the proper time, iron is poured into the upper end of the trough, delivering iron at a constant rate throughout the pour. Following a dwell time sufficient to

Source: Ductile Iron Pipe Research Association

Figure 3-1 Elements of a casting machine

fill the bell cavity, the casting machine moves the mold away from the stationary end of the trough at a controlled rate so that the iron is distributed evenly along the full length of the mold. Following the metal pour, mold rotation is continued until the newly formed pipe has solidified and cooled sufficiently to be removed. Spinning of the mold is then stopped and the pipe extracted either by holding the pipe stationary and pulling the mold away from it or by pulling the pipe out of the stationary mold with a pipe puller.

Preparations for the next casting cycle consist of removing any solid "splinters" from the bottom of the trough, recoating the inner surface of the mold with an inoculating dry spray, and placing a new core in the bell end. The dry spray is fed to the surface of the rotating mold through tubing mounted on the trough and supplied with exit ports near the spout. The spray is applied by running the trough through the mold in a stroke between casting cycles or just in front of the metal stream during the casting cycle.

3.1.4 Annealing

Because of the mold cooling and rapid iron solidification, the pipe must be annealed (heat treated) to produce the necessary proportion of ferrite that imparts strength, ductility, and impact properties required by specifications for water service (Figure 3-2).

The annealing times and temperatures necessary to accomplish carbide breakdown and ferritization depend on metal chemistry, the degree of inoculation the iron received prior to casting, and the influence of the solidification rate during casting. Pipe rapidly cooled by casting directly against the metal mold surface requires austenitizing at approximately 1,750°F (954°C) prior to a ferritizing treatment at approximately 1,400°F (760°C). Pipe cast against other surfaces cools more slowly and may require only the ferritizing treatment.

Source: Ductile Iron Pipe Research Association

Figure 3-2 Pipe entering an annealing oven

Modern annealing ovens are equipped with automatic controls for maintaining correct temperatures in heating and cooling zones and for maintaining correct travel speed through the oven. Except in batch-type ovens in which a pipe stands vertically (Figure 3-3), provisions must usually be made for rotating the pipe or supporting the pipe while at elevated temperatures. Pipe larger than 12 in. (305 mm) in diameter may be passed through a "rounder" that kneads the pipe between full-length rollers while it is still at an elevated temperature from the annealing treatment.

3.2 TESTING

The annealing treatment relieves casting stresses, increases toughness, and improves ductility while maintaining required strength. The effectiveness of the annealing treatment, as well as the spheroidal properties of the graphite, are proven using performance tests. The types of tests required, test frequencies, and acceptance limits are specified in ANSI/AWWA C151/A21.51. The current edition of the standard requires that samples for 70°F (21°C) notched Charpy impact tests (ASTM E23 as modified by AWWA C151) be taken during each operating hour and that the samples be selected to represent the extremes of pipe diameters and thicknesses. Samples for tensile testing are required to be taken at least once during each casting period of approximately 3 h. At least 10 percent of the required samples selected for Charpy testing must be also tested at –40F (–40°C).

Acceptance values for the notched Charpy impact tests are a minimum of 7 ft-lb (9.49 J) at 70°F (21°C and 3 ft-lb (4.07 J) at –40°F (–40°C). If a sound specimen fails to meet the

Source: Ductile Iron Pipe Research Association

Figure 3-3 Vertical annealing furnace

minimum requirement, this indicates that the specimen contains excessive pearlite and/ or carbides (ineffective anneal), excessive silicon, or degenerated graphite. A microscopic examination of the Charpy test bar can indicate which of the three imperfections or combination thereof caused the failure. Pipe with excessive pearlite or carbide may be salvaged by further annealing treatments and retesting. However, pipe with degenerated graphite and excessive silicon content cannot be corrected by heat treatment and must be scrapped.

Acceptance values for tensile tests per ASTM E8 (2022a) are the following:

Tensile strength:	60,000 psi minimum (413.7 × 10^3 kPa)
Yield strength:	42,000 psi minimum (289.6 × 10^3 kPa)
Elongation:	10 percent minimum

A pipe sample that meets the required notched Charpy impact requirement with a sound specimen will rarely fail to meet the tensile requirement. When the results from a test fail to meet acceptance requirements, all pipe back to the last acceptable test are suspect and are bracketed for further investigation.

When a pipe exits the annealing treatment, it may be internally ground to remove protrusions that may have formed on the inside surface by the centrifugal action during casting. In accordance with the requirements of ANSI/AWWA C151/A21.51, each pipe is hydrostatically tested at a minimum of 500 psi (3,450 kPa) for a minimum of 5 s (Figure 3-4). In further compliance with the standard, sufficient pipes are gauged inside the socket and outside the spigot to ensure that dimensional requirements are met.

Source: Ductile Iron Pipe Research Association

Figure 3-4 Hydrostatic testing of pipe to a minimum of 500 psi (3,450 kPa)

Although there are many factors that must be controlled in proper balance, the centrifugal casting process has replaced all alternative methods of making ductile-iron pipe. In addition to its benefits of high productivity and economy, the process produces pipe with close dimensional control and a dense wall structure.

3.3 FITTINGS TESTING

Fittings for ductile-iron pipe are cast by conventional foundry methods for static sand castings. Fittings of the older ANSI/AWWA C110/A21.10 designs are produced of either gray-iron or ductile-iron in sizes 3 in. through 48 in. (76 mm–1,219 mm). Ductile-iron fittings of the newer compact design are produced in accordance with ANSI/AWWA C153/A21.53 in the 3- through 64-in. (76- through 1,600-mm) sizes. Both fittings standards require that the iron used be tested for chemical and mechanical properties and that at least one sample be selected representing each period of approximately 3 h, during which the melting unit is operated continuously.

Chemical limitations are 0.08 percent maximum phosphorus in ductile-iron fittings and 0.15 percent maximum sulfur and 0.90 percent maximum phosphorus in gray-iron fittings. Control of the other chemical constituents is specified to meet the respective physical property requirements of these standards.

The most common acceptance test for the physical characteristics of ductile-iron fittings is a tensile test of specimens from (1) a separately cast coupon, (2) the castings runner bar system, or (3) the casting, conducted in accordance with ASTM A536 (2019). The manufacturer of fittings must have records to demonstrate that such fittings meet the performance test requirements of the standard. The most common grade of iron is 70-50-05, with acceptance values as follows:

Tensile strength:	70,000 psi minimum (482.7 × 10^3 kPa)
Yield strength:	50,000 psi minimum (344.8 × 10^3 kPa)
Elongation:	5 percent minimum

Other permissible grades of iron are 60-42-10 and 65-45-12. The acceptance values of grades are as follows, respectively:

Tensile strength:	60,000 psi minimum (413.7 × 10^3 kPa)
Yield strength:	42,000 psi minimum (289. 10^3 kPa)
Elongation:	10 percent minimum
Tensile strength:	65,000 psi minimum (448.1 x 10^3 kPa)
Yield strength:	45,000 psi minimum (310.2 x 10^3 kPa)
Elongation:	12 percent minimum

For gray-iron fittings, the mechanical properties may be determined by tensile tests conducted per ASTM A48/A48M (2022b). The acceptance values for tensile tests shall be as shown in Table 3-1.

Table 3-1 Acceptance values for tensile tests

Iron Strength, psi 1,000	Fitting Size, in.	Minimum Tensile Strength, psi
25	3–14	25,000
30	14–24	30,000
30	20–48	30,000
35	16–24	35,000
35	20–36	35,000

NOTE: To convert inches (in.) to millimeters (mm), multiply by 25.4; to convert pounds per square inch (psi) to megapascal (MPa), multiply by 6.895 × 10^{-3}.

ANSI/AWWA C153/A21.53 requires that the manufacturer shall have qualified the design of the product by performing pressure tests on 90°bends and crosses of each size to demonstrate a minimum safety factor of 2.5 times the rated working pressure and shall have records to show the results of these tests. Tests of equal opening tees shall be substituted for tests of crosses if the manufacturer produces tees but not crosses.

REFERENCES

American Foundry Society (AFS). 2019. *Ductile Iron Handbook.* AFS FC9200. Schaumburg, Ill.: AFS.

ASTM. 2019. Standard Specification for Ductile Iron Castings. A536. West Conshohocken, PA.: ASTM International.

ASTM. 2022a. Standard Test Methods for Tension Testing of Metallic Materials. E8. West Conshohocken, PA.: ASTM International.

ASTM. 2022b. Standard Specification for Gray Iron Castings. A48/A48M. West Conshohocken, PA.: ASTM International.

ASTM. 2023. Standard Test Methods for Notched Bar Impact Testing of Metallic Materials. E23. West Conshohocken, PA.: ASTM International.

Chapter 4

Design

4.1 BACKGROUND

The design of ductile-iron pipe for most services requiring pressure pipe is described in ANSI/AWWA C150/A21.50, Standard for Thickness Design of Ductile-Iron Pipe. The design procedures published in this standard have evolved through the work of the A21 Committee on Ductile-Iron Pipe and Fittings (A21 Committee) of AWWA. Ductile-iron pipe has been carefully studied since it was introduced in the 1950s, and the A21 Committee has been provided with extensive test data on virtually all aspects of the material that could affect its performance. The A21 Committee meets periodically to review and update ANSI/AWWA C150/A21.50 and other standards concerning ductile-iron pipe and fittings.

The design of ductile-iron pipe is based on the flexible conduit theory developed in the 1940s by Professor M.G. Spangler (1982) and others at Iowa State University and confirmed by tests conducted by E.C. Sears (1964) in 1961 (Figure 4-1). Additional tests conducted in 1971 at Utah State University (Kennedy 1976) further confirmed the conservatism of the design method.

Briefly, the design procedure for ductile-iron pipe is as follows:

1. Design for internal pressures (static or operating pressure plus a transient pressure allowance).
2. Design for ring bending stress resulting from the external load (earth load plus truck or transient load).
3. Use the larger resulting wall thickness.
4. Add a 0.08-in. (2.0-mm) service allowance, check for ring deflection under external load (earth load plus truck or transient load), and increase wall thickness if necessary.
5. Add a standard allowance for casting tolerances.

This design procedure results in the total calculated design thickness from which the appropriate class is chosen.

Figure 4-1 Performance test on 36-in. (914-mm) ductile-iron pipe with push-on joints

4.2 MAJOR DESIGN CRITERIA AND FORMULAS

The A21 Committee carefully chose the following criteria for use in calculating required thicknesses of ductile-iron pipe. Definitions of symbols used in Section 4.2 are provided in Section 4.2.1.

4.2.1 Nomenclature

A = Outside radius of pipe. For A in ft, D in in.: $A = D/24$ (For A in m, D in mm: $A = D/2{,}000$)

a = Conversion factor. For lb/ft^2 to psi: $a = 144$ (for kN/m^2 to kPa: $a = 1$)

B = 1.5 ft (0.457 m)

b = Effective pipe length: 36 in. (0.914 m). This is the length of pipe over which the traffic load is distributed at pipe depth.

C = Surface load factor

D = Outside diameter (OD), in. (mm) (see Table 4-4)

E = Modulus of elasticity: 24×10^6 psi (165.5×10^3 kPa)

E' = Modulus of soil reaction, psi (kPa) (see Table 4-2)

F = Impact factor: 1.5

f = Design bending stress: 48,000 psi (331×10^3 kPa)

H = Depth of cover, ft (m)

K_b = Bending moment coefficient (see Table 4-2)

K_x = Deflection coefficient (see Table 4-2)

P = Wheel load: 16,000 lb (71,170 N)

P_e = Earth load, psi (kPa)

P_i = Design internal pressure, in psi (kPa): $P_i = 2\,(P_w + P_s)$

P_w = Working pressure, psi (kPa)

P_s = Standard transient allowance, 100 psi (689 kPa) based on an assumed flow velocity of 2 ft per s (fps)

P_t = Truck load, in psi (kPa)

P_v = Trench load, in psi (kPa): $P_v = P_e + P_t$

S = Minimum yield strength of ductile-iron pipe in tension: 42,000 psi (289.6×10^3 kPa)

t = Net thickness, in. (mm). This is the thickness resulting from internal pressure or ring bending designs, whichever governs.
t_1 = Minimum manufacturing thickness. In in.: $t_1 = t + 0.08$ (in mm: $t_1 = t + 2.0$). This is the nominal thickness minus the allowance for the casting tolerance.
w = Soil weight: 120 lb/ft^3 (18.85 kN/m^3)
Δx = Design deflection, in./in. (mm/mm): $\Delta x = 0.03D$.

4.2.2 Internal Pressure Design

The hoop stress formula (Equation 4-1) is used to calculate the net thickness, *t*, required for internal pressure. In the formula, the design internal pressure is equal to the sum of the working pressure and transient allowance multiplied by a safety factor of 2.0. The transient allowance used in standard ductile-iron pipe design is 100 psi (689 kPa), which is the surge pressure resulting from a velocity of flow of 2.0 fps and is adequate for most water service applications. (See Chapter 9 for a discussion on water hammer or surge.) If anticipated surge pressures are other than 100 psi, the designer should use the actual anticipated pressure.

$$t = \frac{P_i D}{2S} \qquad \text{(Eq 4-1)}$$

Where
t = net pipe wall thickness, in. (mm)
P_i = design internal pressure, psi (kPa) = $2 (P_w + P_s)$
P_w = working pressure, psi (kPa)
P_s = standard transient allowance, 100 psi (689 kPa)*
D = OD of pipe, in. (mm)
S = minimum yield strength in tension = 42,000 psi (289,590 kPa)

* If anticipated surge pressures are greater than 100 psi (689 kPa), the maximum anticipated pressure must be used.

4.2.3 Earth Load

The magnitude of the earth load is based on the prism load model as shown in formula (Equation 4-2). This formula gives the weight of the column of earth directly above the pipe and assumes no shear support from soils outside of this column. This is the maximum earth load that can be imposed on a flexible conduit in a trench condition and is the simplest version of the modified Marston model for calculating earth loads. In trench conditions, in most cases, this load is reduced by the support of the sidefill soil. The unit weight of the backfill soil is conservatively assumed to be 120 lb/ft^3 (18.85 kN/m^3).

$$P_e = \frac{wH}{a} \qquad \text{(Eq 4-2)}$$

4.2.4 Truck Loads

Truck loads are computed by Equation 4-3 using the surface load factor *C* from Equation 4-4 and the reduction factor *R* from Table 4-1. The wheel load, *P*, is based on the American Association of State Highway and Transportation Officials (AASHTO) H-20 truck on unpaved road or flexible pavement, 16,000-lb (71,170-N), and an impact factor *F* of 1.5 at all depths. The surface load factor is calculated by using Equation 4-4, which assumes a single concentrated wheel load centered over an effective pipe length of 3 ft (0.914 m). The

Table 4-1 Reduction factors, *R*, for truck load calculations

	Depth of Cover, ft			
	<4	4–7	7–10	>10
Nominal Pipe Size, in.		*R*		
3–12	1.00	1.00	1.00	1.00
14	0.92	1.00	1.00	1.00
16	0.88	0.95	1.00	1.00
18	0.85	0.90	1.00	1.00
20	0.83	0.90	0.95	1.00
24–30	0.81	0.85	0.95	1.00
36–64	0.80	0.85	0.90	1.00

NOTE: To convert inches (in.) to millimeters (mm), multiply by 25.4; to convert feet (ft) to meters (m), multiply by 0.3048.

reduction factor *R* takes into account that the part of the pipe directly below the wheels is aided in carrying the truck load by adjacent parts of the pipe that receive little or no load from the wheels.

$$P_t = RF\frac{CP}{bD} \quad \text{(Eq 4-3)}$$

$$C = 1 - \frac{2}{\pi}\text{Arcsin}\left(H\sqrt{\frac{A^2+B^2+H^2}{\left(A^2+H^2\right)\left(B^2+H^2\right)}}\right) + \left(\frac{2}{\pi}\right)\left(\frac{A \times H \times B}{\sqrt{A^2+H^2+B^2}}\right)\left(\frac{1}{A^2+H^2}+\frac{1}{B^2+H^2}\right) \quad \text{(Eq 4-4)}$$

NOTE: Angles are in radians.

4.2.5 External Load Design

The net wall thickness required for the external load is based on two design considerations: limitation of ring bending stress, which addresses the strength of ductile-iron in bending, and ring deflection, which addresses the flexibility of the lining. Most other flexible conduit material designs do not include calculations for bending stress caused by external load (with accompanying ring deflection). Ductile-iron pipe design conservatively includes calculations for bending stress caused by external loads, which in some cases controls the design thickness. When a trench load of sufficient magnitude is applied, ductile-iron pipe can deflect amply to engage the passive resistance of the sidefill soil to movement, thereby transmitting part of the trench load to the sidefill soil. Thus, the load-carrying capacity of ductile-iron pipe is a function of soil properties and ring stiffness. In addition, a reaction equal to the vertical trench load exerted on the pipe develops where the pipe embeds into the bottom of the trench. This reaction is distributed almost uniformly over the width of the bedding of the pipe. The width of bedding is the effective contact area on the pipe bottom where the bedding support develops an equal reaction to the vertical trench load. The greater the width of bedding, the greater the load-carrying capacity of the pipe. Therefore, design criteria dependent on the width of bedding and on the passive resistance of the sidefill soil are used to calculate the resulting ring bending stress and ring deflection of ductile-iron pipe. These criteria have been established from test data for the standard laying conditions shown in Figure 4-2.

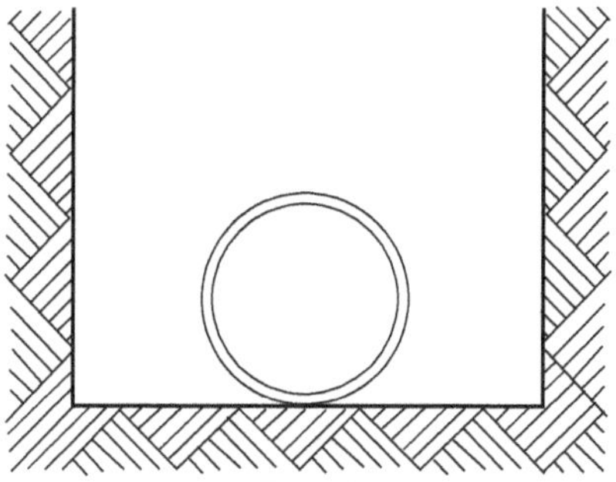

Type 1*

Flat-bottom trench† with loose backfill.

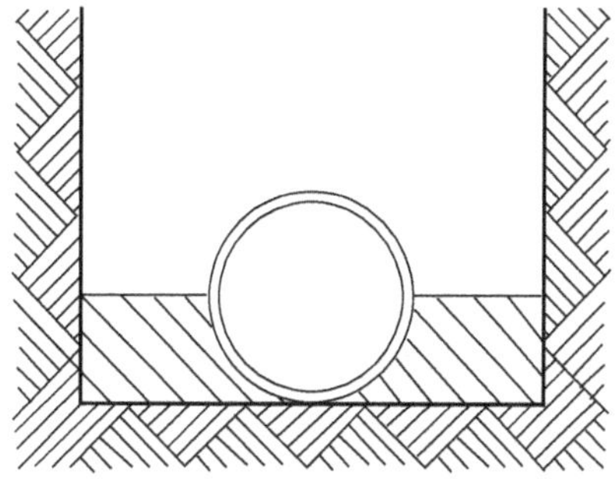

Type 2

Flat-bottom trench† with loose backfill lightly consolidated to centerline of pipe.

Type 3

Pipe bedded in 4-in. (100-mm) minimum loose soil‡ with backfill lightly consolidated to top of pipe.

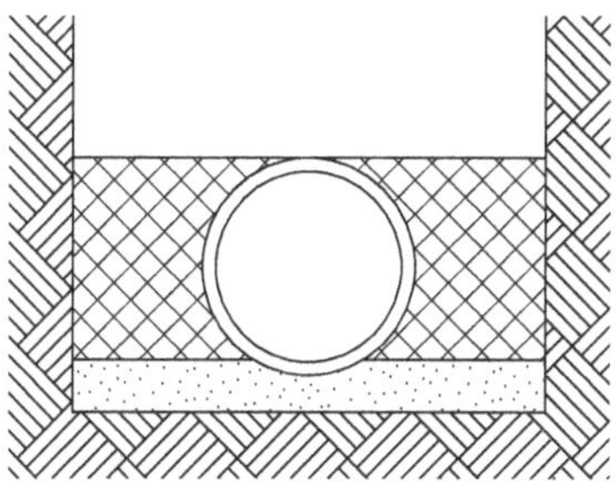

Type 4

Pipe bedded in sand, gravel, or crushed stone to depth of $^1/_8$ pipe diameter, 4-in. (100-mm) minimum, with backfill compacted to top of pipe. (Approximately 80% Standard Proctor, AASHTO§ T-99)

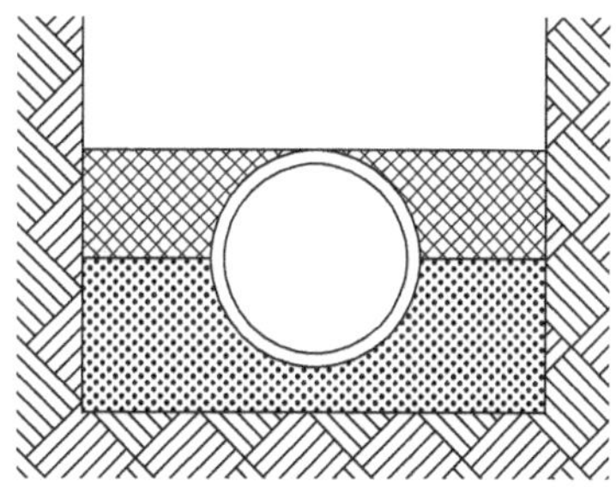

Type 5

Pipe bedded to its centerline in compacted granular material, 4-in. (100-mm) minimum under pipe. Compacted granular** or select material‡ to top of pipe. (Approximately 90% Standard Proctor, AASHTO§ T-99)

* For pipe 14 in. (350 mm) and larger, consideration should be given to the use of laying conditions other than Type 1.

† Flat-bottom trench is defined as "undisturbed earth."

‡ Loose soil or select material is defined as "native soil excavated from the trench free of rocks, foreign material, and frozen earth."

§ AASHTO T-99, Standard Method of Test for Moisture-Density Relations of Soils Using a 2.5-kg (5.5-lb) Rammer and a 305 mm (12 in.) Drop. Available from the American Association of State Highway and Transportation Officials, 444 North Capitol Street N.W., Suite 249, Washington, DC 20001.

** Granular materials are defined per the AASHTO Soil Classification System (ASTM D3282) or the Unified Soil Classification System (ASTM D2487), with the exception that gravel bedding/backfill adjacent to the pipe is limited to 2-in. maximum particle size per ANSI/AWWA C600.

Figure 4-2 Laying conditions for ductile-iron pipe

The magnitude of the external load is the sum of the earth and traffic loads, as shown in Equation 4-5:

$$P_v = P_e + P_t \tag{Eq 4-5}$$

4.2.6 Bending stress Design

Ring bending stress is limited to 48,000 psi (331.0 × 103 kPa). This design value provides ample safety factors on material strength as discussed later in this chapter. Equation 4-6 is used to calculate the trench load that can be supported by the pipe based on pipe size, pipe net wall thickness *t*, and design laying conditions.

$$P_v = \frac{f}{3\left(\frac{D}{t}\right)\left(\frac{D}{t}-1\right)\left(K_b - \frac{K_x}{\left(\frac{8E}{E\left(\frac{D}{t}-1\right)^3} + 0.732\right)}\right)} \tag{Eq 4-6}$$

4.2.7 Deflection Design

The maximum allowable ring deflection is based on the flexibility of the lining being specified. For standard cement–mortar-lined ductile-iron pipe, deflection is limited to 3 percent of the OD of the pipe. This design deflection provides a conservative safety factor of 2.0 against failure of the cement–mortar lining because much larger deflections can be sustained without damage to the pipe wall or cement–mortar lining. Equation 4-7 is derived from Spangler's deflection equation and is used to calculate the magnitude of the trench load that can be sustained based on the deflection design criterion, the minimum manufactured wall thickness t_1, and design laying conditions.

$$P_v = \left(\frac{\left(\frac{\Delta x}{D}\right)}{12K_x}\right)\left(\frac{8E}{\left(\frac{D}{t_1}-1\right)^3} + 0.732E'\right) \tag{Eq 4-7}$$

The conservatism of this design is demonstrated in several ways. Tests in which cement–mortar-lined ductile-iron pipe was loaded to obtain large deflections proved that the cement–mortar lining will not fail by deflections of up to 6 percent and greater; thus, the 3 percent design limit provides ample safety factors, as explained later in this chapter. Furthermore, prism earth loads used in ductile-iron pipe design exceed Spangler's flexible pipe loads, which are used in most other flexible pipe designs.

The maximum ring deflection normally occurs in the relatively short period after installation and before pressurization. After pressurization, deflection is substantially reduced by rerounding. Actual trench tests have shown that no further deflection occurs while the pipe is under pressure; in fact, a slight reduction in deflection occurs with time, probably as a result of a slight yielding of the earth above the pipe as the earth is forced upward by the rerounding force in the pipe. (Internal pressures in most modern pipelines exceed by many magnitudes the external trench load P_v). Tests have also proven that if pressure is relieved in a line after it has been operating, the deflections are unlikely to return to values obtained before initial pressurization. This is because of the settling of soil into voids left at the sides and haunches of the rerounded pipe and a subsequent increase in lateral soil support as the soil consolidates with time.

4.2.8 Laying Conditions

4.2.8.1 Standard laying conditions. As indicated previously, certain factors dependent on the specified type of laying condition are essential to the design of ductile-iron pipe for external loads. Two of these factors, the bending moment coefficient, K_b, and the deflection coefficient, K_x, depend on the width of bedding at the bottom of the pipe. The angle subtended by the width of bedding is commonly referred to as the bedding angle. The remaining factor is the modulus of soil reaction, E', which is a representation of the passive resistance that can be developed in the sidefill soil. To facilitate design calculations, conservative factors have been established from reliable test data for the five standard laying conditions as shown in Table 4-2. In addition, criteria are available that provide for deep bury designs that take into account increased compaction of select soils. Thus, a designer has a great deal of flexibility in selecting the most economical combination of wall thickness and bedding requirements.

It is important to note that the equations for bending stress and deflection cannot be solved for the required wall thickness, *t or* t_1, making the process iterative and potentially

Table 4-2 Design values for standard laying conditions

Laying Condition*	Description	E', psi	Bedding Angle, deg	K_b	K_x
Type 1†	Flat-bottom trench.‡ Loose backfill.	150	30	0.235	0.108
Type 2	Flat-bottom trench. Backfill lightly consolidated to centerline of pipe.	300	45	0.210	0.105
Type 3	Pipe-bedded in 4-in. minimum loose soil.§ Backfill lightly consolidated to top of pipe.	400	60	0.189	0.103
Type 4	Pipe bedded in sand, gravel, or crushed stone to depth of ⅛ pipe diameter, 4-in. minimum. Backfill compacted to top of pipe. (Approximately 80% standard Proctor, AASHTO T-99.)**	500	90	0.157	0.096
Type 5	Pipe bedded to its centerline in compacted granular material, 4-in. minimum under pipe. Compacted granular†† or select material§ to top of pipe. (Approximately 90% standard Proctor, AASHTO T-99.)**	700	150	0.128	0.085

NOTES: 1) Consideration of the pipe-zone embedment conditions included in this table may be influenced by factors other than pipe strength. For additional information, see ANSI/AWWA C600.

2) To convert pounds per square inch (psi) to kilopascals (kPa), multiply by 6.895.

* See Figure 4-2.

† For pipe 14 in. (350 mm) and larger, consideration should be given to the use of laying conditions other than Type 1.

‡ Flat-bottom is defined as "undisturbed earth."

§ Loose soil or select material is defined as "native soil excavated from the trench, free of rocks, foreign material, and frozen earth."

** AASHTO T-99, "Standard Methods of Test for the Moisture-Density Relations of Soils Using a 5.5 lb (2.5 kg) Rammer and a 12-in. (305-mm) Drop," available from the American Association of State Highway and Transportation Officials, 444 N. Capitol St. N.W., Washington, DC 20001.

†† Granular materials are defined per AASHTO Soil Classification System (ASTM D3282) or the Unified Soil Classification System (ASTM D2487), with the exception that gravel bedding/backfill adjacent to the pipe is limited to 2-in. maximum particle size per ANSI/AWWA C600.

time consuming. To expedite calculations, a simple method is described to determine the required class of pipe to be selected based on project design criteria. In addition, design tables that provide diameter–thickness ratios for a wide range of trench loads have been developed from these equations for all five standard laying conditions and are included in ANSI/AWWA C150/A21.50. With these design tables, a designer need only know trench load and desired laying conditions to compute net thickness required for bending stress and deflection designs.

4.2.8.2 Deep bury laying conditions. For gravity sewer installations where the depth of cover may become significant, a deep bury laying condition is defined in ASTM A746. In this laying condition, the pipe is bedded to its top with angular graded stone or well-graded gravel compacted to approximately 95 percent standard Proctor. For a deep bury installation the following design values may be used:

- E′ = 1,500
- Bedding angle = 150
- K_b = 0.128
- K_x = 0.085

4.2.9 Standard Allowances

Once the net thickness requirement *t* has been determined (the larger net thickness from either ring bending stress under external load design or internal pressure design), certain allowances are added to obtain the total calculated thickness that is used to select the appropriate class of pipe. To obtain the minimum manufacturing thickness t_1, a service allowance of 0.08 in. (2.0 mm) for all pipe sizes is added to the net thickness, thereby providing an additional safety factor against possible thickness reduction from surface imperfections or minor damage received during handling and installation. After service allowance is added, check deflection under external load and increase wall thickness if necessary. A standard allowance for the casting tolerance is then added to account for the dynamic aspect of the casting process and to ensure that these tolerances do not compromise the factor of safety used in calculating the required net thickness. The allowance for the casting tolerance depends on the pipe size as shown in Table 4-3.

4.2.10 Standard Pressure Classes

Ductile-iron pipe is manufactured in standard pressure classes varying from Class 150 to Class 350 in 50 psi (344.75 kPa) increments, depending on pipe size. Nominal thicknesses for these standard classes are shown in Table 4-4. Once the total calculated thickness has been determined, a standard pressure class should be selected for ordering and specifying. When the calculated thickness is between two standard thicknesses, the larger of the two should be selected.

Table 4-3 Allowances for casting tolerance

Nominal Pipe Size, in.	Casting Allowance, in.
3–8	0.05
10–12	0.06
14–42	0.07
48	0.08
54–64	0.09

NOTE: To convert inches (in.) to millimeters (mm), multiply by 25.4.

Table 4-4 Standard pressure classes of ductile-iron pipe

Nominal Pipe Size, in.	Outside Diameter, in.	Pressure Class				
		150	200	250	300	350
		Nominal Thickness, in.				
3	3.96	—	—	—	—	0.25*
4	4.80	—	—	—	—	0.25*
6	6.90	—	—	—	—	0.25*
8	9.05	—	—	—	—	0.25*
10	11.10	—	—	—	—	0.26
12	13.20	—	—	—	—	0.28
14	15.30	—	—	0.28	0.30	0.31
16	17.40	—	—	0.30	0.32	0.34
18	19.50	—	—	0.31	0.34	0.36
20	21.60	—	—	0.33	0.36	0.38
24	25.80	—	0.33	0.37	0.40	0.43
30	32.00	0.34	0.38	0.42	0.45	0.49
36	38.30	0.38	0.42	0.47	0.51	0.56
42	44.50	0.41	0.47	0.52	0.57	0.63
48	50.80	0.46	0.52	0.58	0.64	0.70
54	57.56	0.51	0.58	0.65	0.72	0.79
60	61.61	0.54	0.61	0.68	0.76	0.83
64	65.67	0.56	0.64	0.72	0.80	0.87

NOTE: Pressure classes are defined as the rated water working pressure of the pipe in pounds per square inch (psi). The thicknesses shown are adequate for the rated water working pressure plus a transient allowance of 100 psi (689 kPa). Calculations are based on a minimum yield strength in tension of 42,000 psi (289.6 × 103 kPa) and a 2.0 safety factor times the sum of working pressure and 100 psi (689 kPa) transient allowance.

To convert inches (in.) to millimeters (mm), multiply by 25.4.

Thickness can be calculated for rated water working pressure and surges other than those indicated by use of the formula shown in ANSI/AWWA C150/A21.50.

Ductile-iron pipe is available for water working pressures greater than 350 psi (2,413 kPa). Pipe is available with thicknesses greater than pressure class 350. See Table 4-6.

Lowest nominal thicknesses shown in ANSI/AWWA C115/A21.15 for threaded flanged pipe are still required; lowest nominal thicknesses shown in ANSI/AWWA C606 for pipe with grooved and shouldered joints are still required.

* Calculated thicknesses for these sizes and pressure ratings are less than those shown above. (See Table 13 of ANSI/ AWWA C150/A21.50 for actual calculated thicknesses.) Presently these are the lowest nominal thicknesses available in these sizes.

4.2.11 Standard Selection Table

Using the design procedure described, Table 4-5, a standard selection table, was developed to give maximum depth of cover and rated working pressure for each standard pressure class and laying condition. This table has been provided so that a designer may select, rather than calculate, the appropriate pressure class and laying conditions for a given design application.

4.2.12 Special Thickness Classes

Prior to the 1991 revision of AWWA C150/A21.50 and AWWA C151/A21.51, standard thickness classes of ductile-iron pipe were identified as Classes 50 through 56. These are now designated as special thickness classes and continue to be available for design conditions not covered in Tables 4-4 and 4-5. Wall thicknesses for these special thickness classes are shown in Table 4-6, and their rated working pressures and maximum depths of cover are presented in Table 4-7.

Table 4-5 Rated working pressure and maximum depth of cover, standard pressure classes

Nominal Pipe Size, in.	Pressure* Class, psi	Thickness, in.	Laying Conditions: Type 1 Trench	Type 2 Trench	Type 3 Trench	Type 4 Trench	Type 5 Trench
			Maximum Depth of Cover,† ft				
3	350	0.25	78	88	99	100‡	100‡
4	350	0.25	53	61	69	85	100‡
6	350	0.25	26	31	37	47	65
8	350	0.25	16	20	25	34	50
10	350	0.26	11§	15	19	28	45
12	350	0.28	10§	15	19	28	44
14	250	0.28	**	11§	15	23	36
	300	0.30	**	13	17	26	42
	350	0.31	**	14	19	27	44
16	250	0.30	**	11§	15	24	34
	300	0.32	**	13	17	26	39
	350	0.34	**	15	20	28	44
18	250	0.31	**	10§	14	22	31
	300	0.34	**	13	17	26	36
	350	0.36	**	15	19	28	41
20	250	0.33	**	10	14	22	30
	300	0.36	**	13	17	26	35
	350	0.38	**	15	19	28	38
24	200	0.33	**	8§	12	17	25
	250	0.37	**	11	15	20	29
	300	0.40	**	13	17	24	32
	350	0.43	**	15	19	28	37
30	150	0.34	**	—	9	14	22
	200	0.38	**	8§	12	16	24
	250	0.42	**	11	15	19	27
	300	0.45	**	12	16	21	29
	350	0.49	**	15	19	25	33
36	150	0.38	**	—	9	14	21
	200	0.42	**	8§	12	15	23
	250	0.47	**	10	14	18	25
	300	0.51	**	12	16	20	28
	350	0.56	**	15	19	24	32
42	150	0.41	**	—	9	13	20
	200	0.47	**	8	12	15	22
	250	0.52	**	10	14	17	25
	300	0.57	**	12	16	20	27
	350	0.63	**	15	19	23	32

(continued)

Table 4-5 Rated working pressure and maximum depth of cover, standard pressure classes (*Continued*)

Nominal Pipe Size, in.	Pressure* Class, psi	Thickness, in.	Laying Conditions: Type 1 Trench	Type 2 Trench	Type 3 Trench	Type 4 Trench	Type 5 Trench
			Maximum Depth of Cover,† ft				
48	150	0.46	**	—	9	13	20
	200	0.52	**	8	11	15	22
	250	0.58	**	10	13	17	24
	300	0.64	**	12	15	19	27
	350	0.70	**	15	18	22	30
54	150	0.51	**	—	9	13	20
	200	0.58	**	8	11	14	22
	250	0.65	**	10	13	16	24
	300	0.72	**	13	15	19	27
	350	0.79	**	15	18	22	30
60	150	0.54	**	5 §	9	13	20
	200	0.61	**	8	11	14	22
	250	0.68	**	10	13	16	24
	300	0.76	**	13	15	19	26
	350	0.83	**	15	18	22	30
64	150	0.56	**	5§	9	13	20
	200	0.64	**	8	11	14	21
	250	0.72	**	10	13	16	24
	300	0.80	**	12	15	19	26
	350	0.87	**	15	17	21	29

NOTE: To convert inches (in.) to millimeters (mm), multiply by 25.4; to convert feet (ft) to meters (m), multiply by 0.3048; to convert pounds per square inch (psi) to kilopascals (kPa), multiply by 6.895.

* Ductile-iron pipe is adequate for the rated working pressure indicated for each nominal size plus a transient allowance of 100 psi (689 kPa). Calculations are based on a 2.0 safety factor times the sum of working pressure and 100 psi (689 kPa) transient allowance. (See ANSI/AWWA C150/A21.50 for design formulae.) Ductile-iron pipe for working pressures higher than 350 psi (2,413 kPa) is available.

† An allowance for a single H-20 truck with 1.5 impact factor is included for all depths of cover.

‡ Calculated maximum depth of cover exceeds 100 ft (30.5 m).

§ Minimum allowable depth of cover is 3 ft (0.9 m).

** For pipe 14 in. (350 mm) and larger, consideration should be given to the use of laying conditions other than Type 1.

Table 4-6 Special thickness classes of ductile-iron pipe

Nominal Pipe Size in.	OD in.	Special Thickness Class: 50	51	52	53	54	55	56
		Thickness, in.						
3	3.96	—	0.25	0.28	0.31	0.34	0.37	0.40
4	4.80	—	0.26	0.29	0.32	0.35	0.38	0.41
6	6.90	0.25	0.28	0.31	0.34	0.37	0.40	0.43
8	9.05	0.27	0.30	0.33	0.36	0.39	0.42	0.45
10	11.10	0.29	0.32	0.35	0.38	0.41	0.44	0.47

(continued)

Table 4-6 Special thickness classes of ductile-iron pipe (*Continued*)

Nominal Pipe Size in.	OD in.	Special Thickness Class 50	51	52	53	54	55	56
		Thickness, in.						
12	13.20	0.31	0.34	0.37	0.40	0.43	0.46	0.49
14	15.30	0.33	0.36	0.39	0.42	0.45	0.48	0.51
16	17.40	0.34	0.37	0.40	0.43	0.46	0.49	0.52
18	19.50	0.35	0.38	0.41	0.44	0.47	0.50	0.53
20	21.60	0.36	0.39	0.42	0.45	0.48	0.51	0.54
24	25.80	0.38	0.41	0.44	0.47	0.50	0.53	0.56
30	32.00	0.39	0.43	0.47	0.51	0.55	0.59	0.63
36	38.30	0.43	0.48	0.53	0.58	0.63	0.68	0.73
42	44.50	0.47	0.53	0.59	0.65	0.71	0.77	0.83
48	50.80	0.51	0.58	0.65	0.72	0.79	0.86	0.93
54	57.56	0.57	0.65	0.73	0.81	0.89	0.97	1.05

NOTE: To convert inches (in.) to millimeters (mm), multiply by 25.4.

Table 4-7 Rated working pressure and maximum depth of cover, special thickness classes

Nominal Pipe Size in.	Special Thickness Class	Nominal Thickness in.	Rated Water Working Pressure, psi*	Laying Conditions Type 1	Type 2	Type 3	Type 4	Type 5
				Maximum Depth of Cover,* ft				
3	51	0.25	350†	78	88	99	99	100‡
	52	0.28	350†	100‡	100‡	100‡	100‡	100‡
	53	0.31	350†	100‡	100‡	100‡	100‡	100‡
	54	0.34	350†	100‡	100‡	100‡	100‡	100‡
	55	0.37	350†	100‡	100‡	100‡	100‡	100‡
	56	0.40	350†	100‡	100‡	100‡	100‡	100‡
4	51	0.26	350†	62	71	80	98	100‡
	52	0.29	350†	94	100‡	100‡	100‡	100‡
	53	0.32	350†	100‡	100‡	100‡	100‡	100‡
	54	0.35	350†	100‡	100‡	100‡	100‡	100‡
	55	0.38	350†	100‡	100‡	100‡	100‡	100‡
	56	0.41	350†	100‡	100‡	100‡	100‡	100‡
6	50	0.25	350†	26	31	37	47	65
	51	0.28	350†	40	47	53	67	88
	52	0.31	350†	58	66	75	92	100‡
	53	0.34	350†	78	89	100‡	100‡	100‡
	54	0.37	350†	100‡	100‡	100‡	100‡	100‡
	55	0.40	350†	100‡	100‡	100‡	100‡	100‡

(continued)

Table 4-7 Rated working pressure and maximum depth of cover, special thickness classes (*Continued*)

Nominal Pipe Size in.	Special Thickness Class	Nominal Thickness in.	Rated Water Working Pressure, psi*	Laying Conditions Type 1	Type 2	Type 3	Type 4	Type 5
				Maximum Depth of Cover,* ft				
	56	0.43	350+	100‡	100‡	100‡	100‡	100‡
8	50	0.27	350+	21	26	31	40	57
	51	0.30	350+	30	36	42	53	71
	52	0.33	350+	41	48	55	69	90
	53	0.36	350+	55	63	71	88	100‡
	54	0.39	350+	70	79	90	100‡	100‡
	55	0.42	350+	87	99	100‡	100‡	100‡
	56	0.45	350+	100‡	100‡	100‡	100‡	100‡
10	50	0.29	350+	16	21	25	35	51
	51	0.32	350+	23	28	33	43	60
	52	0.35	350+	31	36	42	54	72
	53	0.38	350+	40	46	53	66	87
	54	0.41	350+	50	58	65	81	100‡
	55	0.44	350+	62	71	80	98	100‡
	56	0.47	350+	75	85	96	100‡	100‡
12	50	0.31	350+	15	19	24	33	49
	51	0.34	350+	20	25	30	39	56
	52	0.37	350+	26	31	37	47	65
	53	0.40	350+	33	39	45	57	76
	54	0.43	350+	41	48	54	68	89‡
	55	0.46	350+	50	57	65	81	100‡
	56	0.49	350+	60	68	77	95‡	100‡
14	50	0.33	350+	**	17	21	30	47
	51	0.36	350+	**	21	26	35	52
	52	0.39	350+	**	26	31	41	58
	53	0.42	350+	**	32	38	48	66
	54	0.45	350+	**	39	45	57	76
	55	0.48	350+	**	46	53	66	87
	56	0.51	350+	**	54	62	77	99
16	50	0.34	350+	**	15	20	28	44
	51	0.37	350+	**	19	23	32	49
	52	0.40	350+	**	23	28	37	54
	53	0.43	350+	**	27	33	43	60
	54	0.46	350+	**	33	38	49	67
	55	0.49	350+	**	38	44	56	75
	56	0.52	350+	**	45	51	64	85

(continued)

Table 4-7 Rated working pressure and maximum depth of cover, special thickness classes (*Continued*)

Nominal Pipe Size in.	Special Thickness Class	Nominal Thickness in.	Rated Water Working Pressure, psi*	Laying Conditions Type 1	Type 2	Type 3	Type 4	Type 5
				Maximum Depth of Cover,* ft				
18	50	0.35	330	**	14	18	27	38
	51	0.38	350+	**	17	21	30	46
	52	0.41	350+	**	20	25	34	51
	53	0.44	350+	**	24	29	39	55
	54	0.47	350+	**	28	34	44	61
	55	0.50	350+	**	33	39	50	67
	56	0.53	350+	**	38	44	56	75
20	50	0.36	310	**	13	17	26	35
	51	0.39	350+	**	16	20	29	41
	52	0.42	350+	**	18	23	32	48
	53	0.45	350+	**	22	26	36	52
	54	0.48	350+	**	25	30	40	57
	55	0.51	350+	**	29	34	45	62
	56	0.54	350+	**	33	39	50	68
24	50	0.38	275	**	11	16	22	30
	51	0.41	325	**	14	18	25	34
	52	0.44	350+	**	16	20	29	38
	53	0.47	350+	**	18	23	32	44
	54	0.50	350+	**	21	26	35	51
	55	0.53	350+	**	24	29	38	55
	56	0.56	350+	**	27	32	42	59
30	50	0.39	215	**	9§	13	17	24
	51	0.43	270	**	11	15	20	27
	52	0.47	320	**	13	18	23	31
	53	0.51	350+	**	16	20	27	36
	54	0.55	350+	**	18	23	32	41
	55	0.59	350+	**	21	26	35	48
	56	0.63	350+	**	24	29	39	56
36	50	0.43	205	**	8	12	16	23
	51	0.48	260	**	11	15	18	26
	52	0.53	315	**	13	18	22	30
	53	0.58	350+	**	16	20	26	34
	54	0.63	350+	**	19	23	30	40
	55	0.68	350+	**	21	26	35	46
	56	0.73	350+	**	25	30	39	54

(continued)

Table 4-7 Rated working pressure and maximum depth of cover, special thickness classes (*Continued*)

Nominal Pipe Size in.	Special Thickness Class	Nominal Thickness in.	Rated Water Working Pressure, psi*	Type 1	Type 2	Type 3	Type 4	Type 5
				Laying Conditions — Maximum Depth of Cover,* ft				
42	50	0.47	200	**	8	12	15	22
	51	0.53	260	**	11	14	17	25
	52	0.59	315	**	13	17	21	29
	53	0.65	350⁺	**	16	20	25	33
	54	0.71	350⁺	**	19	23	29	39
	55	0.77	350⁺	**	22	26	35	45
	56	0.83	350⁺	**	25	30	40	53
48	50	0.51	190	**	8	11	14	22
	51	0.58	250	**	10	13	17	24
	52	0.65	305	**	13	16	19	27
	53	0.72	350⁺	**	15	19	23	32
	54	0.79	350⁺	**	18	23	28	37
	55	0.86	350⁺	**	21	26	34	43
	56	0.93	350⁺	**	25	30	39	51
54	50	0.57	190	**	8	11	14	21
	51	0.65	250	**	10	13	16	24
	52	0.73	310	**	13	16	19	27
	53	0.81	350⁺	**	16	19	23	31
	54	0.89	350⁺	**	18	23	28	36
	55	0.97	350⁺	**	22	26	33	43
	56	1.05	350⁺	**	25	30	40	51

NOTE: To convert inches (in.) to millimeters (mm), multiply by 25.4; to convert feet (ft) to meters (m), multiply by 0.3048; to convert pounds per square inch (psi) to kilopascals (kPa), multiply by 6.895.

* Ductile-iron pipe is adequate for the rated water working pressure indicated for each nominal size plus a transient allowance of 100 psi (689 kPa). Calculations are based on a 2.0 safety factor times the sum of working pressure and 100 psi (689 kPa) transient allowance. Calculations are rounded to the nearest 5 psi.

† An allowance for a single H-20 truck with 1.5 impact factor is included for all depths of cover.

⁺ The calculated actual rated water working pressure exceeds 350 psi (2,413 kPa). Ductile-iron pipe for rated water working pressures higher than 350 psi (2,413 kPa) is available.

‡ Calculated maximum depth of cover exceeds 100 ft (30.5 m).

§ Minimum allowable depth of cover is 3 ft (0.9 m).

** For pipe 14 in. (350 mm) and larger, consideration should be given to the use of laying conditions other Type 1.

4.2.13 Safety Factor

The safety factor for internal pressure is 2.0 based on minimum yield strength of ductile iron in tension. For external loads, two safety factors are specified: 1.5 based on minimum ring yield strength and at least 2.0 based on ring ultimate strength. Also, the design ring deflection provides a safety factor of at least 2.0 based on test data regarding deflections required to cause failure in cement–mortar linings.

These safety factors are used to establish minimum design criteria and should not be confused with the total available or actual safety factor of ductile-iron pipe, which is generally much greater than the minimum because of the following:

- Minimum manufacturing thickness for small-size pipes (3-in. through 8-in. diameter) is even greater than that required per the basic design for a 350-psi working pressure rating.
- The design criteria for ductile-iron pipe are not based on performance limits associated with actual failure of the pipe wall.
- Specified safety factors are used to calculate net wall thickness requirements, after which both service allowance and casting allowance are added.
- The yield strength of ductile-iron pipe will consistently exceed the minimum values specified for design.
- Ductile-iron pipe can sustain stresses considerably higher than yield strength determined by standard test methods without damage to the pipe wall.
- Design considerations dependent on laying conditions were established on a conservative basis.

Extensive tests (CIPRA 1964) were conducted in 1964 on ductile-iron pipe to determine average values for tensile strength, ring strength, hardness, and elongation. Test pipe ranged in size from 2 in. to 24 in. (50 mm to 610 mm) and represented five producers. The test results indicated an average bursting tensile strength of 52,320 psi (360.7×10^3 kPa) and an average ring yield strength of 84,880 psi (585.2×10^3 kPa) for all pipes tested. Subsequent testing has confirmed that current quality is consistent with or exceeds these early test results. Using these values, the following example of the total safety factor for internal pressure design can be made.

To determine the total safety factor of 6-in. (152-mm) Class 350 ductile-iron pipe with respect to internal pressure for 350 psi (2,413 kPa) working pressure and a standard transient pressure allowance of 100 psi (689 kPa):

Step 1. Compute the hoop stress developed by using the following:

$$S = \frac{P_i D}{2t_1} \qquad \text{(Eq 4-8)}$$

Where:

P_i = 350 + 100 = 450 psi because total safety factor is desired.

D = 6.90 in.

Nominal thickness of Class 350 = 0.25 in.

Subtract casting allowance to obtain minimum manufacturing thickness t_1.

t_1 = 0.025 – 0.05 = 0.20 in.

$$S = \frac{(450)(6.90)}{(2)(0.20)} = 8,762.5 \text{ psi} \qquad \text{(Eq 4-9)}$$

Step 2. Compare computed hoop stress to average bursting tensile strength to determine a representative total safety factor by using the following:

$$\frac{52,320 \text{ psi average}}{7,762.5 \text{ psi computed}} = 6.74 \qquad \text{(Eq 4-10)}$$

The total safety factor for internal pressure design will vary with pipe size, thickness class, and design working pressure, but the previous example demonstrates that the total available safety factor in ductile-iron pipe can be much greater than the design safety factor of 2.0.

With regard to external load design, tests were conducted on large-diameter ductile-iron pipe at Utah State University for the purpose of verifying the C150/A21.50 procedure. From these test data, safety factors were calculated by dividing the loads at cement–mortar lining failure by allowable loads, as well as by dividing the loads at pipe failure by the allowable loads. Using the C150/A21.50 design procedure for external loads, the allowable loads were calculated. This comparison showed that when cement–mortar lining failure was used, the calculated safety factor of the test pipe averaged 2.98; when pipe failure was used, the calculated safety factor averaged 5.46.

Based on these data, it is apparent that the total available safety factor of ductile-iron pipe with respect to external loads is far greater than the design safety factors of 1.5 and 2.0. Further, the total available safety factors were determined on the basis of a separate stress design; for a combined stress situation (i.e., external load + internal pressure), the total available safety factor would be even greater because internal pressure would tend to reround the pipe, thereby reducing deflection and ring bending stresses created by external load. It is evident, therefore, that the total safety factor for ductile-iron pipe is more than adequate and a thorough analysis of both the pipe material and the design procedure is necessary to properly determine true safety factors.

4.3 TRUCK LOADS ON PIPE BURIED AT SHALLOW DEPTHS

4.3.1 Introduction

Special design considerations should be used for pipe under roads and highways with less than 2.5 ft (0.76 m) of cover because of the possibility of high dynamic loading. This loading could result in damage to the pavements or the pipes. Also, if the impact factor is anticipated to be greater than 1.5, then a higher value should be applied. An impact factor, *F*, of 1.5 is consistent with American Society of Civil Engineers (ASCE) Manual 37 (1972). For any given project, the ultimate responsibility for the proper use of the equations and other data provided rests with the design engineer.

4.3.2 Procedure

The procedure for calculating truck loads on buried ductile-iron pipe is provided in ANSI/AWWA C150/A21.50. This procedure is based on the teachings of Spangler and others, and it employs the same procedures used in the design standard ANSI/AWWA C101/A21.1-1967 for cast-iron pipe. The design method is based on the following two assumptions:

- A single concentrated wheel load at the surface
- Uniform load distribution over an effective pipe length of 3 ft (0.914 m)

The truck load on pipe buried under flexible pavement is calculated using Equation 4-3.

The surface load factor, *C*, is a measure of how the wheel load at the surface is transmitted and distributed through the soil to the pipe. *C* is calculated using Equation 4-4.

This equation is derived from Holl's integration of the Boussinesq formula for vertical unit pressure, assuming the load is to be determined on a 3-ft (0.9-m) section of pipe directly under the point load.

Regarding the point load assumption, the following Boussinesq equation (Equation 4-11) gives the vertical stress at any point in an elastic medium when a point load is exerted at the surface.

$$\sigma_z = \left(\frac{3P}{2\pi}\right)\left(\frac{H^3}{R_1^5}\right) \tag{Eq 4-11}$$

Where

σ_z = vertical stress, psi (kPa)

P = point load at surface, lb (kN)

H = depth, in. (m)

R_1 = distance from the point load to the point at which the stress is to be determined, in. (m) (Figure 4-3)

Integration of the Boussinesq equation (Equation 4-8) over the rectangular area over the pipe (as shown in Figure 4-3) results in the total load on a 3-ft (0.914-m) section of pipe as a result of the point load, P, at the surface. Equation 4-4 is a function of this integration. The quantity bD in the denominator of Equation 4-3 yields the desired units of pounds per square inch in expressing the truck load. The result thus represents an "average" pressure on the 3-ft (0.914-m) length of pipe centered under the load.

The factors discussed previously, as well as other factors—such as the assumed flexible pavement, large wheel load used for design, safety factors in the thickness design procedure, and inherent structural strength of ductile iron (Stroud 1991)—led to the conclusion that the approach to calculating truck loading is adequate at any depth of cover. The actual distributed load of a truck tire "footprint" will produce fewer concentrated effects on a pipe than will the assumed point load. The typical dual truck tire imprint may have a contact area of approximately 200 in.^{2} (1,290.4 cm^2) (The Asphalt Institute 2007). Also, the length of pipe effective in carrying the load may be much greater than that assumed, particularly for large-diameter pipe. Furthermore, in shallow cover situations under highways, the roadbed stability will necessitate well-compacted fill around the pipe, which will increase its load-bearing capacity.

Tables 4-8, 4-9, and 4-10 present values for earth loads, P_e, truck loads, P_t, trench loads, P_v, surface load factors for a single truck on unpaved road, and required pipe wall thicknesses for the sum of these earth and truck loads for the various pipe sizes under shallow cover conditions of 1.0, 1.5, and 2.0 ft (0.30, 0.46, and 0.61 m). These tables can be used in the same manner as described in ANSI/AWWA C150/A21.50 (Tables 1, 6, and 12, respectively).

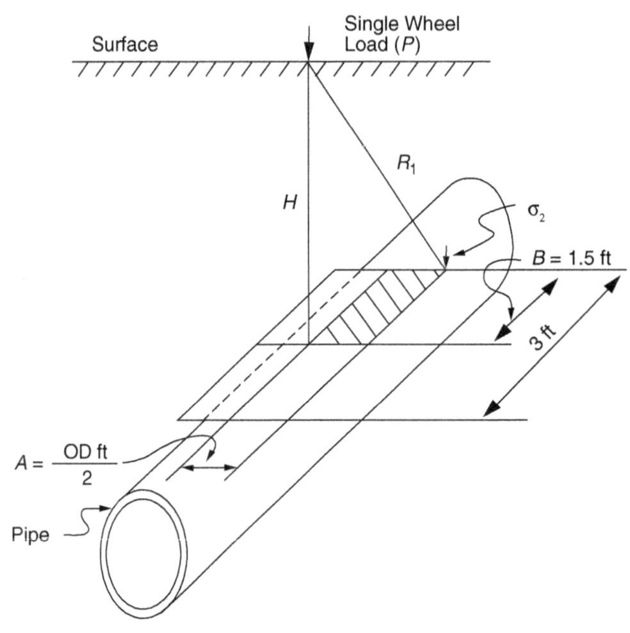

Figure 4-3 Truck loading conditions—single-wheel load

Table 4-8 Earth loads P_e, truck loads P_t, and trench loads P_v (psi)

Depth of Cover,		3-in. Pipe		4-in. Pipe		6-in. Pipe		8-in. Pipe		10-in. Pipe		12-in. Pipe		14-in. Pipe		16-in. Pipe		18-in. Pipe	
ft	P_e	P_t	P_v	P_t	P_v	P_t	P_v	P_t	P_v	P_t	P_v	P_t	P_v	P_t	P_v	P_t	P_v	P_t	P_v
1.0	0.8	33.3	34.1	33.1	33.9	32.2	33.0	31.0	31.8	29.8	30.6	28.4	29.2	24.8	25.6	22.5	23.3	20.6	21.4
1.5	1.3	20.7	22.0	20.6	21.9	20.3	21.6	19.9	21.2	19.5	20.8	19.1	20.4	17.0	18.3	15.8	17.1	14.8	16.1
2.0	1.7	13.9	15.6	13.9	15.6	13.8	15.5	13.6	15.3	13.5	15.2	13.3	15.0	12.0	13.7	11.3	13.0	10.6	12.3
		20-in. Pipe		24-in. Pipe		30-in. Pipe		36-in. Pipe		42-in. Pipe		48-in. Pipe		54-in. Pipe		60-in. Pipe		64-in. Pipe	
		P_t	P_v	P_t	P_v	P_t	P_v	P_t	P_v	P_t	P_v	P_t	P_v	P_t	P_v	P_t	P_v	P_t	P_v
1.0	0.8	19.0	19.8	16.6	17.4	14.2	15.0	12.1	12.9	10.6	11.4	9.4	10.2	8.4	9.2	7.9	8.7	7.4	8.2
1.5	1.3	13.9	15.2	12.6	13.9	11.3	12.6	10.0	11.3	8.9	10.2	8.0	9.3	7.2	8.5	6.8	8.1	6.4	7.7
2.0	1.7	10.2	11.9	9.4	11.1	8.7	10.4	7.9	9.6	7.2	8.9	6.6	8.3	6.0	7.7	5.7	7.4	5.4	7.1

Note: To convert inches (in.) to millimeters (mm), multiply by 25.4; to convert feet (ft) to meters (m), multiply by 0.3048; to convert pounds per square inch (psi) to kilopascals (kPa), multiply by 6.895.

Table 4-9 Surface load factors C for single truck on unpaved road

Depth of Cover, *ft*	Nominal Pipe Size, in.								
	3	4	6	8	10	12	14	16	18
	Surface Load Factor—*C*								
1.0	0.1980	0.2380	0.3329	0.4210	0.4956	0.5623	0.6195	0.6680	0.7087
1.5	0.1227	0.1482	0.2102	0.2708	0.3253	0.3773	0.4252	0.4690	0.5086
2.0	0.0828	0.1001	0.1428	0.1853	0.2244	0.2627	0.2993	0.3338	0.3661
Depth of Cover, *ft*	Nominal Pipe Size, in.								
	20	24	30	36	42	48	54	60	64
	Surface Load Factor—*C*								
1.0	0.7427	0.7944	0.8428	0.8714	0.8881	0.8985	0.9054	0.9082	0.9104
1.5	0.5442	0.6043	0.6700	0.7155	0.7458	0.7667	0.7818	0.7884	0.7936
2.0	0.3964	0.4504	0.5154	0.5656	0.6025	0.6303	0.6520	0.6620	0.6703

Note: To convert inches (in.) to millimeters (mm), multiply by 25.4; to convert feet (ft) to meters (m), multiply by 0.3048.

Table 4-10 Thickness for earth load plus truck load

Pipe Size, in.	Depth of Cover, ft	Laying Condition									
		Type 1 Trench		Type 2 Trench		Type 3 Trench		Type 4 Trench		Type 5 Trench	
		Total Calculated Thickness,* in.	Pressure Class†	Total Calculated Thickness,* in.	Pressure Class†	Total Calculated Thickness,* in.	Pressure Class†	Total Calculated Thickness,* in.	Pressure Class†	Total Calculated Thickness,* in.	Pressure Class†
3	1.0	0.22	350	0.21	350	0.20	350	0.19	350	0.16	350
	1.5	0.20	350	0.19	350	0.18	350	0.17	350	0.15	350
	2.0	0.19	350	0.18	350	0.17	350	0.16	350	0.14	350
4	1.0	0.23	350	0.23	350	0.22	350	0.20	350	0.17	350
	1.5	0.21	350	0.20	350	0.20	350	0.18	350	0.15	350
	2.0	*0.20*	*350*	*0.19*	*350*	*0.18*	*350*	*0.16*	*350*	*0.15*	*350*
6	1.0	0.28	‡	0.27	‡	0.25	350	0.23	350	0.18	350
	1.5	0.25	350	0.24	350	0.22	350	0.20	350	0.16	350
	2.0	0.23	350	0.21	350	0.20	350	0.17	350	0.16	350
8	1.0	0.32	‡	0.30	‡	0.29	‡	0.26	‡	0.20	350
	1.5	0.28	‡	0.27	‡	0.25	350	0.21	350	0.17	350
	2.0	0.26	‡	0.24	350	0.22	350	0.19	350	0.16	350
10	1.0	0.37	‡	0.35	‡	0.33	‡	0.30	‡	0.22	350
	1.5	0.33	‡	0.31	‡	0.29	‡	0.24	350	0.19	350
	2.0	0.30	‡	0.27	‡	0.25	350	0.21	350	0.18	350
12	1.0	0.41	‡	0.38	‡	0.36	‡	0.32	‡	0.24	350
	2.0	0.32	‡	0.30	‡	0.27	350	0.22	350	0.19	350
14	1.0	§	—	0.41	‡	0.38	‡	0.33	‡	0.26	250
	1.5	§	—	0.36	‡	0.33	‡	0.27	250	0.21	250
	2.0	§	—	0.32	‡	0.29	300	0.24	250	0.20	250
16	1.0	§	—	0.43	‡	0.40	‡	0.33	350	0.27	250
	1.5	§	—	0.38	‡	0.34	350	0.27	250	0.22	250
	2.0	§	—	0.33	350	0.30	250	0.24	250	0.21	250

(continued)

Table 4-10 Thickness for earth load plus truck load (*Continued*)

		Laying Condition									
		Type 1 Trench		Type 2 Trench		Type 3 Trench		Type 4 Trench		Type 5 Trench	
Pipe Size, in.	Depth of Cover, ft	Total Calculated Thickness,* in.	Pressure Class†	Total Calculated Thickness,* in.	Pressure Class†	Total Calculated Thickness,* in.	Pressure Class†	Total Calculated Thickness,* in.	Pressure Class†	Total Calculated Thickness,* in.	Pressure Class†
18	1.0	§	—	0.45	‡	0.41	‡	0.33	300	0.27	250
	1.5	§	—	0.39	‡	0.35	350	0.28	250	0.22	250
	2.0	§	—	0.35	350	0.31	250	0.25	250	0.21	250
20	1.0	§	—	0.46	‡	0.42	‡	0.34	300	0.27	250
	1.5	§	—	0.41	‡	0.36	300	0.29	250	0.23	250
	2.0	§	—	0.36	300	0.32	250	0.26	250	0.22	250
24	1.0	§	—	0.49	‡	0.44	‡	0.37	250	0.26	200
	1.5	§	—	0.44	‡	0.38	300	0.31	200	0.24	200
	2.0	§	—	0.39	300	0.34	250	0.27	200	0.23	200
30	1.0	§	—	0.53	‡	0.46	350	0.40	250	0.27	150
	1.5	§	—	0.48	350	0.41	250	0.34	150	0.25	150
	2.0	§	—	0.43	300	0.37	200	0.30	150	0.24	150
36	1.0	§	—	0.56	350	0.48	300	0.41	200	0.28	150
	1.5	§	—	0.51	300	0.43	250	0.33	150	0.27	150
	2.0	§	—	0.46	250	0.40	200	0.31	150	0.26	150
42	1.0	§	—	0.58	350	0.50	250	0.39	150	0.29	150
	1.5	§	—	0.54	300	0.45	200	0.35	150	0.28	150
	2.0	§	—	0.49	250	0.42	200	0.33	150	0.27	150
48	1.0	§	—	0.60	300	0.52	200	0.39	150	0.31	150
	1.5	§	—	0.57	250	0.48	200	0.37	150	0.30	150
	2.0	§	—	0.53	250	0.45	150	0.36	150	0.29	150
54	1.0	§	—	0.64	250	0.53	200	0.41	150	0.33	150
	1.5	§	—	0.60	250	0.51	150	0.40	150	0.32	150
	2.0	§	—	0.57	200	0.48	150	0.39	150	0.31	150

(continued)

Table 4-10 Thickness for earth load plus truck load (*Continued*)

		Laying Condition									
		Type 1 Trench		Type 2 Trench		Type 3 Trench		Type 4 Trench		Type 5 Trench	
Pipe Size, in.	Depth of Cover, ft	Total Calculated Thickness,* in.	Pressure Class†	Total Calculated Thickness,* in.	Pressure Class†	Total Calculated Thickness,* in.	Pressure Class†	Total Calculated Thickness,* in.	Pressure Class†	Total Calculated Thickness,* in.	Pressure Class†
60	1.0	§	—	0.65	250	0.54	150	0.42	150	0.33	150
	1.5	§	—	0.61	200	0.52	150	0.41	150	0.33	150
	2.0	§	—	0.58	200	0.50	150	0.40	150	0.32	150
64	1.0	§	—	0.66	250	0.55	150	0.43	150	0.34	150
	1.5	§	—	0.62	200	0.53	150	0.42	150	0.33	150
	2.0	§	—	0.59	200	0.51	150	0.41	150	0.32	150

Note: To convert inches (in.) to millimeters (mm), multiply by 25.4; to convert feet (ft) to meters (m), multiply by 0.3048.

* Total calculated thickness includes service allowance and casting tolerance added to net thickness.

† For wall thickness of standard pressure class pipe see Table 4-4.

‡ Special thickness classes are available for this application. See Table 4-6.

§ For pipe 14 in. and larger, consideration should be given to the use of laying conditions other than Type 1.

4.4 THICKNESS DESIGN FOR DUCTILE-IRON PIPE UNDER RAILROADS

4.4.1 Introduction

If authorized by the railroad company, ductile-iron pipe can be installed beneath railroads without a casing if the live load caused by passing trains is taken into account in the pipe thickness design. The approach is essentially the same as for normal highway installations, except the truck load should be replaced by the Cooper E-80 railroad loading series as described by the American Railway Engineering and Maintenance-of-Way Association (AREMA 2007).

4.4.2 Railroad Loads

The loading configuration for the Cooper E-80 railroad loading is a uniformly loaded rectangle with dimensions of 20 ft × 8 ft (6.1 m × 2.4 m) (Figure 4-4). The source of these dimensions is the width of the railroad ties (8 ft [2.4 m]) and the spacing between the drive wheels of the typical locomotive. Loading is based on the axle weight exerted on the trackage by two locomotives coupled together in doubleheader fashion (Figure 4-5). For E-80 loading, the axle load is 80,000 lb (335.84 kN), and there are four axles. The entire load is presumed to be uniformly distributed on the rectangle.

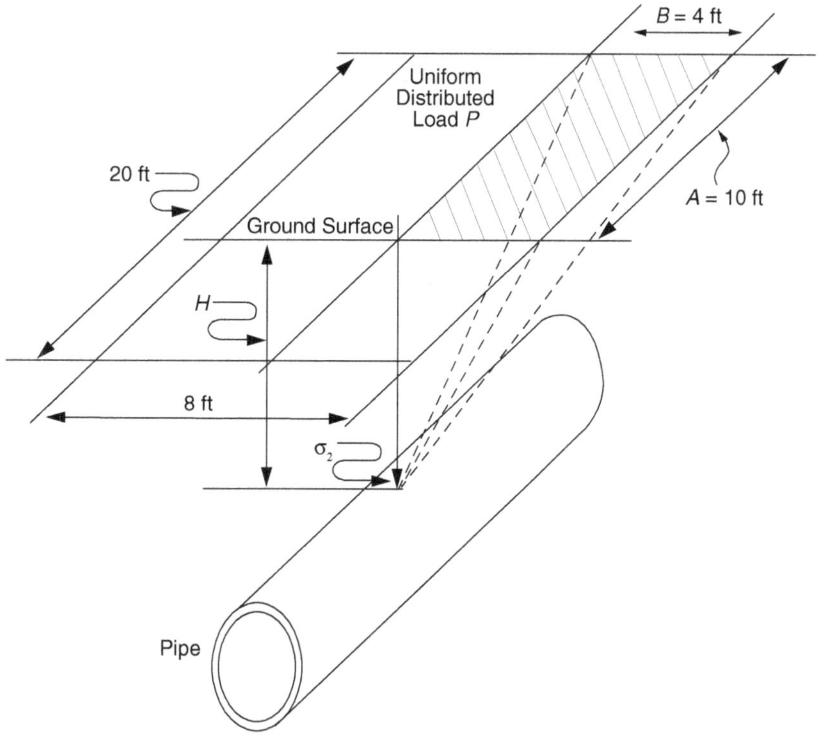

Figure 4-4 Pipe loading from uniformly distributed railroad loading

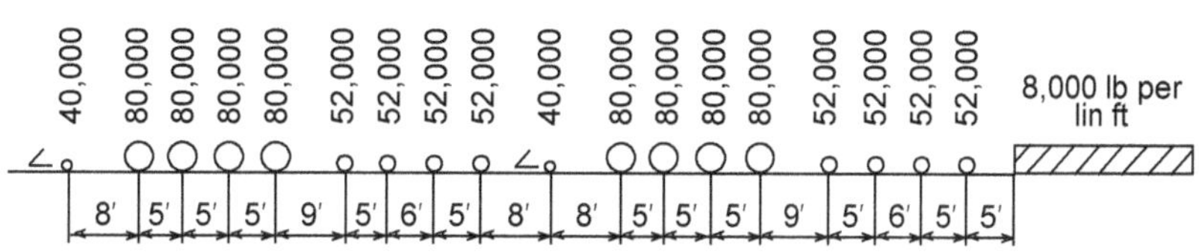

Figure 4-5 Railroad loading for Cooper e-80 locomotives

The trench load with railroad loading taken into account is calculated by using Equation 4-12.

$$P_r = CPF \tag{Eq 4-12}$$

Where:

P_r = railroad locomotive load at pipe depth, psi (kPa)
C = surface load factor (influence coefficient)
P = pressure on the loaded rectangle (wheel load)
= 13.9 psi (95.8 kPa) for E-80 loading
F = impact factor (1.5)

The influence coefficient at depth H for a surface load distributed over a rectangle of dimensions $2A \times 2B$ is given by the integration of the Boussinesq equation:

$$C = 1 - \frac{2}{\pi}\left[\sin^{-1} H \sqrt{\frac{A^2 + B^2 + H^2}{\left(A^2 + H^2\right)\left(B^2 + H^2\right)}} - \frac{ABH}{\sqrt{A^2 + B^2 + H^2}}\left(\frac{1}{A^2 + H^2} + \frac{1}{B^2 + H^2}\right)\right] \tag{Eq 4-13}$$

Table 4-11 gives values of C for various values of H for Cooper railroad loading:

Where:

A = 10 ft (3.05 m)
B = 4 ft (1.22 m)
H = depth of cover, ft (m)

NOTE: All angles are in radians.

Example of thickness design beneath a railroad:

Project criteria: 24-in. (610-mm) ductile-iron pipe under 8 ft (2.44 m) of cover, Type 5 trench, 75-psi (517-kPa) operating pressure; employ Cooper E-80 railroad load.

Step 1. Design for Internal Pressure

Using Equation 4-1,

$$\text{Net } t = \frac{P_i D}{2S} \tag{Eq 4-14}$$

$P_i = 2\,(P_w + 100) = 350$ psi
D (from Table 4-4) = 25.80 in.

Table 4-11 Surface load factors for Cooper railroad loading

Depth, ft	C	Depth, ft	C
2.5	.928	9.00	.454
3.0	.892	10.00	.405
4.0	.809	12.00	.326
5.0	.725	14.00	.265
6.0	.645	16.00	.219
7.0	.573	18.00	.183
8.0	.509	20.00	.155

NOTE: To convert feet (ft) to meters (m), multiply by 0.3048.

Step 2. Design for External Load
Total trench load $Pv = Pe + Pr$

$$\text{Net } t = \frac{350(25.8)}{2(42{,}000)} = 0.107 \text{ in.} \qquad \text{(Eq 4-15)}$$

Using Equation 4-5,
$Pr = CPF = 0.509(13.9)\ (1.5) = 10.61$ psi
The total trench load is Pv
$Pv = 6.67 + 10.61 = 17.3$ psi
From Table 11 in ANSI/AWWA C150/A21.50, the required D/t for bending stress with $P_v = 17.3$ is

$$D/t = 252$$

Therefore,

$$\text{Net } t = \frac{25.8}{252} = 0.102 \qquad \text{(Eq 4-16)}$$

From the same table, the required D/t_1 for deflection is

$$D/t_1 = 137$$

Therefore,

$$t_1 = \frac{25.8}{137} = 0.188 \text{ in.} \qquad \text{(Eq 4-17)}$$

and

$$\text{Net } t = 0.188 - .08 = 0.108 \text{ in.}$$

Because the deflection calculation yields the highest net t, total required thickness is as follows:

Net t	0.11 in.
+ service allowance	0.08 in.
+ casting allowance	0.07 in.
= Total thickness	0.26 in.

Comparison with Table 4-4 reveals that the required total thickness results in the selection of Pressure Class 200.

4.5 THICKNESS DESIGN FOR DUCTILE-IRON PIPE ON SUPPORTS

4.5.1 Introduction

Neither ductile iron nor any other type of pipe is designed specifically as a beam for normal buried service. It is always assumed that the pipe will be uniformly supported along its length by the soil beneath it. Erosion, excessive traffic loading, frost, expansive soils, and poor installation sometimes result in beam loading on buried pipe. In fact, these conditions, individually or in combination, probably are responsible for many failures in buried pipelines made from other materials. However, because of ductile iron's great beam strength and ductility, beam failures in buried ductile-iron pipe are virtually unknown.

Source: Ductile Iron Pipe Research Association

Figure 4-6 Ductile-iron pipe is well suited for pipe-on-supports applications because of its beam strength

In some situations, it is necessary or desirable to use supports at designated intervals along pipelines (Figure 4-6). Aboveground supported pipe is needed to transport water and other fluids within treatment plants and buildings. Also, pipe on piers is used to cross natural or man-made objects. Sometimes, unstable soil conditions or other factors necessitate the installation of pipe on piers or pilings underground.

This section reviews the pertinent design considerations for aboveground and underground ductile-iron pipe-on-supports installations. Specific procedures, recommended design limits, and allowable stresses are outlined in the example problem. Design tables based on ductile-iron pipe data and suggested loads are also provided. Bridge-crossing installations that require special attention because of their unique situations are not specifically addressed.

4.5.2 Beam Span for Ductile-Iron Pipe on Supports

Ductile-iron pipe is normally manufactured in 18- or 20-ft (5.5- or 6.1-m) nominal lengths, depending on the pipe manufacturer and pipe size. (Ductile-iron pipe may be provided in shorter lengths per AWWA C151. If exact lengths are required to fit on prebuilt piers, this should be specified.) The most common joints used with ductile-iron pipe are the push-on type joint and the mechanical joint. Both of these rubber-gasketed joints allow a certain amount of deflection and longitudinal displacement while maintaining their hydrostatic seal. This makes these pipe joints ideally suited for normal underground installation. The flexibility of the joints reduces the occurrence of excessive beam stresses.

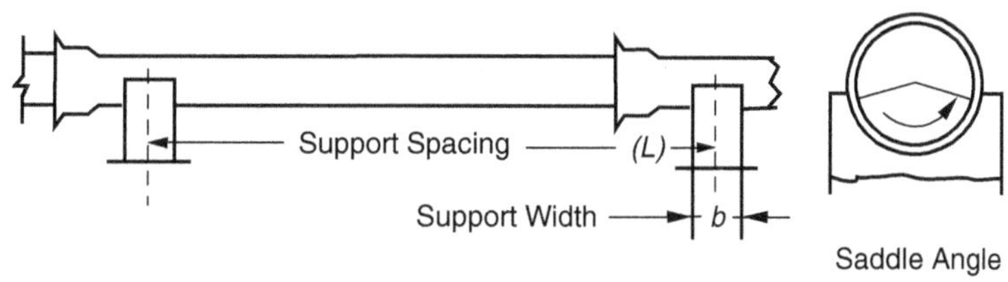

Figure 4-7 Saddle angle and width

For pipe supported at intervals, however, flexible joints usually require that at least one support be placed under each length of pipe for stability.

Various schemes have been successfully used to obtain longer spans where particular installation conditions presented the need, but these are special design situations and are not specifically addressed in this section. The design presented herein is based on one support per length of pipe.

4.5.3 Support Location

System security is maximized by positioning the supports immediately behind the pipe bells. When the support is placed near the bell, the bell section contributes beneficial ring stiffness where it is most needed. This ring stiffness, in turn, reduces the effect of support loads and localized stress. Supports should normally not be placed under spigots adjacent to bells because of possible undesirable effects on joints.

4.5.4 Saddle Angle and Support Width

Pipe supports should cradle the pipe in a saddle (Table 4-7). This cradling, which should follow the contour of the pipe, minimizes stress concentrations at the supports. It is recommended that the saddle angle β of the support is between 90° and 120°. Little or no benefit is gained by increasing the saddle angle to more than 120°. With angles smaller than 90°, the maximum stress tends to increase rapidly with decreasing saddle angle (Evces and O'Brien 1984).

There are some differences among published theories and data regarding the importance of axial support width for saddles. The most accepted formulae are found to be completely independent of saddle width. Some test data, however, show a decrease in measured stresses with an increase in saddle width. There is little effect on the maximum stress when saddle support width is increased more than (Keyes et al 1941). Therefore, for saddle supports, the minimum width b is determined by using Equation 4-12.

$$b = \sqrt{2Dt_e} \qquad \text{(Eq 4-11)}$$

Where:

b = minimum (axial) saddle width, in. (mm)

D = actual OD of pipe, in. (mm)

t_e = nominal pipe wall thickness, in. (mm) (see Table 4-4)

4.5.5 Support Design

Additionally, supports, piles, or foundations should be adequately designed from a structural and soil-engineering standpoint to safely handle any loads transferred from the pipe.

4.5.6 Loads on Pipe

For underground pipe-on-supports design calculations, the total load normally includes the prism earth load plus the weight of the pipe and contents. Usually, buried pipe is installed on supports because of unstable ground conditions. In such cases, vehicle loading would normally not be expected or desirable. Thus, live loads, such as truck loading per ANSI/AWWA C150/A21.50, should be used in design calculations only where they cannot be avoided and are likely to occur. For aboveground design calculations, the total load normally includes the weight of the pipe and contents. If the engineer expects greater loads to occur on aboveground or underground installations, these loads are not in the scope of this procedure but should be incorporated into the design.

4.5.7 Pipe Wall Thickness Calculations

Design calculations include localized stress at supports, hoop stress as a result of internal pressure, and flexural stress and beam deflection at the center of the span.

Because of the conservative approach of this design procedure and in the interest of simplicity, combinations of external load and internal pressure to obtain principal stresses have not been considered. The engineer may elect to investigate principal stresses resulting from extraordinary circumstances: for example, very high internal pressure.

4.5.7.1 Localized stress at supports. The supported pipe is subjected to localized stresses that are a function of the total reaction at the support and the shape (saddle angle) of the support. This maximum stress may be longitudinal or circumferential in nature and is predicted by the equation proposed by Roark (Young and Budynas 2001) (Equation 4-13).

$$f_r = K\left(\frac{wL}{t_n^{\,2}}\right) \text{In}\left(\frac{D}{2t_n}\right) \qquad \text{(Eq 4-12)}$$

Where:

f_r = localized stress caused by support reaction, psi (MPa)
= 48,000 psi (331 MPa) maximum
K = saddle coefficient, see Equation 4-12
w = unit load per linear foot, lb/ft (N/m)
L = span length, ft (m)
t_n = design wall thickness of pipe, in. (mm), see Table 4-12
D = pipe OD, in. (mm), see Table 4-4

For aboveground applications,

t_n = minimum manufacturing thickness of pipe
= nominal pipe wall thickness minus casting allowance

For underground applications,

t_n = net pipe wall thickness
= nominal pipe wall thickness minus casting allowance minus 0.08 in. (2.0 mm) service allowance

Recent research involving ductile-iron pipe has established that the following function

$$K = 0.03 - 0.00017(\beta - 90°) \qquad \text{(Eq 4-13)}$$

provides excellent correlation between the ring stresses predicted by Equation 4-13 and the actual stress as measured when β is between 90° and 120° (DIPRA 2016).

Table 4-12 Pipe plus water weight ($W_p + W_w$) and design wall thickness t_n

Nominal Pipe Size, in.	Pressure Class	$W_p + W_w$, lb/lin ft	t_n, in. Aboveground Applications	t_n, in. Underground Applications
3	350	14	0.20	0.12
4	350	18	0.20	0.12
6	350	31	0.20	0.12
8	350	48	0.20	0.12
10	350	68	0.20	0.12
12	350	92	0.22	0.14
14	250	119	0.21	0.13
	300	122	0.23	0.15
	350	123	0.24	0.16
16	250	151	0.23	0.15
	300	154	0.25	0.17
	350	157	0.27	0.19
18	250	185	0.24	0.16
	300	190	0.27	0.19
	350	193	0.29	0.21
20	250	225	0.26	0.18
	300	230	0.29	0.21
	350	233	0.31	0.23
24	200	306	0.26	0.18
	250	314	0.30	0.22
	300	320	0.33	0.25
	350	326	0.36	0.28
30	150	453	0.27	0.19
	200	462	0.31	0.23
	250	473	0.35	0.27
	300	481	0.38	0.30
	350	491	0.42	0.34
36	150	637	0.31	0.23
	200	650	0.35	0.27
	250	665	0.40	0.32
	300	677	0.44	0.36
	350	693	0.49	0.41
42	150	848	0.34	0.26
	200	869	0.40	0.32
	250	887	0.45	0.37
	300	905	0.50	0.42
	350	927	0.56	0.48

(continued)

Table 4-12 Pipe plus water weight ($W_p + W_w$) and design wall thickness t_n (*Continued*)

Nominal Pipe Size, in.	Pressure Class	$W_p + W_w$, lb/lin ft	t_n, in. Aboveground Applications	t_n, in. Underground Applications
48	150	1,099	0.38	0.30
	200	1,124	0.44	0.36
	250	1,148	0.50	0.42
	300	1,173	0.56	0.48
	350	1,197	0.62	0.54
54	150	1,403	0.42	0.34
	200	1,436	0.49	0.41
	250	1,468	0.56	0.48
	300	1,501	0.63	0.55
	350	1,533	.070	0.62
60	150	1,608	0.45	0.37
	200	1,643	0.52	0.44
	250	1,678	0.59	0.51
	300	1,717	0.67	0.59
	350	1,752	0.74	0.66
64	150	1,817	0.47	0.39
	200	1,860	0.55	0.47
	250	1,902	0.63	0.55
	300	1,945	0.71	0.63
	350	1982	0.78	0.70

NOTE: To convert inches (in.) to millimeters (mm), multiply by 25.4; to convert pounds per linear foot (lb/lin ft) to newtons per meter (N/m), multiply by 14.59.
Approximate pipe weight based on push-on joint cement–mortar-lined pipe. Weight of water based on actual pipe inside diameter.

It is recommended that the maximum calculated localized stress be limited to 48,000 psi (331.0 MPa). This value is equal to the yield strength in bending for ductile iron (72,000 psi) (496.4 MPa) divided by a safety factor of 1.5. It is the same limiting value of bending stress employed in ANSI/AWWA C150/A21.50.

4.5.7.2 Hoop stress caused by internal pressure. The net thickness required for internal pressure can be determined by using Equation 4-1.

4.5.7.3 Flexural stress at center of span. With one support per length of pipe positioned immediately behind the bells, each span can conservatively be treated as a simply supported beam. The joints being slightly offset from the supports causes some distribution of the simple beam moment and stress from the center of the span to the support. This makes the simple beam approach conservative. Equation 4-15 represents the flexural stress at the center of the span of a uniformly loaded, simply supported beam:

$$f_b = \frac{15.28DwL^2}{D^4 - d^4} \qquad \text{(Eq 4-14)}$$

Where:

f_b = allowable flexural stress, psi
= 48,000 psi maximum

D = pipe OD, in.
w = unit load per linear foot, lb/lin ft
L = length of span, ft
d = $D - 2t_n$, in.

In metric units,

$$f_b = \frac{1,270DwL^2}{D^4 - d^4} \tag{Eq 4-15}$$

Where:
f_b = allowable flexural stress, MPa 331
= MPa maximum
D = pipe OD, mm
w = unit load per linear meter, N/m
L = length of span, m
d = $D - 2t_n$, mm

4.5.7.4 Beam deflection at center of span. Computations for beam deflection are also based on the simply supported beam concept. This is likewise conservative because of the reality of offset joints. To prevent damage to the cement–mortar lining, the maximum allowable deflection at midspan is limited to the following:

$$y_r = \frac{L}{10} \tag{Eq 4-16}$$

Where:
y_r = maximum allowable deflection at center of span, in.
L = length of span, ft

In metric units,

$$Y_r = 8.33L \tag{Eq 4-17}$$

Where:
y_r = maximum allowable deflection at center of span, mm
L = length of span, m

Less deflection may be desired. The deflection of the beam may be significant for aesthetic reasons in aboveground installations or possibly for hydraulic reasons in gravity-flow pipelines. Limitations on the deflection, if any, should be determined by the designer as appropriate to a specific installation.

The beam deflection at center span for a uniformly loaded, simply supported beam can be calculated using Equation 4-18:

$$y = \frac{458.4wL^4}{E\left(D^4 - d^4\right)} \tag{Eq 4-18}$$

Where:
y = deflection at center span, in.
w = unit load per linear foot, lb/lin ft
L = length of span, ft
E = modulus of elasticity, 24 × 106 psi
D = pipe OD, in.

d = $D - 2t_n$, in.

In metric units,

$$y = \frac{265.2 \times 10^6 wL^4}{E\left(D^4 - d^4\right)} \qquad \text{(Eq 4-19)}$$

Where:

y = deflection at center span, mm
w = unit load per linear meter, N/m
L = length of span, m
E = modulus of elasticity, 165.5×10^3 MPa
D = pipe OD, mm
d = $D - 2t_n$, mm

4.5.8 Aboveground Installations

For aboveground installations with one support per length of pipe—that is, a span length of 18 or 20 ft (5.5 or 6.1 m)—the minimum pressure class of ductile-iron pipe in all sizes is more than adequate to support the weight of the pipe and water it contains when analyzed in accordance with the suggestions of this procedure.

Other design considerations for pipes supported aboveground may include the carrying capacity of the supports and saddles themselves, the strength of the structure from which a pipe may be suspended, and unusual or additional loads not in the scope of this manual. Bridge crossings normally incorporate a pipe-on-supports design with some special considerations (DIPRA 2016). Such loading may include seismic loads and loads resulting from frequency or resonance of vibrations, wind, water current, and other special design considerations.

It is also necessary to ensure a minimum of lateral and vertical stability at the supports for aboveground piping. Deflected pipe joints in pressure pipelines can result in unbalanced thrust forces of hydrostatic or hydrodynamic origin. If these joints are not laterally and vertically restrained, unbalanced forces may result in additional joint deflection and possible failure of the pipeline.

Thermal expansion of ductile-iron pipelines supported aboveground is not usually of concern in correctly designed and installed systems because of the nature of the push-on or mechanical joint. A 100°F (55.6°C) change in temperature results in expansion or contraction of a 20-ft (6.1-m) length of ductile-iron pipe of approximately 0.15 in. (3.81 mm). This is easily accommodated with correctly installed pipe and joints. Occasionally, where structures from which ductile-iron pipe is to be suspended are expected to have significantly different behavior from the pipeline, special considerations for expansion, contraction, and supports may be necessary. For reference, the following are coefficients of thermal expansion for various materials:

Ductile iron:	6.2×10^{-6} in./in. °F (11.2×10^{-6} mm/mm °C)
Steel:	6.5×10^{-6} in./in. °F (11.7×10^{-6} mm/mm °C)
Concrete:	7.0×10^{-6} in./in. °F (12.6×10^{-6} mm/mm °C)

4.5.9 Design Procedure

Step 1. Select the length of span (18 ft or 20 ft) (5.5 m or 6.1 m), saddle angle (90°–120°), and pipe diameter.

Step 2. Determine the unit load per linear foot, w, based on the minimum pressure class pipe.

a. For aboveground installations, $w = (W_p + W_w)$

b. For underground installations,

- no truck loads

$$w = (W_p + W_w) + 12DP_e$$

- truck loads included

$$w = (W_p + W_w) + 12D\,(P_e + P_t)$$

NOTE: For D, see Table 4-4.
For $(W_p + W_w)$, see Table 4-12.
For P_e and P_t, see Table 4-13.

Table 4-13 Earth loads P_e and truck loads P_t (psi)

Depth of Cover,		3-in. Pipe	4-in. Pipe	6-in. Pipe	8-in. Pipe	10-in. Pipe	12-in. Pipe
ft	P_e	P_t	P_t	P_t	P_t	P_t	P_t
2.5	2.1	9.9	9.9	9.9	9.8	9.7	9.6
3	2.5	7.4	7.4	7.3	7.3	7.2	7.2
4	3.3	4.4	4.5	4.4	4.4	4.4	4.4
5	4.2	3.0	3.0	3.0	3.0	2.9	2.9
6	5.0	2.1	2.1	2.1	2.1	2.1	2.1
7	5.8	1.6	1.6	1.6	1.6	1.6	1.6
8	6.7	1.2	1.2	1.2	1.2	1.2	1.2
9	7.5	1.0	1.0	1.0	1.0	1.0	1.0
10	8.3	0.8	0.8	0.8	0.8	0.8	0.8
12	10.0	0.6	0.6	0.6	0.6	0.5	0.5
14	11.7	0.4	0.4	0.4	0.4	0.4	0.4
16	13.3	0.3	0.3	0.3	0.3	0.3	0.3
20	16.7	0.2	0.2	0.2	0.2	0.2	0.2
24	20.0	0.2	0.1	0.1	0.1	0.1	0.1
28	23.3	0.1	0.1	0.1	0.1	0.1	0.1
32	26.7	0.1	0.1	0.1	0.1	0.1	0.1
Depth of Cover,		14-in. Pipe	16-in. Pipe	18-in. Pipe	20-in. Pipe	24-in. Pipe	30-in. Pipe
ft	P_e	P_t	P_t	P_t	P_t	P_t	P_t
2.5	2.1	8.7	8.2	7.8	7.5	7.1	6.7
3	2.5	6.6	6.2	5.9	5.7	5.4	5.2
4	3.3	4.4	4.1	3.9	3.9	3.6	3.5
5	4.2	2.9	2.8	2.6	2.6	2.4	2.4
6	5.0	2.1	2.0	1.9	1.9	1.7	1.7
7	5.8	1.6	1.5	1.4	1.4	1.3	1.3
8	6.7	1.2	1.2	1.2	1.1	1.1	1.1
9	7.5	1.0	1.0	1.0	0.9	0.9	0.9
10	8.3	0.8	0.8	0.8	0.7	0.7	0.7
12	10.0	0.5	0.5	0.5	0.5	0.5	0.5
14	11.7	0.4	0.4	0.4	0.4	0.4	0.4

(continued)

Table 4-13 Earth loads P_e and truck loads P_t (psi) (*Continued*)

Depth of Cover,		14-in. Pipe	16-in. Pipe	18-in. Pipe	20-in. Pipe	24-in. Pipe	30-in. Pipe
ft	P_e	P_t	P_t	P_t	P_t	P_t	P_t
16	13.3	0.3	0.3	0.3	0.3	0.3	0.3
20	16.7	0.2	0.2	0.2	0.2	0.2	0.2
24	20.0	0.1	0.1	0.1	0.1	0.1	0.1
28	23.3	0.1	0.1	0.1	0.1	0.1	0.1
32	26.7	0.1	0.1	0.1	0.1	0.1	0.1
Depth of Cover,		36-in. Pipe	42-in. Pipe	48-in. Pipe	54-in. Pipe	60-in. Pipe	64-in. Pipe
ft	P_e	P_t	P_t	P_t	P_t	P_t	P_t
2.5	2.1	6.2	5.8	5.4	5.0	4.8	4.5
3	2.5	4.9	4.6	4.4	4.1	3.9	3.8
4	3.3	3.4	3.3	3.1	3.0	2.9	2.8
5	4.2	2.3	2.3	2.2	2.1	2.1	2.1
6	5.0	1.7	1.7	1.6	1.6	1.6	1.5
7	5.8	1.3	1.3	1.2	1.2	1.2	1.2
8	6.7	1.1	1.0	1.0	1.0	1.0	1.0
9	7.5	0.8	0.8	0.8	0.8	0.8	0.8
10	8.3	0.7	0.7	0.7	0.7	0.7	0.7
12	10.0	0.5	0.5	0.5	0.5	0.5	0.5
14	11.7	0.4	0.4	0.4	0.4	0.4	0.4
16	13.3	0.3	0.3	0.3	0.3	0.3	0.3
20	16.7	0.2	0.2	0.2	0.2	0.2	0.2
24	20.0	0.1	0.1	0.1	0.1	0.1	0.1
28	23.3	0.1	0.1	0.1	0.1	0.1	0.1
32	26.7	0.1	0.1	0.1	0.1	0.1	0.1

NOTE: To convert feet (ft) to meters (m), multiply by 0.3048; to convert pound per square inch (psi) to kilopascals (kPa), multiply by 6.895.

Step 3. Determine whether the design thickness, t_n, corresponding to the pipe class selected in Step 2 and found in Table 4-12, results in an acceptable localized stress ≤ 48,000 psi (331.0 MPa).

a. Calculate the saddle coefficient, K, using Equation 4-14.

b. Calculate f_r using Equation 4-13. If f_r exceeds 48,000 psi (331.0 MPa), increase t_n to that corresponding to the next higher pressure class and recalculate starting with Step 2. Repeat until the resulting f_r is less than or equal to 48,000 psi (331.0 MPa).

Step 4. Determine the pipe pressure class required as a result of internal pressure.

a. Calculate the net thickness, t, required for hoop stress caused by internal pressure using Equation 4-1.

b. Determine the total calculated thickness, T, caused by internal pressure.

For aboveground applications,

$$T = t + \text{casting allowance}$$

For underground applications,

$$T = t + \text{casting allowance} + 0.08$$

c. Using Table 4-4, select the standard pressure-class thickness. When the total calculated thickness is between two standard thicknesses, select the larger of the two. (NOTE: For aboveground application, the standard pressure class selected from Table 4-3 may be less than the design working pressure because the 0.08 in. [2.0 mm] service allowance is not required.)

Step 5. Calculate the flexural stress, f_b, at midspan using Equation 4-15 and the largest pressure class pipe required in Step 3 or 4 along with its corresponding t_n and w values. If f_b exceeds 48,000 psi (331.0 MPa), increase t_n to that corresponding to the next higher pressure class and recalculate f_b using the new class thickness and corresponding t_n and w values. Repeat until the resulting f_b is less than or equal to 48,000 psi (331.0 MPa).

Step 6. Check deflection at midspan.

a. Calculate the deflection at midspan y using Equation 4-17 and the largest pressure class pipe required in Step 3, 4, or 5 along with its corresponding t_n and w values.

b. Calculate the maximum allowable deflection at midspan y_r using Equation 4-16. (NOTE: Less deflection may be desired.) If the deflection y is greater than the deflection y_r, increase t_n to that corresponding to the next higher pressure class and recalculate y using the new class thickness and corresponding t_n and w values. Repeat until the resulting y is less than or equal to y_r.

Step 7. Choose the largest pressure class corresponding to the largest t_n required in Step 3, 4, 5, or 6 and calculate the minimum saddle width using Equation 4-12.

4.5.10 Design Example

Find the required pipe class for 24-in. (610-mm) ductile-iron pipe under 3 ft (0.91 m) of earth cover installed on piers spaced 20 ft (6.1 m) on center with 120° saddles and an operating pressure of 150 psi (1,034 kPa). Assume no truck load.

Referring to Step 1 of the design procedure,

20-ft span *(L)*
120°saddle angle (β)
24-in. diameter ductile-iron pipe

Referring to Step 2 of the design procedure,

$$w = (W_p + W_w) + 12\,DP_e$$
$$(W_p + W_w) = 306 \text{ lb/ft (Table 4-12)}$$
$$D = 25.8 \text{ in. (Table 4-4)}$$
$$P_e = 2.5 \text{ psi (Table 4-13)}$$
$$w = 306 + 12(25.8)(2.5) = 1{,}080 \text{ lb/ft}$$

Referring to Step 3 of the design procedure:

$$K = 0.03 - 0.00017\,(\beta - 90^\circ)$$
$$K = 0.03 - 0.00017\,(120^\circ - 90^\circ) = 0.025$$

$$f_r = K\left(\frac{wL}{t_n^{\,2}}\right)\text{In}\left(\frac{D}{2t_n}\right)$$

$$t_n = 0.18 \text{ (Table 4-12)}$$

$$f_r = 0.025\left[\frac{(1{,}080)(20)}{0.18^2}\right]\text{In}\left[\frac{25.8}{2(0.18)}\right] = 71{,}200 \text{ psi}$$

71,200 psi > 48,000 psi ∴ try next higher class (Class 250)

For Class 250 (from Table 4-12),

t_n = 0.22 in.

w = 314 + 12(25.8)(2.5) = 1,088 lb/ft

$$f_r = 0.025\left[\frac{(1,088)(20)}{(0.22^2)}\right] \text{ In }\left[\frac{25.8}{2(0.22)}\right] = 45,761 \text{ psi}$$

45,761 psi < 48,000 psi ∴ OK

Referring to Step 4 of the design procedure:

$$t = \frac{P_i D}{2S}$$

P_i = $2(P_w + P_s)$ = 2(150 + 100) = 500 psi

$$t = \frac{500(25.8)}{2(42,000)} = 0.15 \text{ in.}$$

Total calculated thickness $T = t$ + casting allowance + 0.08

Casting allowance = 0.07 (Table 4-3)

T = 0.15 + 0.07 + 0.08 = 0.30 in.

From Table 4-4, pressure class 200 is adequate for internal pressure design. Referring to Step 5 of the design procedure,

using pressure class 250 determined in Step 3,

$$f_b = \frac{15.28 DwL^2}{D^4 - d^4}$$

d = $D - 2t_n$ – 25.8 – 2(0.22) = 25.36 in.

$$f_b = \frac{(15.28)(25.8)(1,088)(20)^2}{25.8^4 - 25.36^4} = 5,824 \text{ psi}$$

5,284 psi < 48,000 psi ∴ OK

Referring to Step 6 of the design procedure,
using pressure class 250 determined in Step 3,

$$y = \frac{458.4wL^4}{E\left(D^4 - d^4\right)}$$

$$y = \frac{458.4(1,088)(20)^4}{\left(24 \times 10^6\right)\left(25.8^4 - 25.36^4\right)} = 11 \text{ in.}$$

$$y_r = \frac{L}{10} = \frac{20}{10} = 2 \text{ in.}$$

0.11 in. < 2 in. ∴ OK

Referring to Step 7 of the design procedure,
using pressure class 250 determined in Step 3,

$$b = \sqrt{2Dt_e} = \sqrt{2(25.8)(0.37)} = 4.37 \text{ in.}$$

Therefore, use pressure class 250 pipe with minimum saddle width of 4.37 in. (111.0 mm).

4.5.11 Ductile-Iron Pipe Supported on Rollers

Unlike cradle or saddle supports that follow the contour of the pipe and have a support/pipe contact area dependent on the diameter of the pipe, saddle angle, and width of the support, pipe supported on rollers normally has only two points of contact locations at each roller support. Due to smaller contact areas, this point loading results in much higher localized stress concentrations that are dependent on pipe size, pipe wall thickness, distance between rollers, location of rollers along the pipe length, radius of rollers, loading, and so on. Formulas addressing these high-stress concentrations for cylindrical shells and pipes have been published in technical literature. However, the stress analysis is difficult, and the results are rendered uncertain by doubtful boundary conditions; therefore, the ultimate responsibility for this design rests with the design engineer.

Even so, there have been thousands of successful ductile-iron pipe bridge crossing installations utilizing roller supports. While one roller support per length of pipe (normally located fairly close behind bells) has been shown in practice to be adequate for small-diameter ductile-iron pipelines (capitalizing on the innate strength and toughness of this pipe), multiple rollers per length may be appropriate for larger sizes. However, rollers should generally not be placed under spigots closely adjacent to bells because of higher developed stresses and possible undesirable effects on joints.

As with cradled supports, it is of utmost importance that sufficient vertical and lateral stability be provided at roller support behind bells, for both restrained and unrestrained joint pipe systems. A double roller "guide" or bracketed design (roller below and above the pipe) with lateral bracing is one method of accomplishing this.

4.6 SPECIAL USE CONSIDERATIONS

Special use considerations should be given to such situations as expansive soils, earthquakes, sinkholes, and other geologically hazardous conditions. Field studies have shown that certain clay soils can exhibit high swell pressures that can cause significant beam loading on underground pipe. It has been estimated that 50,000 earthquakes occur annually throughout the world that are of sufficient magnitude to be felt or noticed without the aid of instruments. Ductile-iron pipe's inherent strength, flexibility of standard joints, and the availability of specially designed joints make it a good choice for such applications.

REFERENCES

American Association of State Highway and Transportation Officials. 2022. AASHTO T-99, Standard Methods of Test for the Moisture-Density Relations of Soils Using a 5.5 lb (2.5 kg) Rammer and a 12-in. (305-mm) Drop. Washington: AASHTO.

American National Standards Institute/American Water Works Association. 2021. C150-21 Thickness Design of Ductile-Iron Pipe. Denver: AWWA.

ANSI/AWWA. 2022. C606. Grooved and Shouldered Joints. Denver: AWWA.

American Railway Engineering and Maintenance-of-Way Association (AREMA). 2019. Manual for Railway Engineering. Landover, MD: AREMA.

American Society of Civil Engineers. 1969. Manual No. 37, Design and Construction of Sanitary and Storm Sewers. 1st ed. New York: ASCE.

ASTM. 2018. Standard Specification for Ductile Iron Culvert Pipe. ASTM A716. West Conshohocken, PA: ASTM.

ASTM. 2022. Standard Specification for Ductile Iron Gravity Sewer Pipe. 2022. ASTM A746. West Conshohocken, PA: ASTM.

Bonds, R.W. 2017. Design of Ductile Iron Pipe on Supports. Ductile Iron Pipe News. Birmingham, AL: Ductile Iron Pipe Research Association.

Bonds, R.W. 1993. Thickness Design for Ductile-Iron Pipe Under Railroads. Birmingham, AL: Ductile Iron Pipe Research Association.

Cast Iron Pipe Research Association. 1964. Ductile-Iron Pipe Test Report. Birmingham, AL: CIPRA.

Ductile Iron Pipe Research Association. 2016. *Bridge Crossings with Ductile Iron Pipe.* Birmingham, AL: DIPRA.

Evces, C.R., and J.M. O'Brien. 1984. Stresses in Saddle-Supported Ductile Iron-Pipe. *Journal AWWA*, 76:11:54–59.

Kennedy, H. 1976. "The New Ductile Iron Pipe Standards." *Journal AWWA*, 68:11: 622–626.

Keyes D. B., H.F. Johnstone, R.G. Sturm, W.M. Wilson, E. D. Olson, L. Byman, R.L. Pigford, and J.H. Chapin. 1941. *Tests on Cylindrical Shells: A Report of an Investigation Conducted by the Engineering Experiment Station, University of Illinois in Cooperation with the Chicago Bridge and Iron Co.* Urbana, IL: University of Illinois.

Sears, E.C. 1964. "Ductile-Iron Pipe Design." *Journal AWWA*, 56:1:4–22.

Spangler, M.G. 1982. *Soil Engineering*. 4th ed. New York: Harper & Row Publishers.

Stroud, T.F. 1991. Design of Ductile Iron Pipe. Ductile Iron Pipe News. Birmingham, AL: Ductile Iron Pipe Research Association.

The Asphalt Institute. *The Asphalt Handbook*. 2007. 7th ed. College Park, MD: The Asphalt Institute.

Young, W. and R.G. Budynas. 2001. *Roark's Formulas for Stress and Strain*. 7th ed. New York: McGraw-Hill.

Chapter 5

Pipe Joints

5.1 INTRODUCTION

Several types of rubber-gasket joints are available with ductile-iron pipe and fittings to accommodate a wide variety of applications. The most prevalent are the push-on joint, mechanical joint, and flanged joint.

Restrained joints are available to accommodate the thrust forces acting on a pipeline. Variations of restrained joints are available to accommodate special demands on piping in seismic active areas. Other joints available include the ball-and-socket joint, grooved and shouldered joints, and special bolted joints used with sleeves and couplings.

5.2 JOINTS

5.2.1 Push-on Joint

The push-on joint (see Figure 5-1a) uses a single rubber gasket positioned in an annular recess in the socket of the pipe or fitting bell. The joint is assembled by pushing a beveled plain end past the gasket into the socket. The plain end compresses the gasket radially to form a positive seal that is effective against high internal pressure, vacuum, and external water head. The gasket and the annular recess in the bell are designed and shaped so that the gasket is secured against displacement.

Push-on joints are made to conform with ANSI/AWWA C111/A21.11. This standard sets forth joint performance requirements, gasket and lubricant properties, and marking criteria.

There are several designs of push-on joints used with ductile-iron pipe and fittings. These joints use different bell and gasket configurations. Consequently, the gaskets are not interchangeable. However, the outside diameters of all ductile-iron pipe of the same size are manufactured to the same dimensions and tolerances; therefore, the plain ends of ductile-iron pipe are interchangeable. The two principal push-on joints used with ductile-iron pipe produced in the United States are illustrated in Appendix A. A third push-on joint design used in certain fittings is also provided.

Push-on joints must be installed in accordance with the manufacturer's assembly instructions and care exercised to make sure that the correct gasket is used. The plain end of field-cut pipe must be beveled and ground smooth to prevent damage to the gasket. General assembly instructions and information on joint deflection can be found in ANSI/AWWA Standard C600 and in Chapter 11 of this manual.

Push-on joints are designed for conveying water and other liquids and may also be used for air service to 100 psi (689 kPa). Air piping applications sometimes involve high temperatures that may require special care in selecting appropriate joints, gasket materials, and linings. Ductile-iron pipe and its push-on joints are also effective under negative pressure conditions.

5.2.2 Mechanical Joint

A mechanical joint, as illustrated in Figure 5-1b, is a bolted nonrestrained joint of the stuffing-box type. The pipe or fittings bell provides an exterior flange having bolt holes or slots and an internal socket with annular recesses for both the gasket and the mating plain end. Accessories consist of the gasket, a follower gland, and tee-head bolts with hexagonal nuts.

Detailed assembly instructions for the mechanical joint are outlined in AWWA Standard C600. Additional information on proper assembly of this joint can be found in Appendix A of ANSI/AWWA C111/A21.11 and in Chapter 11 of this manual.

The mechanical joint is manufactured in accordance with ANSI/AWWA C111/A21.11. This standard sets forth dimensions and tolerances, performance requirements, gaskets, glands, and tee-head bolts. Recommended bolt torques are shown in Appendix A of the standard.

Because the mechanical-joint bell socket, plain end, gasket, and gland are standardized dimensionally, this joint is interchangeable between pipe and fittings regardless of manufacturer. Different bolt lengths may be used in some sizes for fittings manufactured in accordance with ANSI/AWWA C110/A21.10 and ANSI/AWWA C153/A21.53 standards. Proprietary gasket designs are available for the mechanical joint. The mechanical joints have the same pressure rating as the corresponding pipe or fittings.

5.2.3 Flanged Joint

The flanged-joint (see Figure 5-1c) for ductile-iron pipe is a bolted joint of the type illustrated in Figure 5-2 and depicted in Figure 5-3. A flat or special gasket is inserted between the mating flanges and is compressed by tightening the bolts holding the flanges together.

Pipe with threaded flanges is made in accordance with ANSI/AWWA C115/A21.15. The flanges are individually fitted on threaded pipe, machine tightened, and faced by the manufacturer. Flanges and threaded pipe from different sources are not normally interchangeable in the field. Because it is difficult to tighten large-diameter loose flanges in the field, it is recommended that "hand-tight" flanges be limited to 16-in. (406-mm) and smaller sizes. Ductile-iron and gray-iron fittings with integral flanges are made in accordance with ANSI/AWWA C110/A21.10 and C153/A21.53.

ANSI/AWWA C110/A21.10, C153/A21.53, and C115/A21.15 set forth flange dimensions, material requirements, facing, flange, and bolt hole alignment. AWWA has not published a standard covering flanged joint gaskets, performance requirements, and bolt torques because of the variety of gasket configurations and materials used with the joint.

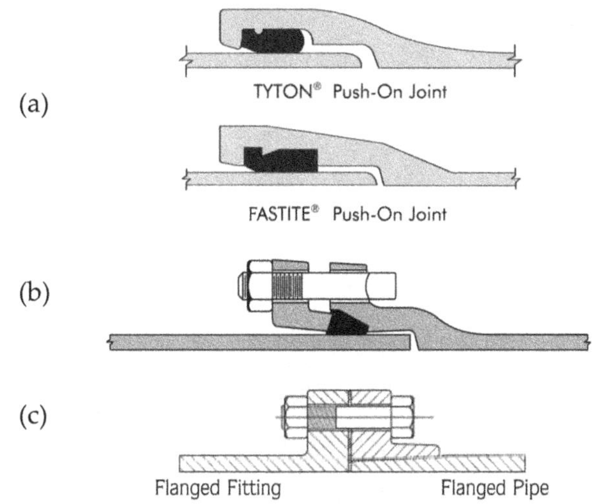

Source: Ductile Iron Pipe Research Association

Figure 5-1 Push-on (a), mechanical (b), and flanged (c) joints

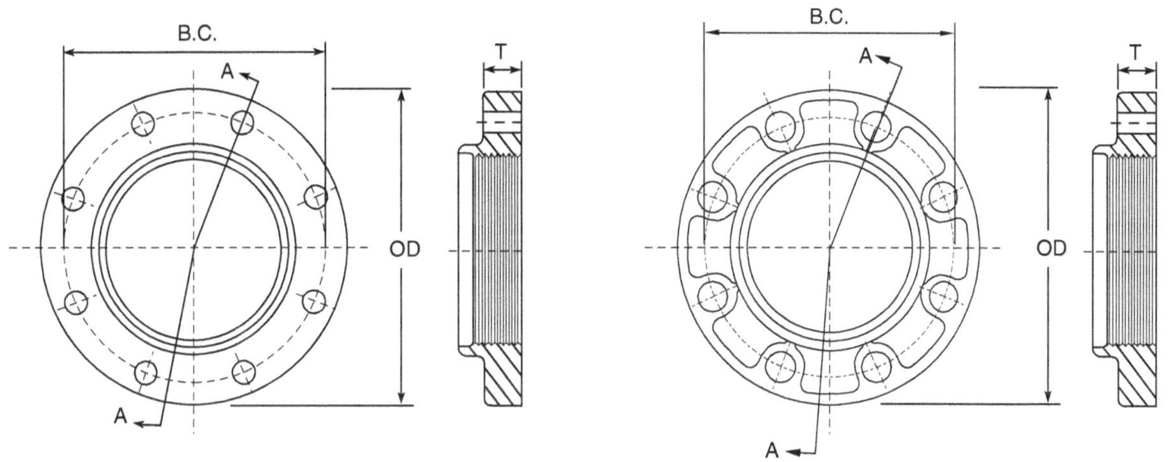

Figure 5-2 Flanged joint details

Information on bolts, gaskets, and installation can be found in Appendix A of ANSI/AWWA C110/A21.10, C153/A21.53, and C115/A21.15. Ring gaskets, ⅛ in. thick, are recommended for 14-in. (356-mm) and larger sizes if flat gaskets are used. Specially designed gaskets, either ring or full face, employing one or more annular rings molded into the gasket to improve joint performance, are also available for all pipe sizes. By using these or other special gaskets, it may be possible to obtain a pressure rating greater than 350 psi (2,413 kPa) in the 24-in. and smaller sizes and greater than 250 psi (1,720 kPa) in the 30-in. and larger sizes.

Underground use of the flanged joint is generally not recommended because of the rigidity of the joint.

The flanges provided in accordance with ANSI/AWWA C110/A21.10, C153/A21.53, and C115/A21.15 are typically rated for water service of 250 psi (1,720 kPa) working pressure. Attention is directed to an apparent conflict between these standards and ANSI/ASME B16.1 (2020), with regard to pressure ratings for flanged pipe and fittings. ANSI/

Source: Ductile Iron Pipe Research Association

Figure 5-3 Flanged ductile-iron pipe and fittings in a water treatment plant

ASME B16.1 flanges carry a pressure–temperature rating (for service at ambient and elevated temperatures), whereas the ANSI/AWWA C110/A21.10, C153/A21.53, and C115/A21.15 flange ratings are for water at ambient temperature. The Class 125 ANSI/ASME B16.1 flange, although having facing and drilling identical to the ANSI/AWWA C110/A21.10, C153/A21.53, and C115/A21.15 flanges, is rated for only 150 to 200 psi (1,030 to 1,380 kPa) service, depending on the flange size, the class or grade of iron, and fluid temperature. ANSI/ASME Standard B16.1 also contains the criteria for a Class 250 flange, which is heavier, has a larger bolt circle, and uses larger bolts than Class 125 B16.1 flanges and the flanges defined in ANSI/AWWA C110/A21.10, C153/A21.53, and C115/A21.15. This Class 250 B16.1 flange will not match either the Class 125 B16.1 flange or the ANSI/AWWA C110/A21.10, C153/A21.53, and C115/A21.15 flanges.

Products may be available that meet the intent of A21 standards, though they may vary in dimensions, characteristics, material properties, or other features. The purchaser and specifier should become knowledgeable of the details of the standards and carefully evaluate the effect deviations from the standards may have on their project. One example of such a deviation is that some manufacturers may supply fittings with thinner flanges than indicated in the standard. While these fittings may have adequate strength for an application, a standard bolt kit may have insufficient threads on a bolt to adequately compress the gasket to prevent leakage.

5.2.4 Special Joints

A variety of special joints have been developed for specific applications involving ductile-iron pipe, fittings, and specialty products. Detailed information on designs, pressure ratings, and assembly procedures can be obtained by contacting the manufacturers of these particular joints; however, most of the information is proprietary. AWWA's *Sourcebook* (2024) provides a listing of manufacturers that supply joints for ductile-iron pipe and accessories. Several of the more commonly used special joints are discussed in the following sections.

5.2.4.1 Ball and socket joint. The ball and socket type joint, illustrated in Appendix A, is available for installations requiring more joint deflection than can be provided by standard push-on or mechanical joints. This joint type uses an enlarged bell and a ball on the plain end of the pipe to provide up to 15° joint deflection. It is commonly used for subaqueous installations (including stream and lake crossings, intakes, outfalls, and so on).

5.2.4.2 Grooved and shouldered joints. The grooved joint illustrated in Appendix A uses a bolted, segmental, clamp-type mechanical coupling having a housing that encloses a U-shaped rubber gasket. The housing locks the pipe ends together and compresses the gasket against the outside of the pipe ends. The ends of the pipe are machine grooved to accept the housing. Grooved joints may be provided as either rigid or flexible joints. The shouldered joint is similar except that the pipe ends are shouldered instead of grooved. These joints are covered in ANSI/AWWA C606 (2022), which defines material requirements, general design criteria, pipe grooving details, and coupling test requirements.

5.2.4.3 Miscellaneous joints. A variety of modified mechanical joints and similar stuffing-box type joints have been developed, and the joints are used with related products, such as tapping sleeves, repair sleeves, couplings, and connectors (see Figure 5-4).

5.2.5 Restrained Joints

The push-on and mechanical joints used with ductile-iron pipe and fittings do not provide significant restraint against longitudinal separation when the joint is subjected to axial forces. Tests have shown that the friction between the gasket and the plain end is unpredictable and that these joints offer no appreciable longitudinal restraint for design purposes.

Restrained push-on and mechanical joints are available, which provide the longitudinal restraint required to accommodate axial thrust forces acting on a pipeline.

- Gaskets with locking segments: Locking segments or wedges include gripping teeth to engage the pipe spigot to prevent joint separation. These gaskets are used in standard push-on or mechanical joint sockets and can be used with factory- or field-cut spigots.
- Extended pipe bell designs: An extension to the bell socket is included to provide an annular recess to house a restraint mechanism. These designs may include locking segments that provide a mechanical lock with a weld bead on the spigot. A variation where restraint of field cuts is needed is a toothed mechanism that fits in the recess to grip the pipe spigot.
- External integral restraint systems: A locking mechanism is located outside of the bell socket and is attached by bolts to the bell. Restraint of the spigot may include engagement with a weld bead or ring attached to the spigot, or engagement may be achieved by gripping elements forced down onto the spigot surface.

Product information, pressure rating, and assembly instructions can be obtained from the manufacturer of the restrained joint. Many of these joints are proprietary and are not interchangeable between manufacturers. Additional discussion of restrained joints

Source: Ductile Iron Pipe Research Association

Figure 5-4 Installation of mechanical-joint-type restrained joints

and notes regarding their use can be found in Chapter 8 of this manual. Some of the many restrained joints available are illustrated in Appendix A.

5.2.6 Seismic Joints

Restrained push-on and mechanical joints are available, which provide for additional axial movement, deflection, and resistance against longitudinal separation. Sockets are extended in length to provide increased axial movement capabilities. The combination of axial movements and joint deflections allows displacement of the pipeline in response to soil movements, increasing the probability of maintaining pipeline integrity during seismic events. The magnitude of the resistance against axial separation may be substantially more than that required to resist the forces generated by internal hydraulic pressures. The pipeline designer may choose to incorporate a combination of special joints and fittings to achieve the desired flexibility in pipeline axial movements and deflections.

5.3 GASKETS

Gasket requirements for push-on joints and the mechanical joint are set forth in ANSI/AWWA C111/A21.11. Dimensions and tolerances, rubber properties, and marking are specified for mechanical joint gaskets. While push-on joint gasket dimensions are provided in accordance with the manufacturer's standard design, C111/A21.11 specifies rubber properties and marking for these gaskets. Gaskets for other types of joints are made in accordance with the manufacturer's design. Information on such gaskets can be obtained from the manufacturer.

The gaskets normally provided with push-on and mechanical joints are produced from a styrene butadiene rubber (SBR), which has excellent properties for its intended applications. Other synthetic rubber materials are available for special service conditions, including polychloroprene (CR), ethylene propylene diene monomer (EPDM), nitrile (NBR), and fluorocarbon (FKM). Natural rubber gaskets are excluded by ANSI/AWWA C111/A21.11 and are not available for ductile-iron pipe.

Table 5-1 lists elastomers that are generally available for gaskets and also maximum recommended service temperatures. This information is provided as a general guide for gasket use. Manufacturers usually do not warrant gaskets to meet special uses.

Table 5-1 Gasket materials used for ductile-iron pipe in water and sewage service

	Maximum Service Temperature, °F *†			
	Water and Sewer	Air§**		
Description	Push-On and Mechanical Joint Gaskets	Push-On Joint Gaskets	Mechanical Joint Gaskets	Uses‡
SBR	150	150	125	Common: Drinking water, seawater, sanitary sewage, reclaimed water, raw water, stormwater
EPDM	212	200	150	Common: Alcohols, dilute acids, dilute alkalis, ketones (methyl ethyl ketone, acetone), vegetable oil. Other acceptable services: Drinking water, seawater, sanitary sewage, reclaimed water, raw water, stormwater
NBR (acrylonitrile butadiene)	150	150	125	Common: Hydrocarbons, fats, oils, greases, chemicals, oils and fluids, refined petroleum. Other acceptable services: Drinking water, sanitary sewage, reclaimed water, raw water, stormwater
CR (Neoprene) (polychloroprene)	200	180	150	Common: Greasy waste. Other acceptable services: Seawater, sanitary sewage, reclaimed water, raw water, stormwater
FKM (Viton; Fluorel)†† (fluorocarbon)	212	300	300	Common: Aromatic hydrocarbons and fuels, acids, vegetable oils, petroleum products, chlorinated hydrocarbons, most chemicals and solvents. Other acceptable services: Drinking water, reclaimed water, raw water, stormwater

* Maximum service temperatures listed are intended as general guidelines for ductile-iron pipe gaskets. For service temperatures greater than those listed, consult manufacturer for specific recommendations.

† Maximum service temperature is not usually a meaningful parameter for piping gaskets; however, low temperatures during pipeline installation may necessitate precautions. Consult manufacturer for pertinent recommendations.

‡ Water, including sewage, with low levels of the listed contaminants.

§ Lubricating oil in the air will adversely affect SBR and EPDM performance.

** SBR, nitrile, or Neoprene® are not recommended for air exposure in wastewater treatment systems.

†† Consult manufacturer for availability of FKM push-on gaskets.

5.4 JOINT ACCESSORIES

Purchase of mechanical joint or flanged joint pipe and fittings without accessories is common because these joints are standardized throughout the industry and joint accessories are available from a number of sources.

Purchase of push-on and restrained joint pipe and fittings without accessories is less common because they are not standardized throughout the industry. If purchased from a source other than the manufacturer of the pipe or fitting, the purchaser must exercise caution to specify joint accessories that are compatible with the particular push-on joint being utilized.

5.5 PERMEATION

The selection of materials is critical for water-service and distribution piping in locations where there is likelihood the pipe will be exposed to significant concentrations of pollutants comprised of low-molecular-weight petroleum products or organic solvents or their vapors. Research has documented that certain pipe materials (such as polyethylene, polybutylene, polyvinyl chloride, and asbestos cement) and elastomers (such as those used in jointing gaskets and packing glands) may be subject to permeation by lower-molecular-weight organic solvents or petroleum products. If a water pipe must pass through such a contaminated area or an area subject to contamination, the manufacturer should be consulted regarding permeation of pipe walls and jointing materials before selecting materials for use in that area.

5.6 CHLORINE AND CHLORAMINE DEGRADATION OF ELASTOMERS

5.6.1 General

The selection of materials is critical for water service and distribution piping in locations where there is a possibility that elastomers will be in contact with chlorine or chloramines. Documented research has shown that elastomers such as gaskets, seals, valve seats, and encapsulations may be degraded when exposed to chlorine or chloramines. The impact of degradation is a function of the type of elastomeric material, chemical concentration, contact surface area, elastomer cross-section, and environmental conditions as well as temperature. Careful selection of and specifications for elastomeric materials and the specifics of their application for each water system component should be considered to provide long-term usefulness and minimum degradation (swelling, loss of elasticity, or softening) of the elastomer specified.

5.6.2 Effects of Chloramines on Ductile-Iron Pipe Gaskets

Evidence has been presented indicating that certain rubber plumbing and waterworks products serving in chloramine-disinfected water environments may deteriorate significantly. One reported example is neoprene rubber flush valves in toilet tanks that have swollen and lost tensile strength. Accordingly, research has been accomplished as to the effect of chloramine disinfected water on synthetic elastomer pipe joint gaskets. In "Effect of Chloramines on Ductile-Iron Pipe Gaskets of Various Elastomer Compounds" (Bonds 2004), the testing of rubber sheet and specific purposed ductile-iron pipe push-on and mechanical joint gaskets exposed to 110 mg/L ± 10% chloramine solution is discussed. The conclusion expressed for this testing effort is that

> Actual assembled push-on and mechanical joint ductile-iron pipe sections subjected to the same exposure showed no visual degradation of the gaskets and only slight detectable Cl_2 penetration of SBR 60, NBR 65, EPDM 65, and NBR 75.
>
> On the basis of these tests, utilities can be assured that the use of chloramines for disinfection in potable water systems will not compromise the life of SBR, NBR, CR, EPDM, or FKM elastomer gaskets with push-on and mechanical joint ductile-iron pipe.

The Bonds testing protocol placed the assembled pipe joints on end with the enclosed volume filled with the test medium. The top of the test specimen was covered with a glass plate sealed with petroleum jelly. This represents a full-strength solution with full exposure on the exposed portion of the gasket. In the test condition, the chloramine-disinfected water medium was stagnant in the pipe joint assembly, realizing constant exposure of the gasket to the testing medium. To assure the test solution concentration was constant, the test solution was replaced approximately twice a week. The exposure period was approximately one year. Depth of penetration analysis found chlorine penetration to be shallow or at nondetectable levels.

Examination of gaskets from disassembled joints in pipelines conveying chloramine-disinfected water revealed traces of chloramine embedded on the exposed gasket surface. After exposure periods of 26 and 28 years, the gaskets had minimal visible imperfections on the surface of the nose of the bulb in contact with the chloraminated water.

In conclusion, examination of gaskets from exposure during laboratory tests and in-service conditions demonstrates minimal surface impact from exposure to chloramines, indicating to utilities that chloramine disinfection will not compromise the life of push-on and mechanical-joint gaskets used in ductile-iron pipe.

REFERENCES

American National Standards Institute/American Water Works Association. 2022. Grooved and Shouldered Joints. ANSI/AWWA C606. Denver: AWWA.

American Society of Mechanical Engineers. Gray Iron Pipe Flanges and Flanged Fittings: Classes 25, 125, and 250. ASME B16.1. New York: ASME.

American Water Works Association. 2024. *AWWA Sourcebook*. Denver: AWWA.

Bonds, R.W. 2004. "Effect of Chloramines on Ductile-Iron Pipe Gaskets of Various Elastomer Compounds." *Journal AWWA*, 96:4:153–160.

This page intentionally blank.

Chapter 6

Ductile- and Gray-Iron Fittings

6.1 INTRODUCTION

The wide range of designs available in ductile-iron pressure piping systems results in part from the ready availability of a great variety of fittings. The ability to go around or bypass unexpected obstacles encountered in the planned course of a line by cutting the pipe in the field and installing the appropriate fittings has long been recognized as an advantage of iron pipe systems.

6.2 FITTINGS

6.2.1 Standard and Special Fittings

Standard fittings for ductile-iron pressure pipe for water and wastewater are produced in accordance with the following standards:

- ANSI/AWWA C104/A21.4 Cement–Mortar Lining for Ductile-Iron Pipe and Fittings.
- ANSI/AWWA C110/A21.10 Ductile-Iron and Gray-Iron Fittings.
- ANSI/AWWA C116/A21.16 Protective Fusion-Bonded Coatings for the Interior and Exterior Surfaces of Ductile-Iron and Gray-Iron Fittings.
- ANSI/AWWA C153/A21.53 Ductile-Iron Compact Fittings.

As indicated in the titles, C110/A21.10 fittings are available in gray or ductile iron, and C153/A21.53 compact fittings are available only in ductile iron.

The available configurations vary between the two standards, with bends, tees, crosses, reducers, caps, plugs, and sleeves available from each; base bends, base tees, offsets,

Source: Ductile Iron Pipe Research Association

Figure 6-1 Welded-on push-on joint outlet

connecting pieces, and tapped tees are covered only by C110/A21.10. Special fittings, such as long radius fittings, reducing elbows, reducing on run tees, side outlet fittings, eccentric reducers, wall pipe, welded-on bosses and outlets (see Figure 6-1), dual purpose and transition sleeves, and lateral and true wyes are also available from some manufacturers. These fittings are provided according to manufacturers' standards, and some have overall dimensions, wall thicknesses, and other provisions as applicable in accordance with ANSI/AWWA C110/A21.10 or C153/A21.53.

For underground applications, fittings are normally coated with asphaltic coating and lined with cement mortar in accordance with ANSI/AWWA C104/A21.4, or they are lined and coated with fusion-bonded epoxy in accordance with ANSI/AWWA C116/A21.16. It is the manufacturer's option which coating is used.

6.2.2 End Combinations

Standard fittings are provided with various end combinations, usually with push-on joint or mechanical joint bells, plain ends, flanges, or special restrained joints, as shown in the tables of the applicable standards. All-bell fittings, such as those shown in Figure 6-2, are more commonly available. Fittings with flanged joints in 3-in. through 48-in. sizes are not covered in ANSI/AWWA C153/A21.53 because of the importance of standardized center-to-flange dimensions already covered in ANSI/AWWA C110/A21.10.

Flanged fittings in 54-in. through 64-in. sizes with standardized center-to-flange dimensions are included in ANSI/AWWA C153/A21.53.

Source: Ductile Iron Pipe Research Association

Figure 6-2 Ductile-iron fittings stacked in storage yard

6.2.3 Pressure Rating and Safety Factor

The rated working pressure of standard fittings depends on the material (gray iron or ductile iron), the fitting size and configuration, and the wall thickness; for flanged fittings, the rated working pressure depends on the flange. These working pressures are shown in the tables in the standards.

Flanges conforming to the dimensions shown in ANSI/AWWA C110/A21.10 are adequate for service of 250 psi (1,724 kPa) or 350 psi (2,413 kPa), although for some flanged fittings, the rating of the fitting may differ as a result of the material or the fitting wall thickness.

NOTE: Certain ductile-iron fittings 24 in. (610 mm) and smaller may be rated for 350 psi with the use of special gaskets as noted in Section 5.2.3 of this manual. The rating of C110/A21.10 flanges for water service at ambient temperature should not be confused with the designated class 125-lb and 250-lb flanges covered by ANSI/ASME B16.1 (2020).

6.2.3.1 ANSI/ASME B16.1 flanges. The bolt circle and the bolt holes of ANSI/AWWA C110/A21.10 flanges match those of the Class 125 flanges shown in ASME/ANSI B16.1 and can be joined with Class 125 B16.1 flanges. Flanges on standard C110/A21.10 fittings cannot be joined with ANSI/ASME B16.1 Class 250 flanges.

The ratings of the standard fittings in ANSI/AWWA C110/A21.10 and C153/A21.53 provide for a safety factor of at least 2.5 times the rated working pressure at ambient temperature. The ratings were developed on the basis of hydrostatic testing of fittings to bursting.

6.2.4 Casting and Processing

Because of their irregular shapes, fittings are statically cast. Before the iron is poured, molds forming the outside contour of the fittings are assembled with cores that form the inside shape of the casting. Iron is poured into the assembled mold and flows into the void surrounding the core. Another casting method, known as "the lost foam molding process," consists of production of a foam pattern to the image of the desired fitting. This pattern is then surrounded by molding sand, inside and out. When iron is poured into the flask holding this encapsulated pattern, it consumes and replaces the foam to form the casting. After cooling, the fittings are removed from the molds, cleaned, inspected, gauged for dimensional accuracy, and lined and coated as required.

6.2.5 Iron Strength

Standard C110/A21.10 gray-iron fittings are produced of iron of minimum 25,000-psi (172.4×10^3 kPa) or 30,000-psi (206.9×10^3 kPa) iron strength as listed in the appropriate tables. The most common grade of ductile iron used in fittings is 70-50-05, with acceptance values as follows:

- A minimum tensile strength of 70,000 psi (482.7×10^3 kPa)
- A minimum yield strength of 50,000 psi (344.8×10^3 kPa)
- A minimum elongation of 5 percent

Another permissible acceptance value is the 60-42-10 grade of ductile iron.

6.2.6 Use of Fittings and Unbalanced Thrust Forces

Piping systems are subject to unbalanced thrust forces resulting from static and dynamic fluid action on the pipe main. These forces must be absorbed or balanced if the piping system is to maintain its integrity. Unbalanced thrust forces occur at changes in direction of flow, such as at bends, tees, wyes, crosses, and at reducers, valves, and dead ends. Reactive forces to balance these thrust forces can be provided by restrained fitting and pipe joints or by concrete thrust blocks to allow transfer of the thrust force to the soil surrounding the pipe. Methods of handling these thrust forces, together with useful data, are covered in Chapter 8 of this manual.

6.3 SPECIAL SERVICE REQUIREMENTS

6.3.1 Orientation of Mechanical Joint Bolt Holes

Although AWWA standards do not specify the orientation of bolt holes in the bells of mechanical joint fittings, it is at times convenient and sometimes necessary to have bolt holes specially oriented. The normal (but not universal) practice is to have the bolt holes straddle the vertical centerline of the fittings, valves, and hydrants. (The vertical centerline of a fitting is determined when the fitting is in the position to change the direction of fluid flowing in a horizontal plane. With standard base bends and standard base tees, the vertical centerline is determined when the fitting is in a position to change the fluid flowing in a vertical plane.) If a specific or special bolt hole orientation is required by the application, the orientation should be noted in the purchaser's specification.

6.3.2 Orientation of Flanged Bolt Holes on Reducers

ANSI/AWWA C110/A21.10 and C153/A21.53 require that flanged bolt holes straddle the centerline of the fitting. Because the two flanges on most flanged reducers have a different number of bolt holes, all adjacent bolt holes will not straddle a common centerline on both flanges. In some cases, only one common centerline will be straddled by bolt holes on both flanges. This must be considered during installation so that the reducers can be properly oriented to prevent misalignment of adjoining flanged material.

6.3.3 Use of Push-On and Mechanical Joints with Older Existing Pipe

Care should be exercised when attempting to connect fittings with push-on or mechanical joints to cast-iron pipe in old systems. The outside diameter of pipe in old systems should be measured before cutting because some pipe made in past years was manufactured with different diameters than are presently specified in ANSI/AWWA standards. Special mechanical joint sleeves and adapters are available to provide transition from these larger diameters to the present A21 diameters.

REFERENCE

American Society of Mechanical Engineers 2020. Gray Iron Pipe Flanges and Flanged Fittings: Classes 25, 125, and 250. ASME B16.1. New York: ASME.

This page intentionally blank.

Chapter 7

Valves and Hydrants

7.1 INTRODUCTION

Valve applications in a water system cover a variety of types and functions, depending on the size and complexity of the system. The most common functions are isolation, backflow prevention, and flow control. There are specialized valves to perform the less common functions; however, only the most common valves and hydrants are covered in this manual. AWWA M44, *Distribution Valves: Selection, Installation, Field Testing, and Maintenance*, and M17, *Fire Hydrants: Installation, Field Testing, and Maintenance*, and relevant standards provide a more expanded guide in the use of valves and hydrants.

7.2 TYPES OF VALVES AND HYDRANTS

7.2.1 Resilient-Seated Gate Valves

The resilient-seated gate valve consists of a closure mechanism activated by the rotation of a threaded valve stem or screw. It is by far the most common of all valves installed in the utility system. The closure member moves into the waterway to shut off flow when it is turned to the closed position. A resilient seat seals against a corrosion-resistant surface. Resilient seats may be bonded or mechanically attached to either the gate or valve body, although most designs attach them to the gates. The resilient-seated gate valve is manufactured for zero leakage across the seat. When the valve is opened, the closure member moves into the bonnet cavity, leaving the waterway unobstructed. ANSI/AWWA Standards C509, Resilient-Seated Gate Valves for Water Supply Service, and C515, Reduced-Wall, Resilient-Seated Gate Valves for Water Supply Service, cover resilient-seated gate valves. C509 valves may be constructed of ductile iron or gray iron and are rated for 200 psig (1,380 kPa [gauge]) for 3- through 12-in. (75- through 300-mm) size and 150 psig (1,034 kPa [gauge]) for 14- through 36-in. (350- through 900-mm) sizes. C515 valves are reduced-wall ductile iron only and are rated for 200 psig (1,380 kPa [gauge]) minimum for the 3- through 54-in. (75- through 1,350-mm) sizes included in the standard. Larger sizes as well as higher pressure ratings are available for both valves meeting the design criteria established in C509 and C515.

7.2.2 Metal-Seated Gate Valves

The metal-seated gate valve, like the resilient-seated gate valve, has a closure member that moves into and out of the waterway. Designs of the sealing member include double disc gates that have parallel or inclined seats, as well as solid wedge gates. The gates, however, seal against a bronze surface mounted in the valve body. Because of the metal-to-metal sealing surfaces, a small amount of leakage is allowed past the seats. The standard covering this product in sizes 3 through 72 in. (75 through 1,650 mm) is ANSI/AWWA Standard C500, Metal-Seated Gate Valves for Water Supply Service. These valves have a pressure rating of 200 psig (1,380 kPa [gauge]) for the 12-inch and smaller sizes and 150 psig (1,034 kPa [gauge]) for the larger sizes (Figure 7-1). Valves with higher pressure ratings are available; however, they are not covered by the AWWA C500 standard.

7.2.3 Rubber-Seated Butterfly Valves

The rubber-seated butterfly valve consists of a movable disc closure piece that remains in the waterway and rotates on a shaft. A rubber seat is applied to the disc or valve body and seals against a corrosion-resistant surface. Such valves are covered by ANSI/AWWA

Source: American Cast Iron Pipe Company

Figure 7-1 Photograph of a 16-inch and a 60-inch resilient seated gate valve

C504, Rubber-Seated Butterfly Valves. This standard describes 3-in. through 72-in. (75-mm through 1,800-mm) valves with maximum pressure ratings of 250 psig (1,723 kPa [gauge]). The size ranges covered in the standard depend on the body types and classes. Higher pressure-rated valves are available but not described in the AWWA standard.

7.2.4 Swing-Check Valves

There are various types of valves to control and prevent reverse flow in piping systems. Depending on the requirements of the installation, owner and operator preferences, transient concerns, and numerous other factors, the designer must consider and specify the appropriate control device. Because this document is focused on ductile iron and its related products, not all options for reverse flow control are discussed.

Swing check valves are commonly used to permit flow in only one direction (prevent reverse flow). Swing-check valves have a disc attached to a hinge that pivots on a hinge pin or shaft. These valves are described in ANSI/AWWA C508, Swing-Check Valves for Waterworks Service, 2-In. Through 24-In. (50-mm Through 600-mm) NPS. The seating can be resilient to coated, metal to metal or composition to metal. The swing-check valve is activated by flow through the valve. Some swing-check valves are manufactured with levers, weights, and springs to control opening and closing (Figure 7-2); however, these special condition valves are not described in ANSI/AWWA C508, which details standard swing-check valves for waterworks service. This standard describes 2-in. through 48-in. (50-mm through 1,200-mm) valves rated for a water working pressure of 200 psi (1,370 kPa [gauge]) and 14-in. through 48-in. (350-mm through 1,200-mm) valves rated for 150 psig (1,034 kPa [gauge]). Again, higher pressure-rated valves and larger sizes are available but are not described in the AWWA standard.

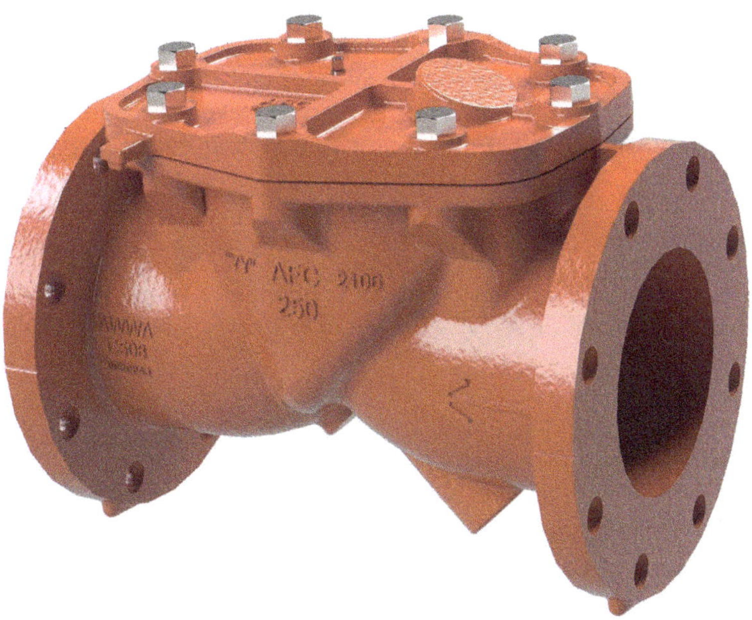

Source: American Cast Iron Pipe Company

Figure 7-2 Common resilient seated check valve

7.2.5 Dry-Barrel Fire Hydrants

Dry-barrel fire hydrants are post-type hydrants, with compression (opening against or with the pressure) or gate shutoff, used in water supply systems in all climates, including those where freezing occurs. They are described by ANSI/AWWA C502, Dry-Barrel Fire Hydrants, and are rated for a minimum of 150 psig (1,034 kPa [gauge]). The hydrant is designed so that no excavation is required to remove the main valve, the operating mechanism, or the movable parts of the drain valve. In the event of accidental damage or breaking of the hydrant above or near the grade level, the main valve is designed to remain closed and reasonably tight against leakage. These hydrants are generally provided with an outlet and valve for draining the hydrant as soon as the main valve is closed.

To reduce the probability of damaging the hydrant sealing surfaces as a result of debris trapped in the hydrant valve during the flushing of a newly installed line, it is important to make sure the hydrant is fully open. After flushing and prior to closing the hydrant, the gate valve in front of the hydrant should be closed slowly to reduce the likelihood of any remaining rocks and debris being trapped in the hydrant seat. Care should be taken to ensure that the connections provided on the fire hydrant comply with the requirements established by the local governing authorities. The use of fire hydrants as source of water for street cleaning, construction projects, or any purpose other than firefighting is beyond the primary purpose for which the hydrant is installed. The use of fire hydrants in this manner should be strictly restricted and controlled in the interest of maintaining the equipment in satisfactory working condition for use at times of emergencies. AWWA M17, *Fire Hydrants: Installation, Field Testing, and Maintenance,* provides an excellent guide for owners of hydrants.

7.2.6 Wet-Barrel Fire Hydrants

In the United States, the wet-barrel fire hydrants, described in ANSI/AWWA C503, Wet-Barrel Fire Hydrants, can be used in water supply systems where freezing temperatures are not a concern. The barrel is pressurized at all times. Independently controlled compression-type (opening against or with the pressure) valves are at each outlet, and all valve working parts are mounted in the hydrant top section above the ground line. As with dry barrel hydrant installations, care should be taken to ensure that the proper connections are provided and that their use is restricted to their intended purpose.

7.3 INSTALLATION

Proper installation will ensure lower maintenance costs. See Chapter 11 of this manual and other appropriate AWWA valve and hydrant manuals and standards for installation recommendations.

7.4 OPERATION

7.4.1 Valves

Each valve should be operated through a full cycle and then returned to its normal position on a regular schedule. The time interval between operations should be determined by the manufacturer's recommendations, size of the valve, severity of the operating conditions, and the importance of the installation. The number of turns should be recorded and compared with permanent installation records to ensure full valve travel.

When using portable auxiliary power actuators with input torque capacities exceeding the maximum recommended operating torques, extreme care should be exercised to avoid applying excessive torque to the valve. A torque-limiting device should be used where possible. If no torque-limiting device is available, the last few turns of travel should be by manual operation.

7.4.2 Hydrants

Fire hydrants should be inspected and operated through a full cycle on a regular schedule. As stated in AWWA M17, hydrants should be inspected regularly, at least once a year, to ensure their satisfactory operation. Most hydrant manufacturers recommend inspections twice per year. The hydrant should be flushed to prevent sediment buildup in the hydrant or connecting piping. Any damage or malfunction should be immediately reported to the authority responsible for maintenance. The wrench used for operation should be a standard-length operating wrench. Wrench handle extensions should never be used.

If the hydrant cannot be operated with the standard wrench, it should be checked by maintenance personnel.

7.5 GOOD PRACTICE

Select the correct type of valve or hydrant and joint for the application. Follow recommended installation procedures for a longer, maintenance-free service life of the piping system. Care should be taken to prevent damage to the hydrant main valve if the hydrant is used for line flushing. Refer to the applicable AWWA standards and manual of practice, as well as the manufacturer's instructions. Clean the interior of all valves and hydrants before installation. Where possible, test joints before backfilling. Valves and hydrants should be operated periodically to check for proper operation and shutoff. Lubrication should be in accordance with the manufacturer's recommendations.

REFERENCES

American National Standards Institute/American Water Works Association. 2023. Resilient-Seated Gate Valves for Water Supply Service. ANSI/ AWWA C509. Denver: AWWA.

American National Standards Institute/American Water Works Association. 2020. Reduced-Wall, Resilient-Seated Gate Valves for Water Supply Service. ANSI/AWWA C515. Denver: AWWA.

American National Standards Institute/American Water Works Association. 2019. Metal-Seated Gate Valves for Water Supply Service. ANSI/AWWA C500. Denver: AWWA.

American National Standards Institute/American Water Works Association. 2023. Rubber-Seated Butterfly Valves. ANSI/AWWA C504. Denver: AWWA.

American National Standards Institute/American Water Works Association. 2017. Swing-Check Valves for Waterworks Service, 2-In. Through 24-In. (50-mm Through 600-mm) NPS. ANSI/AWWA C508. Denver: AWWA.

American National Standards Institute/American Water Works Association. 2018. Dry-Barrel Fire Hydrants. ANSI/AWWA C502. Denver: AWWA.

American National Standards Institute/American Water Works Association. 2018. Wet-Barrel Fire Hydrants. ANSI/AWWA C503. Denver: AWWA.

American Water Works Association. 2016. AWWA M17: *Installation, Field Testing, and Maintenance of Fire Hydrants*. Denver: AWWA.

Chapter 8

Thrust Restraint Design for Ductile-Iron Pipe

8.1 INTRODUCTION

NOTE: Information in this chapter is used with permission from *Thrust Restraint Design for Ductile Iron Pipe*, 7th edition (DIPRA 2016).

Ductile-iron pipe and fittings are most often joined with push-on (Figure 8-1) or mechanical joints. Neither of these joints provides significant restraint against longitudinal separation other than friction between the gasket and the plain end of the pipe or fitting. Tests have shown that this frictional resistance in the joint is unpredictable and varies widely with installation conditions and other factors that are insignificant in other respects. Thus, these joints should be considered as offering no longitudinal restraint for design purposes.

At many locations in underground or aboveground pipelines, the configuration of the pipeline results in unbalanced forces of hydrostatic or hydrodynamic origin that can result in joint separation unless restrained.

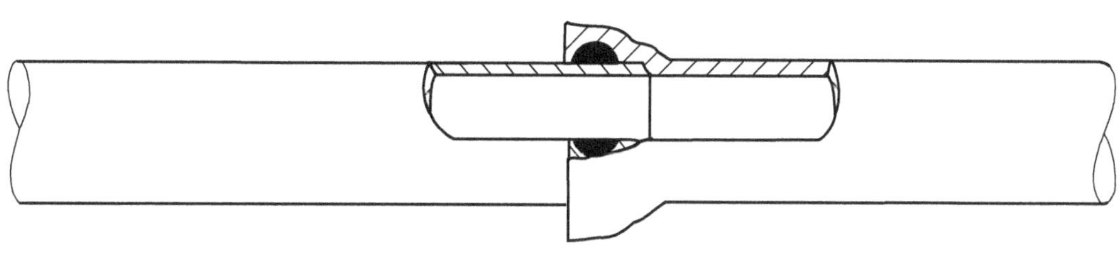

Figure 8-1 Push-on joint detail

Generically, these unbalanced hydrostatic and hydrodynamic forces are called *thrust forces*. In the range of pressures and fluid velocities found in drinking water or wastewater piping, the hydrodynamic thrust forces are generally insignificant in relation to the hydrostatic thrust forces and are usually ignored. Simply stated, thrust forces occur at any point in the piping system where the direction or cross-sectional area of the waterway changes. Thus, there will be thrust forces at bends, reducers, offsets, tees, wyes, dead ends, and valves.

Balancing thrust forces in underground pipelines is usually accomplished with bearing or gravity thrust blocks, restrained joint systems, or a combination of these methods. Presented in this chapter is a general discussion of the nature of thrust forces as well as suggested approaches to the design of both thrust block and restrained joint systems for balancing these forces. The suggested design approaches are conservatively based on accepted principles of soil mechanics.

8.1.1 Nomenclature

A = cross-sectional area of pipe, in.^{2} (m^2) (see Table 8-4)

A_p = surface area of pipe exterior, ft^2/ft (m^2/m)

B = thrust block width, ft (m)

C = pipe cohesion, lb/ft^2 (kN/m^2)

C_s = soil cohesion, lb/ft^2 (kN/m^2) (see Table 8-2)

D' = outside diameter of pipe, ft (m) (see Table 8-4)

f_c = ratio of pipe cohesion to soil cohesion (see Table 8-2)

F_f = unit frictional resistance lb/ft (kN/m)

F_s = unit frictional force assuming half the pipe circumference bears against the soil, lb/ft (kN/m)

$(F_s)_b$ = unit frictional force assuming the entire pipe circumference contacts the soil, lb/ft (kN/m)

f_ϕ = ratio of pipe friction angle to soil friction angle (see Table 8-2)

h = thrust block height, ft (m)

H = depth of cover to top of pipe, ft (m)

H_c = depth of cover to pipe centerline, ft (m)

K_n = trench condition modifier (see Table 8-2)

H_t = depth to bottom of thrust block, ft (m)

L = minimum required restrained pipe length, ft (m)

N_ϕ = $\tan^2\left(45° + \frac{\varphi}{2}\right)$

P = design pressure, psi (kN/m^2)

P_p = passive soil pressure, lb/ft^2 (kN/m^2)

R_s = unit bearing resistance, lb/ft (kN/m)

S_b = horizontal bearing strength of soil, lb/ft^2 (kN/m^2) (see Table 8-1)

S_f = safety factor (usually 1.5)

T = resultant thrust force, lb (kN)

V_g = volume of thrust block, ft^3 (m^3)

W = unit normal force on pipe = $2W_e + W_p + W_w$, lb/ft (kN/m)

W_e = earth prism load = $\gamma H D'$, lb/ft (kN/m)

W_m = density of thrust block material, lb/ft^3 (kN/m^3)

$W_p + W_w$ = unit weight of pipe and water, lb/ft (kN/m) (see Table 8-4)

γ = backfill soil density, lb/ft^3 (kN/m^3) (see Table 8-2)

δ	=	pipe friction angle, degrees
θ	=	bend angle, degrees
φ	=	internal friction angle of the soil, degrees (see Table 8-2)

8.2 THE THRUST FORCE

The internal hydrostatic pressure acts perpendicularly on any plane with a force equal to the pressure, *P*, times the area, *A*, of the plane. All components of these forces acting radially within a pipe are balanced by circumferential tension in the wall of the pipe. This tension in the pipe wall is controlled by proper selection of pipe class during design, which includes calculation of the wall thickness based on the sum of working and surge pressures (See Chapter 4). Axial components acting on a plane perpendicular to the pipe through a straight section of the pipe are balanced internally by the force acting on each side of the plane, as indicated in Figure 8-2.

However, in the case of a bend as shown in Figure 8-3, the forces, *PA*, acting axially along each leg of the bend, are not balanced. The vector sum of these forces is shown as the thrust force, *T*. To prevent separation of the joints, a reaction equal to and in the opposite direction of *T* must be established.

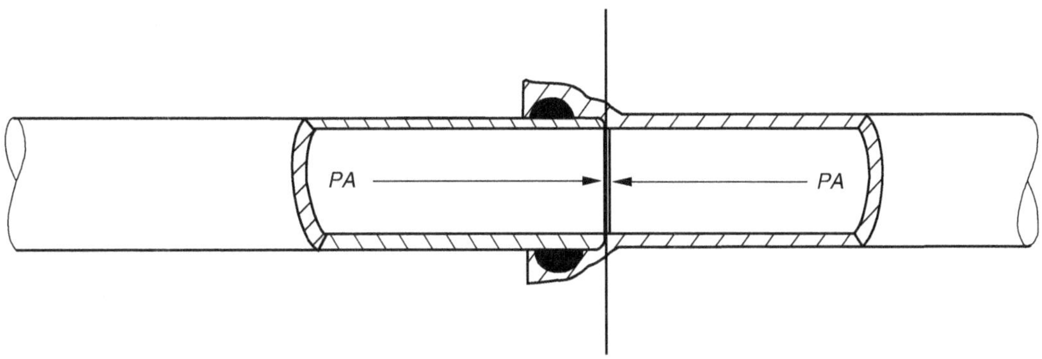

Figure 8-2 Thrust forces at joint of straight pipe

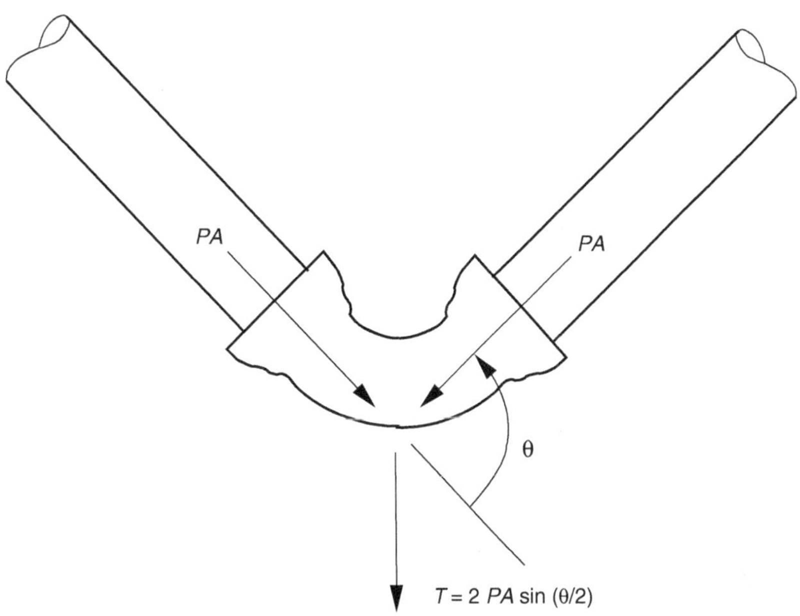

Figure 8-3 Thrust forces at bend

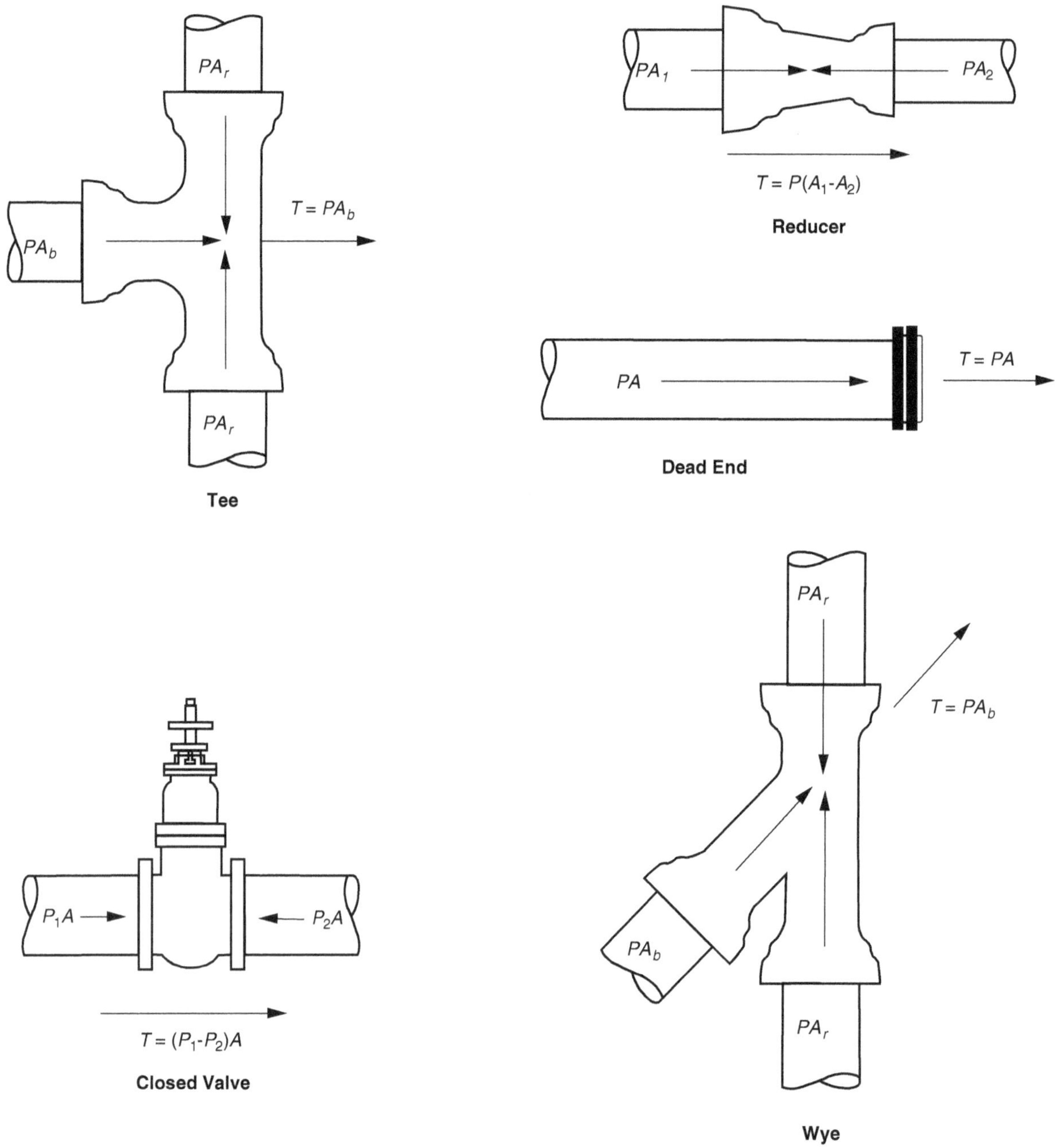

Figure 8-4 Thrust forces on other pipe configurations or appurtenances

Figure 8-4 depicts the net thrust force for various other configurations. In each case, the expression for *T* can be derived by the vector addition of the axial forces.

8.2.1 Design Pressure

The design pressure, *P*, is the maximum pressure to which the pipeline will be subjected, with consideration given to the vulnerability of the pipe-soil system when the pressure is expected to be applied. In most cases, this will be the test pressure of the pipe, applied shortly after installation when the pipe–soil system is normally the most vulnerable.

8.2.2 Pipe–Soil Structure

For buried pipelines, thrust restraint is achieved by transferring the thrust force to the soil structure outside the pipe. The objective of the design is to distribute the thrust forces to the soil structure in such a manner that damage does not occur to the restrained pipe system and joint separation does not occur in unrestrained joints. The thrust force is distributed as a pressure that is less than the soil's ability to resist that pressure. The soil is, therefore, a part of the restraining design, which must be considered if the soil structure is subject to future disturbance (see Section 8.11).

8.3 THRUST BLOCKS

One of the most common methods of providing resistance to thrust forces is the use of thrust blocks. Figure 8-5 depicts a typical bearing thrust block on a horizontal bend. Resistance is provided by transferring the thrust force through the bearing block to the soil. The bearing block is designed such that the bearing area of the block is large enough to result in a bearing pressure against the soil that does not exceed the bearing strength of the soil. Design of thrust blocks consists of determining the appropriate bearing area of the block for a particular set of conditions. The parameters involved in the design include pipe size, design pressure, angle of the bend (or configuration of the fitting involved), and the horizontal bearing strength of the soil.

The following are general criteria for bearing block design (Figure 8-5).

- Bearing surface should, where possible, be placed against undisturbed soil. Where it is not possible, the fill between the bearing surface and undisturbed soil must be compacted to at least 90 percent standard Proctor density.
- Block height h in ft (m) should be equal to or less than one-half the total depth to the bottom of the block H_t in ft (m) but not less than the pipe diameter D' in ft (m).
- Block height h should be chosen such that the calculated block width b in ft (m) varies between one and two times the height.

The required bearing block area is

$$A_b = h_b = \frac{S_f T}{S_b}$$

Then, for a horizontal bend,

$$b = \frac{S_f 2PA \sin\left(\frac{\theta}{2}\right)}{hS_b} \quad \text{(Eq 8-1)}$$

where S_f is a safety factor (usually 1.5 for thrust block design). A similar approach may be used to design bearing blocks to resist the thrust forces at tees and dead ends. Typical values for horizontal bearing strengths of various soil types are listed in Table 8-1.

In lieu of the values for soil-bearing strength shown in Table 8-1, a designer might choose to use the calculated Rankine passive pressure, P_p, or other determination of soil-bearing strength based on actual soil properties.

Gravity thrust blocks may be used to resist the thrust force at vertical down-bends (see Figure 8-6). In a gravity block, the weight of the block is the force providing equilibrium with the thrust force. Therefore, the design is to calculate the required volume of the thrust block of known density. The vertical component of the thrust force in Figure 8-6 is balanced by the weight of the block.

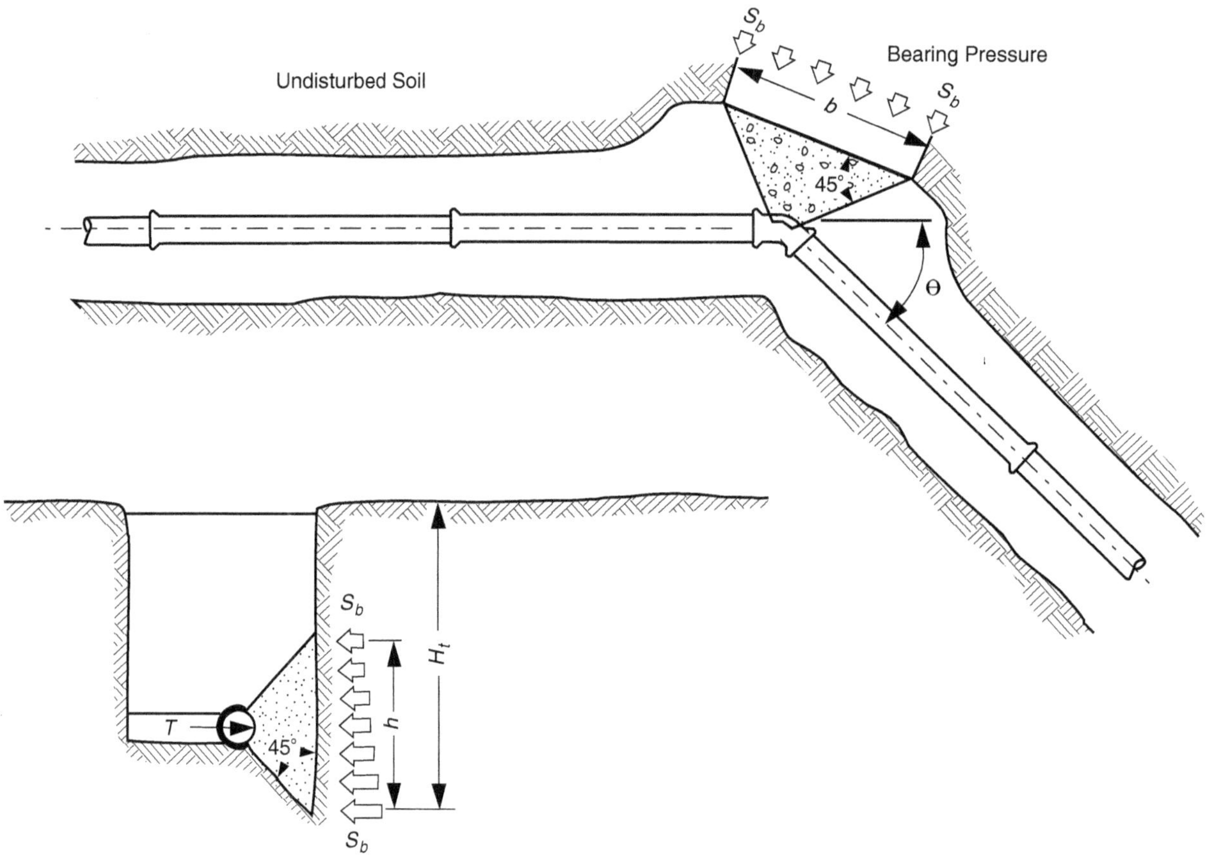

Figure 8-5 Bearing block details

Table 8-1 Horizontal soil-bearing strength

Soil	Bearing Strength, S_b* (lb/ft²)
Muck	0
Soft clay	1,000
Silt	1,500
Sandy silt	3,000
Sand	4,000
Sandy clay	6,000
Hard clay	9,000

NOTE: To convert pounds per square foot (lb/ft2) to kilonewtons per square meter (kN/m2), multiply by 0.0479.

* Although the listed bearing strength values have been used successfully in the design of thrust blocks, their accuracy is totally dependent on accurate soil identification and evaluation. The ultimate responsibility for selecting the proper bearing strength of a particular soil type must rest with the engineer.

It can easily be shown that $T_y = PA \sin \theta$. Then the required volume of the block is

$$V_g = \frac{S_f PA \sin\theta}{W_m} \quad \text{(Eq 8-2)}$$

where W_m = density of the block material. In this case, the horizontal component of the thrust force

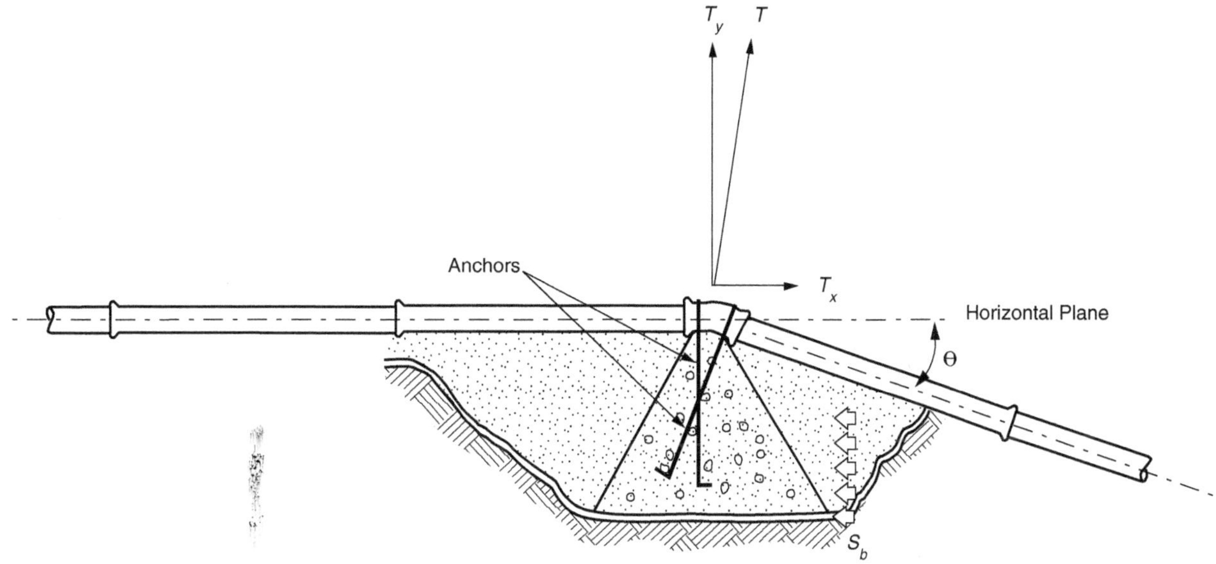

Figure 8-6 Gravity thrust block

$$T_x = PA\left(1 - \cos\theta\right)$$

must be resisted by the bearing of the right side of the block against the soil. Analysis of this aspect will follow like the previous description on bearing blocks.

Calculations of V_g and T_x for orientations other than when one leg is horizontal should reflect that specific geometry.

8.4 RESTRAINED JOINTS

An alternative method of providing thrust restraint is the use of restrained joints. A restrained joint is a special type of push-on or mechanical joint that is designed to provide longitudinal restraint against separation. Restrained-joint systems function in a manner similar to thrust blocks, insofar as the reaction of the entire restrained unit of piping balances the thrust forces against the resisting forces that develop in the soil (Figure 8-7).

The objective in designing a thrust restraint system using restrained pipe joints is to determine the length of pipe that must be restrained on each side of the focus of a thrust force. This restrained length will be a function of the pipe size, internal pressure, depth of cover, the characteristics of the soil surrounding the pipe, and whether the pipe is polyethylene encased. The following is a method of accomplishing the design objective.

As with most engineering problems, the exact nature of the interaction of the restrained pipe unit and the soil is extremely complex. Limitations of the ability to measure the actual parameters involved and limitations on available knowledge of the precise nature of the interaction require that a practical design procedure be based on various assumptions. The assumptions employed in the following design procedure are, in each case, intended to be conservative. This conservatism, together with the explicit safety factor employed in the procedure, should result in a conservative design with an adequate overall safety factor. However, each thrust restraint problem is a site-specific condition. Therefore, it is the design engineer's responsibility to select the appropriate design parameters for the specific case involved.

The design equation for horizontal bends (Equation 8-7) and the soil parameters presented in Table 8-2 are the outgrowths of a design procedure originally proposed by

Figure 8-7 Restrained joints in a buried piping system

Carlsen (1975). Carlsen's design procedure was based solely on theoretical considerations and was conservatively limited to well-compacted trench conditions. The modification of Carlsen's design procedure is the result of full-scale tests of 12-in. (305-mm) ductile-iron pipe with 45° and 90° bends buried in clay (Conner 1988). The data generated by these tests and data available from model studies with 2-in. (50-mm) pipe in sand confirm the conservatism of the present design procedure.

The thrust force must be restrained or balanced by the reaction of the restrained pipe unit with the surrounding soil. The source of the restraining forces is twofold: first, the static friction between the pipe unit and the soil, and second, the resistance to movement of the pipe as it bears against the sidefill soil along each leg of the bend. Both forces are presumed to be functions of the restrained length L on each side of the bend, and they are presumed to act in the direction opposing the thrust force (i.e., directly opposing the impending movement of the bend).

8.4.1 Horizontal Bends

Figure 8-8 is a free body diagram of a restrained pipe unit where L is the length of the restrained pipe on each side of the bend. The unit frictional resistance is shown as a uniformly distributed force of the unit value F_f. The total frictional resistance on each side of the bend is then $F_f L \cos(\theta/2)$.

Figure 8-8 does not represent the actual pipe–soil behavior with all trench types and various restrained joint designs available. Variation in the way different restrained joints respond to loadings, along with soil and installation variables, makes this a situation that defies precise theoretical representation. The approach presented, which includes safety factors, is a practical and conservative general thrust restraint design that has been verified by available test data and the success of numerous installed systems.

Table 8-2 Suggested values for soil parameters and reduction constant k_n

Soil Designation*	Soil Description‡	φ deg.	fφ	Cs, lb/ft²	fc	γ† lb/ft†	Kn A21.50 Laying Condition 2	3	4	5
Clay 1	Clay of medium to low plasticity, LL** <50, <25% coarse particles†† (CL & CL-ML)‡	0	0	300	0.80 (0.50§)	90	.20	.40	.60	.85
Silt 1	Silts of medium to low plasticity, LL** <50, <25% coarse particles†† (ML & ML-CL) ‡	29	0.75 (0.50§)	0	0	90	.20	.40	.60	.85
Clay 2	Clay of medium to low plasticity w/sand or gravel, LL** <50, 25–50% coarse particles†† (CL) ‡	0	0	300	0.80 (0.75§)	90	.40	.60	.85	1.0
Silt 2	Silt of medium to low plasticity w/sand or gravel, LL** <50, 25–50% coarse particles†† (ML)‡	29	0.75 (0.50§)	0	0	90	.40	.60	.85	1.0
Coh-gran	Cohesive granular soils, >50% coarse particles†† (GC & SC)‡	20	0.65 (0.40§)	200	0.40	90	.40	.60	.85	1.0
Sand silt	Sand or gravel w/silt, >50% coarse particles†† (GM & SM)‡	30	0.75 (0.50§)	0	0	90	.40	.60	.85	1.0
Clean sand or clean gravel	Clean sand or clean gravel, >95% coarse particles†† (SW, SP & GW)‡	36	0.80 (0.75§)	0	0	100	.40	.60	.85	1.0

NOTE: To convert pounds per square foot (lb/ft²) to kilonewtons per square meter (kN/m²), multiply by 0.0479; to convert pounds per cubic foot (lb/ft³) to kilonewtons per cubic meter (kN/m³), multiply by 0.1571.

* See Section 8.8 Select Backfill Considerations.

† Values for γ shown in the table and used in this procedure are lower than the soil weight values used to calculate earth loads in ANSI/AWWA C150/A21.50. All other values assume saturated soil conditions. These values were selected for conservatism.

‡ See Table 8-3 for more detailed soil descriptions.

§ These values to be used for laying conditions Type 2.

** LL-liquid limit

†† "Coarse particles" are those particles held on a No. 200 sieve.

The bearing resistance is shown as a triangular distributed force with a maximum unit value of R_s at the bend, diminishing linearly to 0 at L. This assumption is based on the fact that the bearing resistance (passive resistance in the soil) is proportional to deformation or movement. As the restrained joints take load, maximum movement will occur at the bend. The total assumed bearing resistance on each side of the bend is ½ $R_s L \cos(\theta/2)$.

The equilibrium equation for the free body is then

$$PA \sin\left(\frac{\theta}{2}\right) = F_f L \cos\left(\frac{\theta}{2}\right) + \frac{1}{2} R_s L \cos\left(\frac{\theta}{2}\right)$$

Employing a safety factor and solving for L,

$$L = \frac{S_f PA \tan\left(\frac{\theta}{2}\right)}{F_f + \frac{1}{2} R_s} \qquad \text{(Eq 8-3)}$$

Table 8-3 Soil classification chart—ASTM Standard D2487*

Soil Classification Chart—ASTM Standard D2487

Major Divisions			Group Symbols	Typical Names	Classification Criteria		
Coarse-grained Soils More than 50% retained on No. 200 sieve*	Gravels 50% or more of coarse fraction retained on No. 4 sieve	Clean Gravels	GW	Well-graded gravels and gravel-sand mixtures, little or no fines	Classification on basis of percentage of fines Less than 5% Pass No. 200 sieve GW, GP, SW, SP More than 12% Pass No. 200 sieve GM, GC, SM, SC 5% to 12% Pass No. 200 sieve Borderline Classification requiring use of dual symbols	$C_u = D_{60}/D_{10}$ Greater than 4 $C_z = \frac{(D_{30})^2}{D_{10} \times D_{60}}$ Between 1 and 3	
			GP	Poorly graded gravels and gravel-sand mixtures, little or no fines		Not meeting both criteria for GW	
		Gravels With Fines	GM	Silty gravels, gravel-sand-silt mixtures		Atterberg limits plot below "A" line or plasticity index less than 4.	Atterberg limits plotting in hatched area are borderline classifications requiring use of dual symbols.
			GC	Clayey gravels, gravel-sand-clay mixtures		Atterberg limits plot above "A" line and plasticity index greater than 7.	
	Sands More than 50% of coarse fraction passes No. 4 sieve	Clean Sands	SW	Well-graded sands and gravelly sands, little or no fines		$C_u = D_{60}/D_{10}$ Greater than 6 $C_z = \frac{(D_{30})^2}{D_{10} \times D_{60}}$ Between 1 and 3	
			SP	Poorly graded sands and gravelly sands, little or no fines		Not meeting both criteria for SW	
		Sands With Fines	SM	Silty sands, sand-silt mixtures		Atterberg limits plot below "A" line or plasticity index less than 4.	Atterberg limits plotting in hatched area are borderline classifications requiring use of dual symbols.
			SC	Clayey sands, sand-clay mixtures		Atterberg limits plot above "A" line and plasticity index greater than 7.	
Fine-grained Soils 50% or more passes No. 200 sieve*	Silts And Clays Liquid limit 50% or less		ML	Inorganic silts, very fine sands, rock flour, silty or clayey fine sands			
			CL	Inorganic clays of low to medium plasticity, gravelly clays, sandy clays, silty clays, lean clays			
			OL	Organic silts and organic silty clays of low plasticity			
	Silts And Clays Liquid limit greater than 50%		MH	Inorganic silts, micaceous or diatomaceous fine sands or silts, elastic silts			
			CH	Inorganic clays of high plasticity, fat clays			
			OH	Organic clays of medium to high plasticity			
Highly Organic Soils			PT	Peat, muck, and other highly organic soils			

Plasticity Chart

For classification of fine-grained soils and fine fraction of coarse-grained soils.
Atterberg limits plotting in hatched area are borderline classifications requiring use of dual symbols.
Equation of A-line:

$$PI = 0.73\,(LL - 20)$$

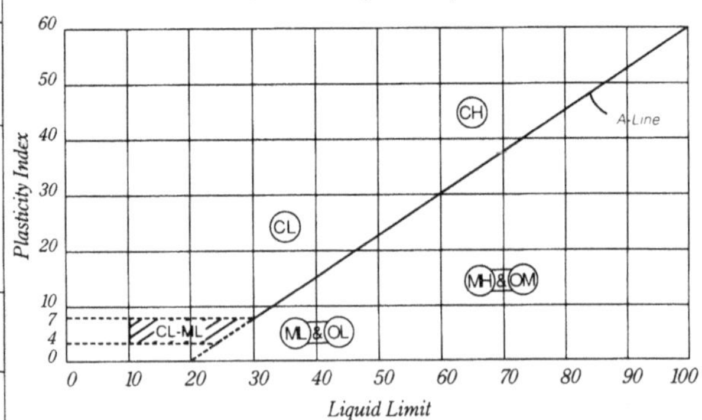

* Based on the material passing the 3-in. (75-mm) sieve.

NOTE: Adapted with permission from D2487-93, Standard Classification of Soils for Engineering Purposes (DIPRA 2016).

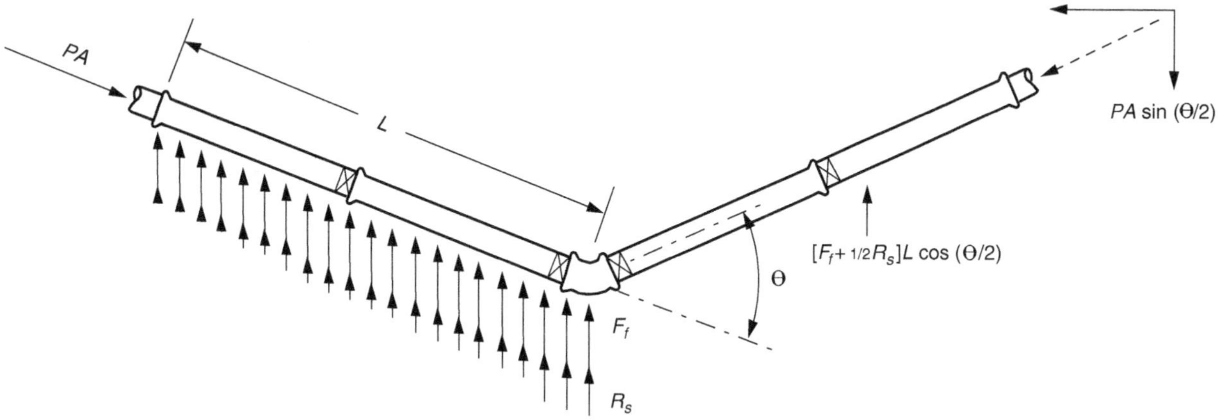

Figure 8-8 Schematic of pipe–soil behavior at horizontal bend (or vertical up-bend)

8.4.2 The Unit Frictional Force F_f

A static frictional force acting on a body is equal in magnitude to the applied force up to a maximum value. In the conventional analysis, the maximum static friction is proportional to the normal force between the surfaces that provide the friction. The constant of proportionality, in this case called the coefficient of friction, depends on the nature of the surfaces. The empirical work from Potyondy (1961) indicates that for friction between pipe and soils, the force is also dependent on the cohesion of the soil.

Thus

$$F_f = A_p C + W \tan \delta$$

Where

A_p = surface area of the pipe bearing on the soil, ft²/ft (m²/m)

= $\frac{\pi D'}{2}$ (for bends, assume ½ the pipe circumference bears against the soil)

= $\pi D'$ (for tee branches, dead-end conditions, and reducers, assume the full pipe circumference bears against the soil)

C = $f_c C_s$, lb/ft² (kN/m²)

f_c = ratio of pipe cohesion to soil cohesion

W = unit normal force, lb/ft (kN/m)

C_s = soil cohesion, lb/ft² (kN/m²)

δ = $f_\varphi \varphi$, degrees

f_φ = ratio of pipe friction angle to soil friction angle

φ = internal friction angle of the soil, degrees

Note in the above that the surface area of the pipe bearing on the soil depends on the direction of the thrust force compared to the direction of the centerline of the restrained pipe. For bends, the direction of the thrust force is at an angle to the centerline of the pipe length being restrained and the impending movement of the pipe is at a lateral direction. Therefore, the friction force only develops on the leading face of the pipe surface, and the surface area is calculated using ½ of the pipe circumference. The resultant friction force is denoted as F_s. For tee branches, dead ends and reducers, the direction of the thrust force is parallel to the centerline of the pipe to be restrained, so the full circumference of the pipe is used to calculate the surface area of the pipe bearing on the soil. For these fittings, the friction force is denoted as $(F_s)_b$.

Values of soil cohesion C_s and internal friction angle of the soil φ must be known or conservatively estimated for the soil at a particular installation. Values for f_c and f_φ are related to soil conditions and the pipe material. Conservative values of these parameters for ductile-iron pipe in seven general classifications of saturated soil types are presented in Table 8-2.

The unit normal force W is given by

$$W = 2W_e + W_p + W_w$$

where the earth load W_e is taken as the prism load on the pipe in pounds. The earth load W_e is doubled to account for the forces acting on both the top and bottom of the pipe (Figure 8-9). The total unit weight of the pipe and water ($W_p + W_w$) for each size of pipe is given in Table 8-4.

Then, for bends,

$$F_s = \frac{\pi D'C}{2} + \left(2W_e + W_p + W_w\right)\tan\delta \qquad \text{(Eq 8-4)}$$

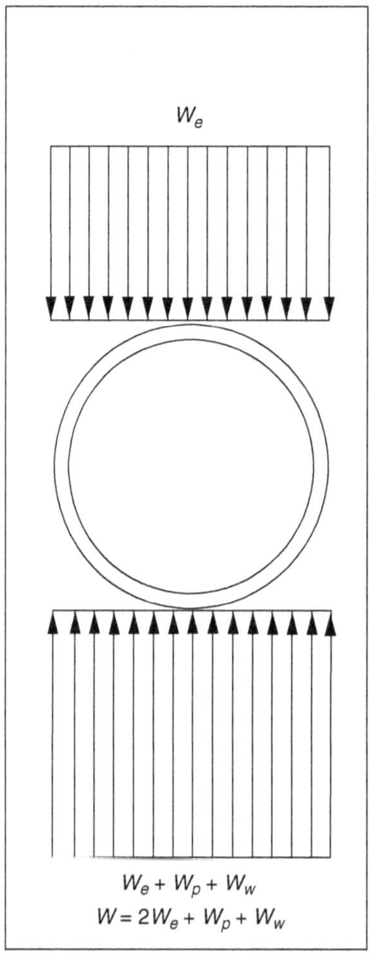

Figure 8-9 Unit normal forces on pipe

Table 8-4 Values for D', A, W_p, W_w, and $(W_p + W_w)$

Nominal Pipe Size, in.	Pressure Class	Pipe Outside Diameter D', ft	Cross-sectional Area of Pipe A, in.^{2}	Wp,* lb/ft	Ww,* lb/ft	$Wp + Ww$,* lb/ft
3	350	0.33	12.3	10	4	14
4	350	0.40	18.1	12	6	18
6	350	0.58	37.3	18	13	31
8	350	0.75	64.3	24	24	48
10	350	0.93	96.7	30	37	67
12	350	1.10	136.8	39	53	92
14	250	1.28	183.8	47	72	119
16	250	1.45	237.7	57	94	151
18	250	1.63	298.6	66	119	185
20	250	1.80	366.4	78	147	225
24	200	2.15	522.7	93	212	305
30	150	2.67	804.2	123	329	452
36	150	3.19	1,152.0	163	473	636
42	150	3.71	1,555.2	206	642	848
48	150	4.23	2,026.8	261	838	1,099
54	150	4.80	2,602.1	325	1,078	1,403
60	150	5.13	2,981.2	371	1,237	1,608
64	150	5.47	3,387.0	410	1,407	1,817

NOTE: To convert inches (in.) to millimeters (mm), multiply by 25.4; to convert feet (ft) to meters (m), multiply by 0.3048; to convert square inches (in.^{2}) to square meters (m^2), multiply by 6.45×10^{-4}; to convert pounds per foot (lb/ft) to kilonewtons per meter (kN/m), multiply by 0.0146.

The difference in $Wp + Ww$ for other pipe pressure classes is not normally significant in relation to these calculations, and these values may be used conservatively regardless of pipe pressure class. However, the designer may use actual pipe weights for optimum design, if desired. To obtain the values tabulated in Tables 8-5 through 8-18, it may be necessary to use more precise numbers than those listed in this table.

* Based on minimum pressure class pipe with standard cement–mortar lining.

For tee branches, dead-end conditions, and reducers,

$$(F_s)_b = \pi D'C + \left(2W_e + W_p + W_w\right)\tan\delta \qquad \text{(Eq 8-5)}$$

Extraordinary installations might result in lesser loads and frictional resistance on the pipes than those calculated by these equations and as shown in Figure 8-8. When such conditions exist, this must be provided for in the design.

8.4.3 Frictional Resistance with Polyethylene Encasement

Limited experimental data suggest that the frictional resistance terms, F_s and $(F_s)_b$, should be multiplied by a factor of 0.7 for pipe encased in polyethylene film to determine the appropriate value of F_f to use in the equations.

8.4.4 The Unit Bearing Resistance R_s

The maximum unit lateral resistance, R_s, at the bend is limited so as not to exceed a rectangular distribution of the Rankine passive soil pressure, P_p, which is generally less than the ultimate capacity of the soil to resist pipe movement. Passive soil pressure is a term generally defined as the maximum horizontal pressure that will be resisted by the soil structure without shearing failure of the soil. Horizontal subgrade pressure will result in a deformation of the soil structure. The resistance offered by the subgrade soil increases with this deformation or strain for pressures less than the passive soil pressure. In a soil with density exceeding the critical void ratio (this condition is usually obtained in stable undisturbed soil and in backfill compacted to approximately 80 percent or more of the standard Proctor density), the movement or deformation that occurs in developing the full passive soil pressure is very small in relation to the allowable or available movement at the bend in restrained push-on or mechanical joint systems used with ductile-iron pipe.

The passive soil pressure for a particular soil is given by the Rankine formula:

$$P_p = \gamma H_c N_\phi + 2C_s\sqrt{N_\phi} \qquad \text{(Eq 8-6)}$$

Where

P_p = passive soil pressure, lb/ft^2 (kN/m^2)

γ = backfill soil density, lb/ft^3 (kN/m^3)

H_c = mean depth from surface to the plane of resistance (centerline of a pipe or center of bearing area of a thrust block), ft (m)

N_φ = $\tan^2 (45° + \varphi/2)$

C_s = soil cohesion, lb/ft^2 (kN/m^2)

φ = internal friction angle of the soil, degrees

As discussed previously, the full Rankine passive soil pressure, P_p, can be developed with insignificant movement in well-compacted soils. For some of the standard laying conditions (Figure 4-2) for ductile iron, the design value of passive soil pressure should be modified by a factor K_n to ensure that excessive movement will not occur. (Empirically determined values for K_n can be found in Table 8-2.)

Therefore,

$$R_s = K_n P_p D' \qquad \text{(Eq 8-7)}$$

In this context, the value chosen for K_n depends on the compaction achieved in the trench, the backfill materials, and the undisturbed earth.

For convenience of the designer, computed values for F_s (Equation 8-4), $(F_s)_b$ (Equation 8-5), and R_s (Equation 8-7) are included in Tables 8-5 through 8-11. Additionally, values of solutions to Equation 8-3 are included in Tables 8-12 through 8-18. In Tables 8-5 through 8-18, the values shown are based on the seven soil types and suggested parameters from Table 8-2.

The suggested values of the parameters listed in Table 8-2 are believed to be very conservative; however, AWWA cannot assume responsibility for these values corresponding to actual conditions at any particular jobsite.

8.5 RESTRAINT DESIGN FOR VERTICAL BENDS, TEES, REDUCERS, AND DEAD ENDS

The following design equations for vertical bends, tees, reducers, and dead ends were derived with assumptions similar to those used in the derivation of the horizontal bend equation (Equation 8-3). Space does not permit full discussion of the derivations, nor does it allow discussion of all possible fittings and thrust configurations.

8.5.1 Vertical Down Bends

Because the bend system will move upward in the direction of the thrust force, it will bear against backfilled (disturbed) soil. For conservatism, the weight of the earth, pipe, and water directly opposing the thrust force is ignored, and the unit bearing resistance, R_s, is not relied upon to calculate the length of pipe to restrain. However, the weight of the earth, pipe, and water is used in calculating the unit frictional force, F_f.

Referring to Figure 8-10 and summing forces in the *Y* direction,

$$\sum Fy = 0, \text{ yields}$$

$$2PA \sin\left(\frac{\theta}{2}\right) - 2F_f\, L \cos\left(\frac{\theta}{2}\right) = 0$$

Employing a safety factor and solving for *L*,

$$L = \left(\frac{S_f\, PA \tan\left(\frac{\theta}{2}\right)}{F_f}\right) \qquad \text{(Eq 8-8)}$$

Where

S_f = safety factor (usually 1.5)

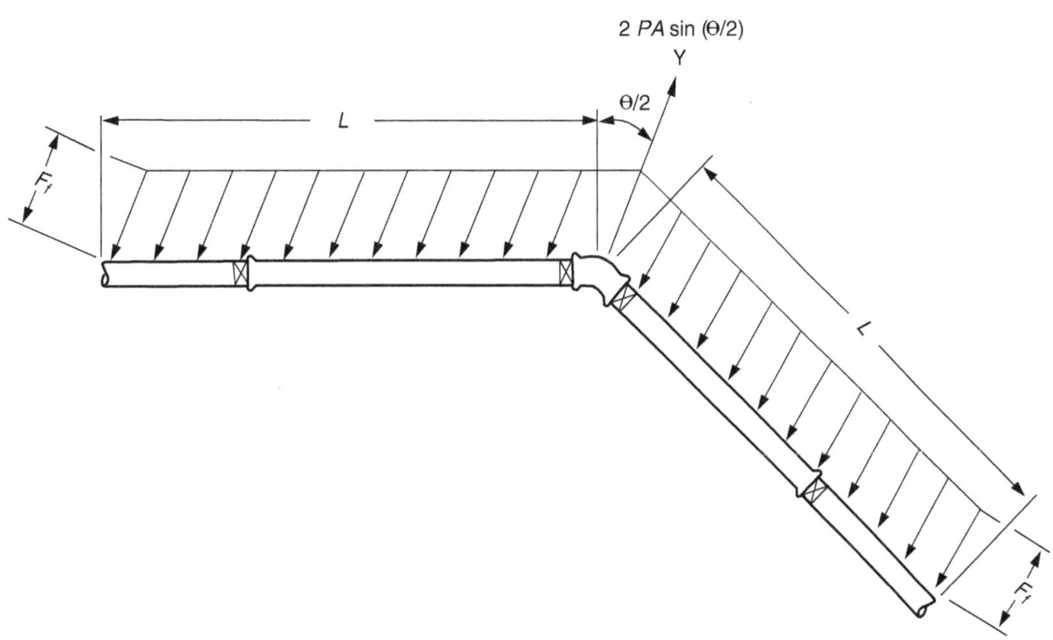

Figure 8-10 Vertical down-bend diagram

8.5.2 Vertical Up bends

Referring to Figure 8-8 for force diagram,

$$L = \left(\frac{S_f\, PA \tan\left(\frac{\theta}{2}\right)}{F_f + \frac{1}{2} R_s} \right) \qquad \text{(Eq 8-9)}$$

The bend system in this case will attempt to move downward in the direction of thrust, against the bottom of the trench, and therefore transfer to undisturbed soil. Consequently, the values of K_n in this case should be chosen to reflect the conditions of the trench bottom on which the pipe rests, assuming adequate bell holes are provided. In most cases, values representing those of type 4 or 5 trench conditions may be used, as the trench bottom is normally relatively undisturbed.

8.5.3 Tees

Referring to Figure 8-11, the friction force that develops is along the branch of the tee, while the unit soil resistance is developed over the first lengths of pipe on either side of the run of the tee.

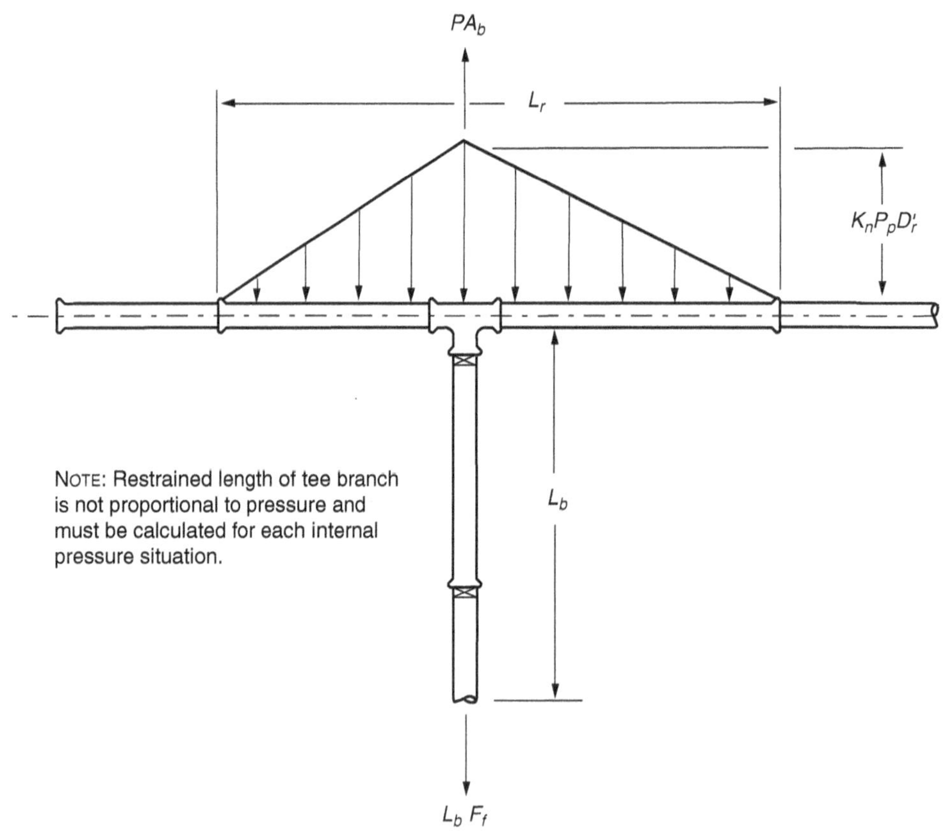

Figure 8-11 Reaction forces at tees

$$PA_b = L_b F_f + \left(\frac{1}{2} R_s L_r\right)$$

Employing a safety factor and solving for L_b,

$$L_b = \left(\frac{S_f PA_b - \left(\frac{1}{2} R_s L_r\right)}{F_f}\right) \quad \text{(Eq 8-10)}$$

$$R_s = K_n P_p D_r'$$

Where

- A_b = cross-sectional area of branch, in.^{2} (m^2)
- L_b = length of branch to be restrained, ft (m)
- F_f = $(F_s)_b$; for standard asphaltic-coated pipe
- F_f = 0.7 $(F_s)_b$; for polyethylene-encased pipe
- $(F_s)_b$ = unit frictional force on branch, lb/ft (kN/m)
- = $\pi D'C + (2 W_e + W_p + W_w) \tan \delta$ (used for tee branches, reducers, and dead ends)
- D'_r = diameter of run, ft (m)
- L_r = total length between first joints on either side of tee on the run, ft (m)
- S_f = safety factor (usually 1.5)

Note that the restrained length of the tee branch is not proportional to the internal pressure of the pipeline and must be calculated for each internal pressure situation.

8.5.4 Reducers

Referring to Figure 8-12, the direction of the thrust force is toward the smaller-sized pipe, and the restrained joints are calculated for the larger-sized pipe.

$$L_1 = \left(\frac{S_f P(A_1 - A_2)}{F_{f1}}\right) \quad \text{(Eq 8-11)}$$

Where

- S_f = safety factor (usually 1.5)
- A_1 = cross-sectional area of larger pipe, in.^{2} (m^2)
- A_2 = cross-sectional area of smaller pipe, in.^{2} (m^2)

NOTE: If the straight run of pipe on the small side of the reducer exceeds the length L_2 determined from Equation 8-12, then no restrained joints are necessary.

$$L_2 = \left(\frac{S_f P(A_1 - A_2)}{F_{f2}}\right) \quad \text{(Eq 8-12)}$$

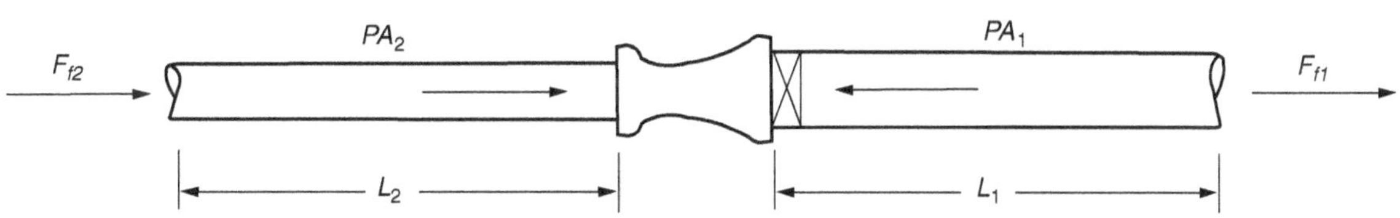

Figure 8-12 Reaction forces for reducers

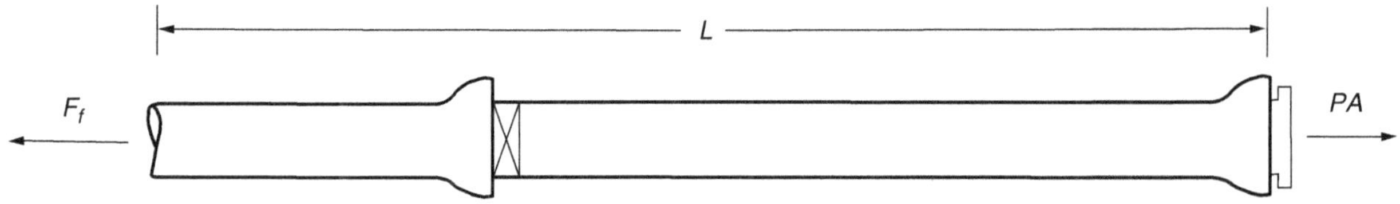

Figure 8-13 Reaction forces at dead ends

8.5.5 Dead ends

Referring to Figure 8-13, the only force resisting the thrust is the friction force that develops around the full circumference of the pipe approaching the dead end.

$$L = \left(\frac{S_f PA}{F_f}\right) \quad \text{(Eq 8-13)}$$

Where

S_f = safety factor (usually 1.5)

8.6 ENCROACHING RESTRAINED LENGTHS

Horizontal and vertical offsets are commonly encountered in restrained sections of a line. These offsets should be made with as small a degree bend as possible to minimize the thrust loads and restrained length required. Also, in these configurations, an increase in line segment length could be detrimental to the pipeline or surrounding structures due to overdeflection of the joints. Therefore, the restrained joints should be fully engaged during installation. For some restrained joints, this involves extending the joint to ensure there would be no longitudinal movement of the pipe when the internal pressure is applied.

In certain configurations, fittings may be close enough to one another that adjacent calculated restrained lengths overlap. In situations of this type, one approach is to

- restrain all pipe between the two fittings,
- assume half the restrained pipe length between the two fittings will act to resist the thrust force of each fitting, and
- calculate the additional restrained length required on the outer legs of the fitting using the appropriate equations.

For each encroaching restrained length condition, it is noted that as the degree of bend approached 90°, the lateral movement of the outer legs is restricted to zero. For this condition, restrain all pipe between the fittings and restrain the outer legs as dead ends.

Note: Calculated restrained joint length for outer legs at encroaching combined bends should not exceed that required for dead ends. If the calculated additional restrained length exceeds the length that would be obtained if the elbow were treated as a dead end, you may use the dead-end length.

8.6.1 Equal Angle Vertical Offset (θ°)

Referring to Figure 8-14, this assumes the pipeline is changing elevation through the use of two equal bends.

For L_1,

$\Sigma F = 0$

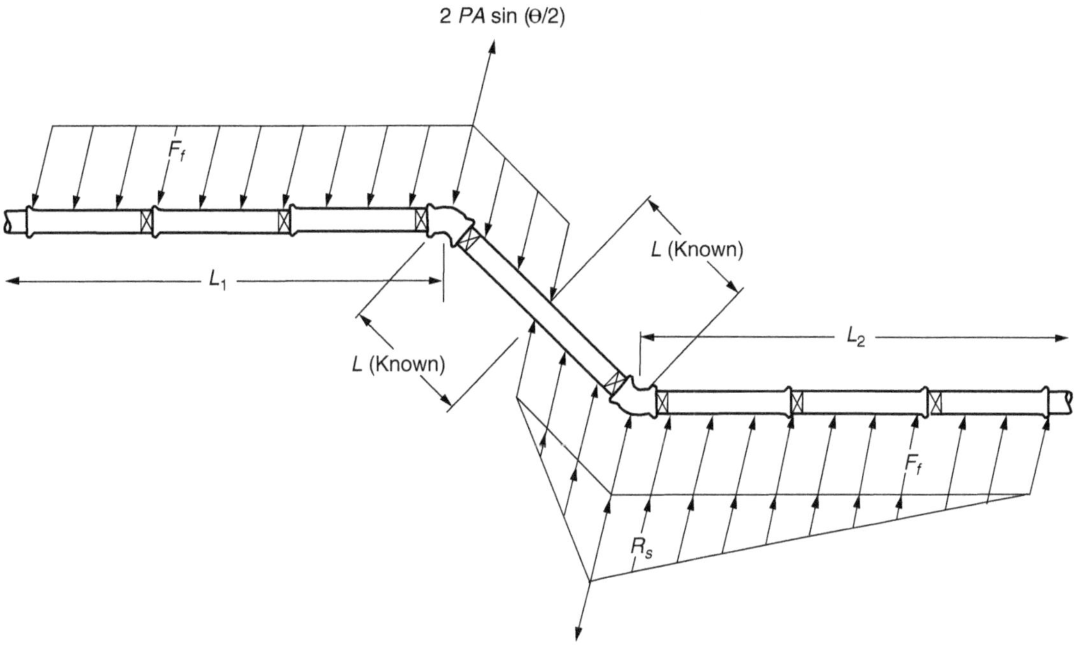

NOTE: As the bend angle approaches 90°, lateral movement of the outer legs approaches zero. For this condition, restrain all pipe between the fittings and restrain the outer legs as dead ends.

Figure 8-14 Equal angle vertical offset (θ°)

$$\left[2PA \sin\left(\frac{\theta}{2}\right)\right] = \left[F_f L \cos\left(\frac{\theta}{2}\right)\right] + \left[F_f L_1 \cos\left(\frac{\theta}{2}\right)\right]$$

Employing a safety factor and solving for L_1,

$$L_1 = \frac{S_f 2PA \tan\left(\frac{\theta}{2}\right)}{F_f} - L \qquad \text{(Eq 8-14)}$$

For L_2,

$\Sigma F = 0$

$$\left[2PA \sin\left(\frac{\theta}{2}\right)\right] = \left[F_f L \cos\left(\frac{\theta}{2}\right)\right] + \left[\frac{1}{2} R_s L \cos\left(\frac{\theta}{2}\right)\right] + \left[F_f L_2 \cos\left(\frac{\theta}{2}\right)\right] + \left[\frac{1}{2} R_s L_2 \cos\left(\frac{\theta}{2}\right)\right]$$

Employing a safety factor and solving for L_2,

$$L_2 = \frac{S_f 2PA \tan\left(\frac{\theta}{2}\right)}{F_f + \frac{1}{2} R_s} - L \qquad \text{(Eq 8-15)}$$

S_f = Safety factor (usually 1.5)

8.6.2 Combined Horizontal Equal Angle Bends (θ°)

Referring to Figure 8-15, this configuration involves a change in the horizontal direction of the pipe through two equal angle horizontal bends:

For L_1,

$\Sigma F = 0$

$$\left[2PA\sin\left(\frac{\theta}{2}\right)\right] = \left[F_f L\cos\left(\frac{\theta}{2}\right)\right] + \left[\frac{1}{2}R_s L\cos\left(\frac{\theta}{2}\right)\right] + \left[F_f L_1\cos\left(\frac{\theta}{2}\right)\right] + \left[\frac{1}{2}R_s L_1\cos\left(\frac{\theta}{2}\right)\right]$$

Employing a safety factor and solving for L_1,

$$L_1 = \frac{S_f 2PA\tan\left(\frac{\theta}{2}\right)}{F_f + \frac{1}{2}R_s} - L \qquad \text{(Eq 8-16)}$$

S_f = Safety factor (usually 1.5)

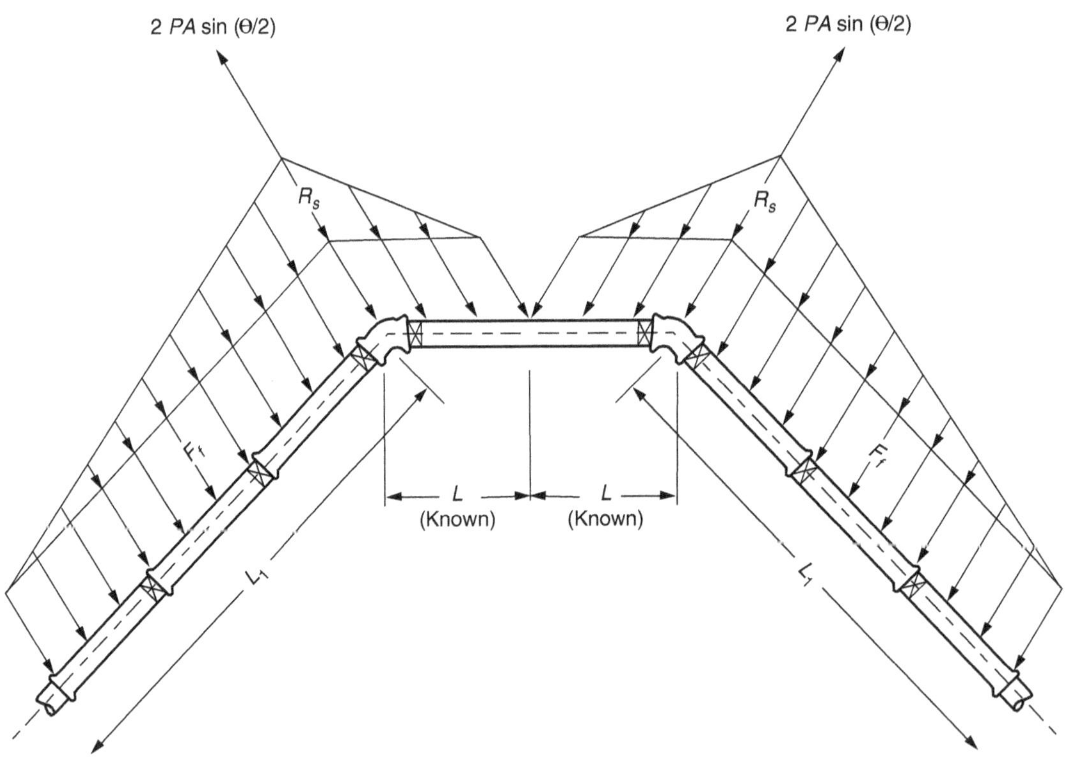

NOTE: As the bend angle approaches 90°, lateral movement of the outer legs approaches zero. For this condition, restrain all pipe between the fittings and restrain the outer legs as dead ends.

Figure 8-15 Combined horizontal equal angle bends (θ°)

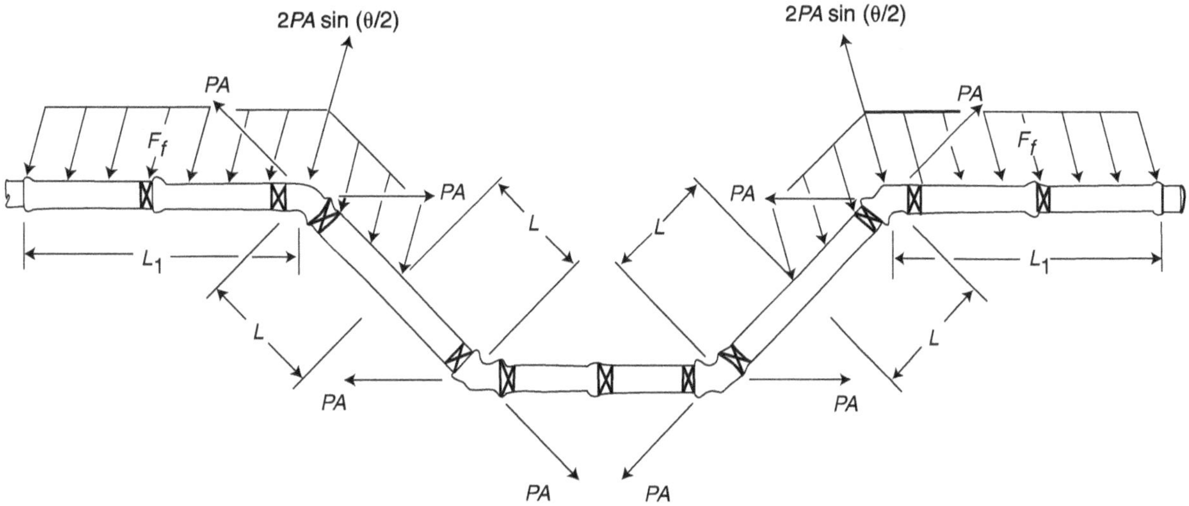

Figure 8-16 Combined equal angle vertical offsets (θ°)

8.6.3 Combined Vertical Equal Angle Offsets (θ°)—Under Obstruction

Vertical offsets are often combined to route a pipeline under an obstruction or existing utility (Figure 8-16). If the required restrained length of the vertical up bends does not overlap, the system may be treated as two individual vertical offsets (see Section 8.6.1). If the required restrained lengths do overlap, the following steps may be taken.

1. Restrain all pipe between the outermost two fittings. Because of opposing forces, the thrust forces of the middle two fittings (vertical up-bends) are counteracted.
2. Assume that one-half of the restrained pipe length between the vertical down bends and vertical up bends acts to resist the thrust force of the vertical down-bends.
3. Using the appropriate equations, calculate the additional restrained length required on the outermost legs of the offset system (vertical down bends). The resulting equation is the same as for the vertical down bend in the single vertical offset (Equation 8-14).

$$L_1 = \frac{S_f 2PA \tan\left(\frac{\theta}{2}\right)}{F_f} - L \qquad \text{(Eq 8-17)}$$

S_f = Safety factor (usually 1.5)

Note: As the bend angle approaches 90°, lateral movement of the outer legs approaches zero. For this condition, restrain all pipe between the fittings and restrain the outer legs as dead ends.

8.6.4 Combined Equal Angle Vertical Offsets (θ°)—Over Obstruction

NOTE: As the bend angle approaches 90°, lateral movement of the outer legs approaches zero. For this condition, restrain all pipe between the fittings and restrain the outer legs as dead ends.

This can be analyzed in the same manner as Figure 8-16, but the outermost bends are vertical up bends, and the following equation applies.

$$L_1 = \frac{S_f 2PA \tan\left(\frac{\theta}{2}\right)}{F_f + \frac{1}{2}R_s} - L \qquad \text{(Eq 8-18)}$$

S_f = Safety factor (usually 1.5)

NOTE: Equation 8-18 also applies to combined horizontal equal angle offsets—around an obstruction.

8.7 RESTRAINED LENGTH

In practice, the actual restrained length attained will generally be in multiples of the length of an individual piece of pipe (usually 18 or 20 ft [5.5 or 6.1 m]). The length calculated indicates the minimum required restrained length for each side of the bend. Thus, calculated lengths of 0 to 18 or 20 ft (5.5 or 6.1 m) normally call for one restrained joint at the fitting, and 18 to 36 ft or 20 to 40 ft (5.5 to 11.0 m or 6.1 to 12.2 m) normally call for two restrained joints. An exception would be joint deflection on buried full lengths of pipe, which does not usually require joint restraint. This is because the lateral frictional and passive resistance developed along two full lengths of pipe is generally more than adequate to restrain thrust forces resulting from the maximum recommended joint deflection.

8.8 SELECT BACKFILL CONSIDERATIONS

If restrained joint pipe is laid in trench backfill with markedly different support characteristics than the native soil, special considerations may be required. As the pipe is pressurized, it will transmit passive pressure to the backfill that will in turn transmit the pressure to the native soil. Therefore, the material that results in the smaller unit bearing resistance, R_s, should be used for the passive resistance, and the unit frictional force, F_s, should be based on the backfill material surrounding the pipe.

If restrained joints are used in swamps or marshes where the soil is unstable or in other situations where the bearing strength of the soil is extremely poor, the entire pipeline should be restrained to provide adequate thrust restraint.

8.9 COMBINING THRUST BLOCKS AND RESTRAINED JOINTS

Combining thrust blocks and restrained joints by designing each system independently of the other and then incorporating both into the piping system normally yields the greatest degree of security.

It is poor practice to mix systems if each system is designed to resist a percentage of the resultant thrust force. Thrust blocks and restrained joint pipe systems require slight movement before their respective thrust restraint capability can be developed. Those movements must be compatible for the combination to be successful. Because of the uncertainties of the degree of these movements being compatible, this design approach must be given special consideration.

8.10 PIPE IN A CASING

It is often necessary to install restrained joint pipe through a casing pipe. The function of restrained joint pipe is to transfer thrust forces to the soil structure. Therefore, if the annular space between the two pipes is not grouted in, the length of restrained pipe inside the casing should not be considered as part of the restrained length to balance the thrust force. When restrained joint pipe is installed through a casing pipe, the restrained joints should normally be fully extended.

8.11 FUTURE EXCAVATIONS

One particular concern of infrastructure pipeline design, installation, and maintenance is the possibility of substantial excavation in the close vicinity of previously installed restrained pipe and fittings, including parallel excavations. Bearing in mind the usual function of restrained pipes in transmitting thrust forces to the soil structure, if this structure is removed or significantly disturbed with the pipeline under pressure, the safety and stability of the system may be compromised. In this regard, it would seem reasonable to temporarily shut down existing restrained lines to do such work or to conduct such operations during lowest-pressure service conditions. Where this is not practical or possible, alternate provisions might be safely employed. These precautions might include supplementary thrust blocking, restraint with laterally loaded piles or batter piles at the thrust focus, special pipe anchors, or other careful, sequential, and innovative engineering and construction procedures. Proper engineering and construction judgment must be exercised in these conditions.

8.12 DEFLECTED UNRESTRAINED DUCTILE-IRON PIPE JOINTS

Unrestrained push-on and mechanical joints for ductile-iron pipe are capable of deflections up to 8° (depending on joint type and pipe size). These joints are well suited for diverting pipelines from obstructions or when following the curvature of streets and roads. To keep thrust forces to a minimum, joint deflections should be utilized whenever possible rather than fittings. In pressurized systems, thrust forces develop at these joint deflections. In the vast majority of installations, the soil–pipe interaction will result in reasonable security and stability of the joints. Only in extraordinary circumstances—e.g., exposed pipe or future excavations, unstable soils, high internal pressure in combination with very shallow cover—is security threatened. In these situations, soil–pipe thrust resisting principles, not unlike those presented in this chapter, may be applied to analyze these unrestrained joint situations.

8.13 SUBMERGED SOIL CONDITIONS

When the pipeline may be subjected to submerged soil conditions (below the water table), some parameters should be adjusted for reduced soil backfill density and buoyant force on the pipe or blocking. These parameters include S_b (horizontal bearing strength) and W_m (thrust block density for gravity thrust block) for blocking and γ (backfill soil density), W_e (earth prism load), and W_p (unit weight of pipe) for restrained joints.

8.14 COMPUTER PROGRAM

For the convenience of the designer, a computer program has been developed based upon the procedures and equations of this chapter. It can be used to assist with calculations of both unit frictional force and unit bearing resistance. Additionally, the computer program may be used to facilitate calculations in determining the required length of restrained piping. This program can be accessed on DIPRA's website at www.dipra.org.

8.15 RESTRAINED LENGTH CALCULATION PROCEDURE

The following is an example of a restrained length calculation procedure.

A 30-in. (762-mm) ductile-iron pipeline is to be buried under 6 ft (1.83 m) of cover in a cohesive granular backfill that will be compacted to 80 percent standard Proctor density to the top of the pipe (laying condition 4). The thrust restraint design pressure is 150 psi (1,034 kPa). The length of restrained piping required should be determined at a 90° horizontal bend.

Step 1. Using Equation 8-3, establish known values for

$$L = \frac{S_f PA \tan\left(\frac{\theta}{2}\right)}{F_f + \frac{1}{2}R_s}$$

Where

R_s = $K_n P_p D'$
S_f = safety factor, 1.5
P = design pressure, 150 psi
A = cross-sectional area of pipe, $36\,\pi\,(D')^2 = 806.3$ in.^{2}
θ = bend angle, 90°
K_n = trench condition modifier, 0.85 (from Table 8-2)
D' = outside diameter of pipe, 32/12 = 2.67 ft
F_f = F_s

Step 2. Determine unit frictional resistance, F_s

$$F_s = \frac{\pi(2.67)(80)}{2} + \left[2(1{,}442) + 452\right] \tan 13°$$

Where

C = $f_c C_s = (0.40)(200) = 80$ lb/ft^2
f_c = 0.40 (from Table 8-2)
C_s = 200 lb/ft^2 (from Table 8-2)

$W_e = H\gamma D' = 6 \times 90 \times 2.67 = 1{,}442$ lb/ft
$H = 6$ ft of cover (given)
$\gamma = 90$ lb/ft^3 (from Table 8-2)
$W_p + W_w = 452$ lb/ft (from Table 8-4)
$\delta = f_\varphi \varphi = (0.65)(20) = 13°$
$f_\varphi = 0.65$ (from Table 8-2)
$\varphi = 20°$ (from Table 8-2)

Then

$$Fs = 335.5 + 770.2 = 1{,}105.7 \text{ lb/ft}$$

Step 3. Using Equation 8-6, determine passive soil resistance P_p

$$P_p = \gamma H_c N_\varphi + 2C_s\sqrt{N_\phi}$$

Where

$$H_c = H + \frac{1}{2}D' = 6 + \frac{2.67}{2} = 7.33 \text{ ft}$$

$$N_\phi = \tan^2\left(45 + \frac{\phi}{2}\right) = \tan^2\left(45 + \frac{20}{2}\right) = 2.04$$

Then

$P_p = (90)(7.33)(2.04) + (2)(200)\sqrt{2.04}$
$P_p = 1{,}345.8 + 571.3 = 1{,}917.1$ lb/ft^2

Step 4. Substitute known and determined values into Equation 8-3 listed in Step 1 to determine required restrained length:

$$L = \frac{(1.5)(150)(806.3)\tan\left(\frac{90}{2}\right)}{1{,}105.7 + \left[\frac{1}{2}(0.85)(1{,}917.1)(2.67)\right]} = \frac{181{,}417.5}{1{,}105.7 + 2{,}175.4}$$

$$L = 55.3 \text{ ft}$$

Note: Tables 8-5 through 8-18 may be used to facilitate calculations to determine the required length of restrained piping. In the example above, a value of 37 ft would have been selected from Table 8-16. Adjusting for 150 psi, 1.5 × 37 = 55.5 ft ≈ 55.3 ft.

8.16 TABULAR VALUES FOR F_s, $(F_s)_b$, AND R_s

Tables 8-5 through 8-11 provide values for F_s, $(F_s)_b$, and R_s for the convenience of the designer of restrained joint systems. The tables are based on the seven general soil types described in Table 8-2. The suggested values of soil parameters for the general soil types are conservative within the range of actual values that may be measured for a given soil. In providing the suggested values in Table 8-2 and the computed values of F_s, $(F_s)_b$, and R_s in the following tables, AWWA cannot assume responsibility that these values accurately reflect the actual values of the soil parameters in a particular installation.

Table 8-5 Thrust restraint design for ductile-iron pipe for clay 1* soil type

Soil Parameters:	A21.50—Laying Conditions:	2	3	4	5
φ = 0 degrees					
C_s = 300 lb/ft^2 (14.37 kN/m^2)	f_φ =	0.00	0.00	0.00	0.00
γ = 90 lb/ft^3 (14.14 kN/m^3)	f_c =	0.50	0.80	0.80	0.80
	K_n =	0.20	0.40	0.60	0.85

		A21.50—Laying Conditions											
Nominal		2			3			4			5		
Pipe Size, *in.*	Depth, *ft*	F_s, *lb/ft*	$(F_s)_b$, *lb/ft*	R_s, *lb/ft*	F_s, *lb/ft*	$(F_s)_b$, *lb/ft*	R_s, *lb/ft*	F_s, *lb/ft*	$(F_s)_b$, *lb/ft*	R_s, *lb/ft*	F_s, *lb/ft*	$(F_s)_b$, *lb/ft*	R_s, *lb/ft*
3	2.5	78	156	55	124	249	111	124	249	166	124	249	236
3	3.0	78	156	58	124	249	117	124	249	175	124	249	248
3	4.0	78	156	64	124	249	129	124	249	193	124	249	273
3	6.0	78	156	76	124	249	152	124	249	229	124	249	324
3	8.0	78	156	88	124	249	176	124	249	264	124	249	374
3	10.0	78	156	100	124	249	200	124	249	300	124	249	425
4	2.5	94	188	67	151	302	135	151	302	202	151	302	187
4	3.0	94	188	71	151	302	142	151	302	213	151	302	302
4	4.0	94	188	78	151	302	156	151	302	235	151	302	333
4	6.0	94	188	93	151	302	185	151	302	278	151	302	394
4	8.0	94	188	107	151	302	214	151	302	321	151	302	455
4	10.0	94	188	121	151	302	243	151	302	364	151	302	516
6	2.5	135	271	98	217	434	196	217	434	294	217	434	416
6	3.0	135	271	103	217	434	206	217	434	309	217	434	438
6	4.0	135	271	113	217	434	227	217	434	340	217	434	482
6	6.0	135	271	134	217	434	268	217	434	402	217	434	570
6	8.0	135	271	155	217	434	310	217	434	464	217	434	658
6	10.0	135	271	175	217	434	351	217	434	526	217	434	746
8	2.5	178	355	130	284	569	259	284	569	389	284	569	551
8	3.0	178	355	136	284	569	273	284	569	409	284	569	579
8	4.0	178	355	150	284	569	300	284	569	450	284	569	637
8	6.0	178	355	177	284	569	354	284	569	531	284	569	753
8	8.0	178	355	204	284	569	408	284	569	613	284	569	868
8	10.0	178	355	231	284	569	463	284	569	694	284	569	983
8	2.5	178	355	130	284	569	259	284	569	389	284	569	551
10	3.0	218	436	169	349	697	337	349	697	506	349	697	717
10	4.0	218	436	185	349	697	371	349	697	556	349	697	788
10	6.0	218	436	219	349	697	437	349	697	656	349	697	929
10	8.0	218	436	252	349	697	504	349	697	756	349	697	1,071
10	10.0	218	436	285	349	697	570	349	697	856	349	697	1,212
12	2.5	259	518	192	415	829	385	415	829	577	415	829	818
12	3.0	259	518	202	415	829	405	415	829	607	415	829	860

(continued)

Table 8-5 Thrust restraint design for ductile-iron pipe for clay 1* soil type (*Continued*)

Soil Parameters:	A21.50—Laying Conditions:				
		2	3	4	5
φ = 0 degrees	f_φ =	0.00	0.00	0.00	0.00
C_s = 300 lb/ft^2 (14.37 kN/m^2)	f_c =	0.50	0.80	0.80	0.80
γ = 90 lb/ft^3 (14.14 kN/m^3)	K_n =	0.20	0.40	0.60	0.85

Nominal Pipe Size, *in.*	Depth, *ft*	A21.50—Laying Conditions 2 F_s, *lb/ft*	2 $(F_s)_b$, *lb/ft*	2 R_s, *lb/ft*	3 F_s, *lb/ft*	3 $(F_s)_b$, *lb/ft*	3 R_s, *lb/ft*	4 F_s, *lb/ft*	4 $(F_s)_b$, *lb/ft*	4 R_s, *lb/ft*	5 F_s, *lb/ft*	5 $(F_s)_b$, *lb/ft*	5 R_s, *lb/ft*
12	4.0	259	518	222	415	829	444	415	829	666	415	829	944
12	6.0	259	518	262	415	829	523	415	829	785	415	829	1,112
12	8.0	259	518	301	415	829	603	415	829	904	415	829	1,280
12	10.0	259	518	341	415	829	682	415	829	1,023	415	829	1,499
14	2.5	300	601	225	481	961	450	481	961	675	481	961	956
14	3.0	300	601	236	481	961	473	481	961	709	481	961	1,005
14	4.0	300	601	259	481	961	519	481	961	778	481	961	1,103
14	6.0	300	601	305	481	961	611	481	961	916	481	961	1,298
14	8.0	300	601	351	481	961	702	481	961	1,054	481	961	1,493
14	10.0	300	601	397	481	961	794	481	961	1,191	481	961	1,688
16	2.5	342	683	258	547	1,093	516	547	1,093	775	547	1,093	1,097
16	3.0	342	683	271	547	1,093	542	547	1,093	814	547	1,093	1,153
16	4.0	342	683	297	547	1,093	595	547	1,093	892	547	1,093	1,264
16	6.0	342	683	350	547	1,093	699	547	1,093	1,049	547	1,093	1,485
16	8.0	342	683	402	547	1,093	803	547	1,093	1,205	547	1,093	1,707
16	10.0	342	683	454	547	1,093	908	547	1,093	1,362	547	1,093	1,929
18	2.5	383	766	292	613	1,225	584	613	1,225	876	613	1,225	1,241
18	3.0	383	766	307	613	1,225	613	613	1,225	920	613	1,225	1,303
18	4.0	383	766	336	613	1,225	672	613	1,225	1,007	613	1,225	1,427
18	6.0	383	766	394	613	1,225	789	613	1,225	1,183	613	1,225	1,676
18	8.0	383	766	453	613	1,225	906	613	1,225	1,358	613	1,225	1,924
18	10.0	383	766	511	613	1,225	1,023	613	1,225	1,534	613	1,225	2,173
20	2.5	424	848	326	679	1,357	652	679	1,357	978	679	1,357	1,386
20	3.0	424	848	342	679	1,357	685	679	1,357	1,027	679	1,357	1,455
20	4.0	424	848	375	679	1,357	750	679	1,357	1,124	679	1,357	1,593
20	6.0	424	848	440	679	1,357	879	679	1,357	1,319	679	1,357	1,868
20	8.0	424	848	504	679	1,357	1,009	679	1,357	1,513	679	1,357	2,144
20	10.0	424	848	569	679	1,357	1,138	679	1,357	1,707	679	1,357	2,419
24	2.5	507	1,013	396	811	1,621	793	811	1,621	1,189	811	1,621	1,684
24	3.0	507	1,013	416	811	1,621	831	811	1,621	1,247	811	1,621	1,767
24	4.0	507	1,013	454	811	1,621	909	811	1,621	1,363	811	1,621	1,931

(continued)

Table 8-5 Thrust restraint design for ductile-iron pipe for clay 1* soil type (*Continued*)

Soil Parameters:		A21.50—Laying Conditions:			
			2	3	4	5
φ = 0 degrees		f_φ =	0.00	0.00	0.00	0.00
C_s = 300 lb/ft^2 (14.37 kN/m^2)		f_c =	0.50	0.80	0.80	0.80
γ = 90 lb/ft^3 (14.14 kN/m^3)		K_n =	0.20	0.40	0.60	0.85

		A21.50—Laying Conditions											
		2			3			4			5		
Nominal Pipe Size, *in.*	Depth, *ft*	F_s, *lb/ft*	$(F_s)_b$, *lb/ft*	R_s, *lb/ft*	F_s, *lb/ft*	$(F_s)_b$, *lb/ft*	R_s, *lb/ft*	F_s, *lb/ft*	$(F_s)_b$, *lb/ft*	R_s, *lb/ft*	F_s, *lb/ft*	$(F_s)_b$, *lb/ft*	R_s, *lb/ft*
24	6.0	507	1,013	532	811	1,621	1,064	811	1,621	1,595	811	1,621	2,260
24	8.0	507	1,013	609	811	1,621	1,218	811	1,621	1,828	811	1,621	2,589
24	10.0	507	1,013	687	811	1,621	1,373	811	1,621	2,060	811	1,621	2,918
30	2.5	628	1,257	504	1,005	2,011	1,008	1,005	2,011	1,512	1,005	2,011	2,142
30	3.0	628	1,257	528	1,005	2,011	1,056	1,005	2,011	1,584	1,005	2,011	2,244
30	4.0	628	1,257	576	1,005	2,011	1,152	1,005	2,011	1,728	1,005	2,011	2,448
30	6.0	628	1,257	672	1,005	2,011	1,344	1,005	2,011	2,016	1,005	2,011	2,856
30	8.0	628	1,257	768	1,005	2,011	1,536	1,005	2,011	2,304	1,005	2,011	3,264
30	10.0	628	1,257	864	1,005	2,011	1,728	1,005	2,011	2,592	1,005	2,011	3,672
36	2.5	752	1,504	618	1,203	2,406	1,237	1,203	2,406	1,855	1,203	2,406	2,628
36	3.0	752	1,504	647	1,203	2,406	1,294	1,203	2,406	1,941	1,203	2,406	2,750
36	4.0	752	1,504	704	1,203	2,406	1,409	1,203	2,406	2,113	1,203	2,406	2,994
36	6.0	752	1,504	819	1,203	2,406	1,639	1,203	2,406	2,458	1,203	2,406	3,482
36	8.0	752	1,504	934	1,203	2,406	1,869	1,203	2,406	2,803	1,203	2,406	3,971
36	10.0	752	1,504	1,049	1,203	2,406	2,098	1,203	2,406	3,148	1,203	2,406	4,459
42	2.5	874	1,748	736	1,398	2,796	1,471	1,398	2,796	2,207	1,398	2,796	3,126
42	3.0	874	1,748	769	1,398	2,796	1,538	1,398	2,796	2,307	1,398	2,796	3,268
42	4.0	874	1,748	836	1,398	2,796	1,672	1,398	2,796	2,507	1,398	2,796	3,552
42	6.0	874	1,748	969	1,398	2,796	1,939	1,398	2,796	2,908	1,398	2,796	4,119
42	8.0	874	1,748	1,103	1,398	2,796	2,206	1,398	2,796	3,308	1,398	2,796	4,687
42	10.0	874	1,748	1,236	1,398	2,796	2,473	1,398	2,796	3,709	1,398	2,796	5,254
48	2.5	997	1,995	860	1,596	3,192	1,720	1,596	3,192	2,579	1,596	3,192	3,654
48	3.0	997	1,995	898	1,596	3,192	1,796	1,596	3,192	2,694	1,596	3,192	3,816
48	4.0	997	1,995	974	1,596	3,192	1,948	1,596	3,192	2,922	1,596	3,192	4,140
48	6.0	997	1,995	1,126	1,596	3,192	2,253	1,596	3,192	3,379	1,596	3,192	4,788
48	8.0	997	1,995	1,279	1,596	3,192	2,558	1,596	3,192	3,837	1,596	3,192	5,435
48	10.0	997	1,995	1,431	1,596	3,192	2,863	1,596	3,192	4,294	1,596	3,192	6,083
54	2.5	1,130	2,260	999	1,808	3,617	1,997	1,808	3,617	2,996	1,808	3,617	4,244
54	3.0	1,130	2,260	1,042	1,808	3,617	2,083	1,808	3,617	3,125	1,808	3,617	4,427
54	4.0	1,130	2,260	1,128	1,808	3,617	2,256	1,808	3,617	3,384	1,808	3,617	4,794
54	6.0	1,130	2,260	1,301	1,808	3,617	2,601	1,808	3,617	3,902	1,808	3,617	5,528

(continued)

Table 8-5 Thrust restraint design for ductile-iron pipe for clay 1* soil type (*Continued*)

Soil Parameters:		A21.50—Laying Conditions:			
			2	3	4	5
φ = 0 degrees						
C_s = 300 lb/ft^2 (14.37 kN/m^2)		f_φ =	0.00	0.00	0.00	0.00
γ = 90 lb/ft^3 (14.14 kN/m^3)		f_c =	0.50	0.80	0.80	0.80
		K_n =	0.20	0.40	0.60	0.85

Nominal Pipe Size, *in.*	Depth, *ft*	A21.50—Laying Conditions 2			3			4			5		
		F_s, *lb/ft*	$(F_s)_b$, *lb/ft*	R_s, *lb/ft*	F_s, *lb/ft*	$(F_s)_b$, *lb/ft*	R_s, *lb/ft*	F_s, *lb/ft*	$(F_s)_b$, *lb/ft*	R_s, *lb/ft*	F_s, *lb/ft*	$(F_s)_b$, *lb/ft*	R_s, *lb/ft*
54	8.0	1,130	2,260	1,473	1,808	3,617	2,947	1,808	3,617	4,420	1,808	3,617	6,262
54	10.0	1,130	2,260	1,646	1,808	3,617	3,292	1,808	3,617	4,938	1,808	3,617	6,996
60	2.5	1,210	2,419	1,084	1,936	3,871	2,169	1,936	3,871	3,253	1,936	3,871	4,609
60	3.0	1,210	2,419	1,131	1,936	3,871	2,261	1,936	3,871	3,392	1,936	3,871	4,805
60	4.0	1,210	2,419	1,223	1,936	3,871	2,446	1,936	3,871	3,669	1,936	3,871	5,198
60	6.0	1,210	2,419	1,408	1,936	3,871	2,816	1,936	3,871	4,223	1,936	3,871	5,983
60	8.0	1,210	2,419	1,593	1,936	3,871	3,185	1,936	3,871	4,778	1,936	3,871	6,769
60	10.0	1,210	2,419	1,777	1,936	3,871	3,555	1,936	3,871	5,332	1,936	3,871	7,554
64	2.5	1,289	2,579	1,172	2,063	4,126	2,345	2,063	4,126	3,517	2,063	4,126	4,983
64	3.0	1,289	2,579	1,222	2,063	4,126	2,443	2,063	4,126	3,665	2,063	4,126	5,192
64	4.0	1,289	2,579	1,320	2,063	4,126	2,641	2,063	4,126	3,961	2,063	4,126	5,611
64	6.0	1,289	2,579	1,517	2,063	4,126	3,035	2,063	4,126	4,552	2,063	4,126	6,448
64	8.0	1,289	2,579	1,714	2,063	4,126	3,429	2,063	4,126	5,143	2,063	4,126	7,286
64	10.0	1,289	2,579	1,911	2,063	4,126	3,823	2,063	4,126	5,734	2,063	4,126	8,123

NOTE: To convert inches (in.) to millimeters (mm), multiply by 25.4; to convert feet (ft) to meters (m), multiply by 0.3048; to convert pounds per foot (lb/ft) to kilonewtons per meter (kN/m), multiply by 0.0146.
* See Tables 8-2 and 8-3 for soil description.

Table 8-6 Thrust restraint design for ductile-iron pipe for silt 1* soil type

Soil Parameters:	A21.50—Laying Conditions:				
		2	3	4	5
φ = 29 degrees	f_φ =	0.50	0.75	0.75	0.75
C_s = 0 lb/ft^2	f_c =	0.00	0.00	0.00	0.00
γ = 90 lb/ft^3 (14.14 kN/m^3)	K_n =	0.20	0.40	0.60	0.85
NOTE: $F_s = (Fs)_b$ in this soil type.					

		A21.50—Laying Conditions							
		2		3		4		5	
Nominal Pipe Size, *in.*	Depth, *ft*	F_s, *lb/ft*	R_s, *lb/ft*	F_s, *lb/ft*	R_s, *lb/ft*	F_s, *lb/ft*	R_s, *lb/ft*	F_s, *lb/ft*	R_s, *lb/ft*
3	2.5	42	46	65	91	65	137	65	194
3	3.0	50	54	77	108	77	163	77	230
3	4.0	65	71	100	143	100	214	100	303
3	6.0	96	106	148	211	148	317	148	449
3	8.0	127	140	195	280	195	419	195	594
3	10.0	157	174	243	348	243	522	243	740
4	2.5	51	56	79	112	79	168	79	238
4	3.0	61	66	93	133	93	199	93	282
4	4.0	79	87	122	174	122	261	122	370
4	6.0	116	129	180	257	180	386	180	547
4	8.0	154	170	237	340	237	510	237	723
4	10.0	191	212	294	423	294	635	294	900
6	2.5	75	83	116	166	116	249	116	353
6	3.0	88	98	136	196	136	294	136	417
6	4.0	115	128	178	256	178	384	178	544
6	6.0	169	188	260	375	260	563	260	797
6	8.0	222	247	343	494	343	742	343	1,051
6	10.0	276	307	425	614	425	921	425	1,304
8	2.5	100	113	155	225	155	338	155	478
8	3.0	118	132	182	264	182	396	182	562
8	4.0	153	171	236	342	236	514	236	728
8	6.0	223	249	344	499	344	748	344	1,060
8	8.0	293	328	452	655	452	983	452	1,393
8	10.0	363	406	561	812	561	1,218	561	1,725
10	2.5	125	142	193	284	193	426	193	604
10	3.0	147	166	226	332	226	498	226	706
10	4.0	190	214	292	428	292	642	292	910
10	6.0	276	310	425	620	425	930	425	1,318
10	8.0	362	406	558	812	558	1,218	558	1,726
10	10.0	448	502	691	1,004	691	1,506	691	2,134

(continued)

Table 8-6 Thrust restraint design for ductile-iron pipe for silt 1* soil type (*Continued*)

Soil Parameters:	A21.50—Laying Conditions:	2	3	4	5
φ = 29 degrees		2	3	4	5
C_s = 0 lb/ft^2	f_φ =	0.50	0.75	0.75	0.75
γ = 90 lb/ft^3 (14.14 kN/m^3)	f_c =	0.00	0.00	0.00	0.00
NOTE: $F_s = (Fs)_b$ in this soil type.	K_n =	0.20	0.40	0.60	0.85

		A21.50—Laying Conditions							
		2		3		4		5	
Nominal Pipe Size, *in.*	Depth, *ft*	F_s, *lb/ft*	R_s, *lb/ft*	F_s, *lb/ft*	R_s, *lb/ft*	F_s, *lb/ft*	R_s, *lb/ft*	F_s, *lb/ft*	R_s, *lb/ft*
12	2.5	152	174	234	348	234	522	234	740
12	3.0	177	203	274	405	274	608	274	861
12	4.0	229	260	353	519	353	779	353	1,103
12	6.0	331	374	511	748	511	1,121	511	1,589
12	8.0	433	488	669	976	669	1,464	669	2,074
12	10.0	536	602	827	1,204	827	1,806	827	2,559
14	2.5	179	208	276	415	276	623	276	882
14	3.0	209	241	322	481	322	722	322	1,023
14	4.0	268	307	414	613	414	920	414	1,304
14	6.0	387	439	597	878	597	1,317	597	1,866
14	8.0	506	571	780	1,143	780	1,714	780	2,428
14	10.0	624	704	963	1,407	963	2,111	963	2,990
16	2.5	208	243	321	485	321	728	321	1,031
16	3.0	242	280	373	560	373	841	373	1,191
16	4.0	309	355	477	711	477	1,066	477	1,511
16	6.0	444	506	685	1,012	685	1,518	685	2,150
16	8.0	579	656	893	1,313	893	1,969	893	2,789
16	10.0	714	807	1,102	1,614	1,102	2,420	1,102	3,429
18	2.5	237	279	366	558	366	838	366	1,187
18	3.0	275	321	424	643	424	964	424	1,366
18	4.0	350	406	541	811	541	1,217	541	1,724
18	6.0	502	574	774	1,149	774	1,723	774	2,441
18	8.0	653	743	1,007	1,486	1,007	2,229	1,007	3,157
18	10.0	804	911	1,241	1,823	1,241	2,734	1,241	3,874
20	2.5	268	317	413	635	413	952	413	1,349
20	3.0	310	364	478	728	478	1,093	478	1,548
20	4.0	393	458	607	915	607	1,373	607	1,945
20	6.0	561	644	865	1,289	865	1,933	865	2,738
20	8.0	729	831	1,124	1,662	1,124	2,493	1,124	3,532
20	10.0	896	1,018	1,382	2,036	1,382	3,053	1,382	4,326

(continued)

Table 8-6 Thrust restraint design for ductile-iron pipe for silt 1* soil type (*Continued*)

Soil Parameters:	A21.50—Laying Conditions:	2	3	4	5
φ = 29 degrees		2	3	4	5
C_s = 0 lb/ft^2	f_φ =	0.50	0.75	0.75	0.75
γ = 90 lb/ft^3 (14.14 kN/m^3)	f_c =	0.00	0.00	0.00	0.00
NOTE: F_s = $(Fs)_b$ in this soil type.	K_n =	0.20	0.40	0.60	0.85

		A21.50—Laying Conditions							
		2		3		4		5	
Nominal Pipe Size, *in.*	Depth, *ft*	F_s, *lb/ft*	R_s, *lb/ft*	F_s, *lb/ft*	R_s, *lb/ft*	F_s, *lb/ft*	R_s, *lb/ft*	F_s, *lb/ft*	R_s, *lb/ft*
24	2.5	329	399	508	797	508	1,196	508	1,695
24	3.0	379	455	585	909	585	1,364	585	1,932
24	4.0	479	566	739	1,132	739	1,698	739	2,406
24	6.0	679	789	1,048	1,578	1,048	2,367	1,048	3,354
24	8.0	880	1,012	1,357	2,024	1,357	3,037	1,357	4,302
24	10.0	1,080	1,235	1,666	2,471	1,666	3,706	1,666	5,250
30	2.5	427	530	659	1,061	659	1,591	659	2,254
30	3.0	489	599	755	1,199	755	1,798	755	2,548
30	4.0	613	738	946	1,476	946	2,213	946	3,136
30	6.0	862	1,014	1,329	2,029	1,329	3,043	1,329	4,312
30	8.0	1,110	1,291	1,712	2,582	1,712	3,873	1,712	5,487
30	10.0	1,358	1,568	2,095	3,136	2,095	4,704	2,095	6,663
36	2.5	536	678	827	1,356	827	2,034	827	2,882
36	3.0	610	761	941	1,522	941	2,283	941	3,234
36	4.0	759	927	1,171	1,853	1,171	2,780	1,171	3,938
36	6.0	1,056	1,258	1,629	2,515	1,629	3,773	1,629	5,345
36	8.0	1,353	1,589	2,087	3,178	2,087	4,766	2,087	6,753
36	10.0	1,650	1,920	2,546	3,840	2,546	5,760	2,546	8,160
42	2.5	651	838	1,004	1,675	1,004	2,513	1,004	3,560
42	3.0	737	934	1,137	1,868	1,137	2,801	1,137	3,969
42	4.0	910	1,126	1,404	2,252	1,404	3,379	1,404	4,786
42	6.0	1,255	1,511	1,936	3,022	1,936	4,533	1,936	6,422
42	8.0	1,600	1,896	2,469	3,791	2,469	5,687	2,469	8,057
42	10.0	1,946	2,280	3,001	4,561	3,001	6,841	3,001	9,692
48	2.5	777	1,014	1,198	2,028	1,198	3,042	1,198	4,309
48	3.0	875	1,124	1,350	2,247	1,350	3,371	1,350	4,776
48	4.0	1,072	1,343	1,654	2,687	1,654	4,030	1,654	5,709
48	6.0	1,467	1,783	2,262	3,565	2,262	5,348	2,262	7,576
48	8.0	1,861	2,222	2,871	4,444	2,871	6,665	2,871	9,442
48	10.0	2,255	2,661	3,479	5,322	3,479	7,983	3,479	11,309

(continued)

Table 8-6 Thrust restraint design for ductile-iron pipe for silt 1* soil type (*Continued*)

Soil Parameters:	A21.50—Laying Conditions:	2	3	4	5
φ = 29 degrees					
C_s = 0 lb/ft^2	f_φ =	0.50	0.75	0.75	0.75
γ = 90 lb/ft^3 (14.14 kN/m^3)	f_c =	0.00	0.00	0.00	0.00
NOTE: $F_s = (Fs)_b$ in this soil type.	K_n =	0.20	0.40	0.60	0.85

Nominal Pipe Size, *in.*	Depth, *ft*	2 F_s, *lb/ft*	2 R_s, *lb/ft*	3 F_s, *lb/ft*	3 R_s, *lb/ft*	4 F_s, *lb/ft*	4 R_s, *lb/ft*	5 F_s, *lb/ft*	5 R_s, *lb/ft*
54	2.5	921	1,219	1,421	2,438	1,421	3,657	1,421	5,180
54	3.0	1,033	1,343	1,593	2,687	1,593	4,030	1,593	5,709
54	4.0	1,256	1,592	1,938	3,184	1,938	4,776	1,938	6,767
54	6.0	1,703	2,090	2,627	4,180	2,627	6,269	2,627	8,882
54	8.0	2,149	2,587	3,315	5,175	3,315	7,762	3,315	10,997
54	10.0	2,596	3,085	4,004	6,170	4,004	9,255	4,004	13,112
60	2.5	1,013	1,350	1,563	2,699	1,563	4,049	1,563	5,736
60	3.0	1,133	1,483	1,748	2,966	1,748	4,448	1,748	6,302
60	4.0	1,372	1,749	2,116	3,498	2,116	5,247	2,116	7,434
60	6.0	1,850	2,282	2,854	4,564	2,854	6,845	2,854	9,698
60	8.0	2,328	2,814	3,591	5,629	3,591	8,443	3,591	11,962
60	10.0	2,806	3,347	4,329	6,694	4,329	10,042	4,329	14,226
64	2.5	1,107	1,487	1,707	2,973	1,707	4,460	1,707	6,318
64	3.0	1,234	1,629	1,904	3,257	1,904	4,886	1,904	6,921
64	4.0	1,489	1,912	2,297	3,825	2,297	5,737	2,297	8,128
64	6.0	1,998	2,480	3,083	4,960	3,083	7,441	3,083	10,541
64	8.0	2,508	3,048	3,869	6,096	3,869	9,144	3,869	12,954
64	10.0	3,017	3,616	4,655	7,232	4,655	10,847	4,655	15,367

NOTE: To convert inches (in.) to millimeters (mm), multiply by 25.4; to convert feet (ft) to meters (m), multiply by 0.3048; to convert pounds per foot (lb/ft) to kilonewtons per meter (kN/m), multiply by 0.0146.
* See Tables 8-2 and 8-3 for soil description.

Table 8-7 Thrust restraint design for ductile-iron pipe for clay 2* soil type

Soil Parameters:	A21.50—Laying Conditions:	2	3	4	5
φ = 0 degrees					
C_s = 300 lb/ft^2 (14.37 kN/m^2)	f_φ =	0.00	0.00	0.00	0.00
γ = 90 lb/ft^3 (14.14 kN/m^3)	f_c =	0.50	0.80	0.80	0.80
	K_n =	0.40	0.60	0.85	1.00

		A21.50—Laying Conditions											
Nominal		2			3			4			5		
Pipe Size, *in.*	Depth, *ft*	F_s, *lb/ft*	$(F_s)_b$, *lb/ft*	R_s, *lb/ft*	F_s, *lb/ft*	$(F_s)_b$, *lb/ft*	R_s, *lb/ft*	F_s, *lb/ft*	$(F_s)_b$, *lb/ft*	R_s, *lb/ft*	F_s, *lb/ft*	$(F_s)_b$, *lb/ft*	R_s, *lb/ft*
3	2.5	78	156	111	124	249	166	124	249	236	124	249	277
3	3.0	78	156	117	124	249	175	124	249	248	124	249	292
3	4.0	78	156	129	124	249	193	124	249	273	124	249	322
3	6.0	78	156	152	124	249	229	124	249	324	124	249	381
3	8.0	78	156	176	124	249	264	124	249	374	124	249	441
3	10.0	78	156	200	124	249	300	124	249	425	124	249	500
4	2.5	94	188	135	151	302	202	151	302	287	151	302	337
4	3.0	94	188	142	151	302	213	151	302	302	151	302	355
4	4.0	94	188	156	151	302	235	151	302	333	151	302	391
4	6.0	94	188	185	151	302	278	151	302	394	151	302	463
4	8.0	94	188	214	151	302	321	151	302	455	151	302	535
4	10.0	94	188	243	151	302	364	151	302	516	151	302	607
6	2.5	135	271	196	217	434	294	217	434	416	217	434	489
6	3.0	135	271	206	217	434	309	217	434	438	217	434	515
6	4.0	135	271	227	217	434	340	217	434	482	217	434	567
6	6.0	135	271	268	217	434	402	217	434	570	217	434	670
6	8.0	135	271	310	217	434	464	217	434	658	217	434	774
6	10.0	135	271	351	217	434	526	217	434	746	217	434	877
8	2.5	178	355	259	284	569	389	284	569	551	284	569	648
8	3.0	178	355	273	284	569	409	284	569	579	284	569	682
8	4.0	178	355	300	284	569	450	284	569	637	284	569	750
8	6.0	178	355	354	284	569	531	284	569	753	284	569	885
8	8.0	178	355	408	284	569	613	284	569	868	284	569	1,021
8	10.0	178	355	463	284	569	694	284	569	983	284	569	1,157
10	2.5	218	436	321	349	697	481	349	697	681	349	697	802
10	3.0	218	436	337	349	697	506	349	697	717	349	697	843
10	4.0	218	436	371	349	697	556	349	697	788	349	697	927
10	6.0	218	436	437	349	697	656	349	697	929	349	697	1,093
10	8.0	218	436	504	349	697	756	349	697	1,071	349	697	1,260
10	10.0	218	436	570	349	697	856	349	697	1,212	349	697	1,426

(continued)

Table 8-7 Thrust restraint design for ductile-iron pipe for clay 2* soil type (*Continued*)

Soil Parameters:	A21.50—Laying Conditions:	2	3	4	5
φ = 0 degrees					
C_s = 300 lb/ft^2 (14.37 kN/m^2)	f_φ =	0.00	0.00	0.00	0.00
γ = 90 lb/ft^3 (14.14 kN/m^3)	f_c =	0.50	0.80	0.80	0.80
	K_n =	0.40	0.60	0.85	1.00

		A21.50—Laying Conditions											
Nominal Pipe		2			3			4			5		
Size, *in.*	Depth, *ft*	F_s, *lb/ft*	$(F_s)_b$, *lb/ft*	R_s, *lb/ft*	F_s, *lb/ft*	$(F_s)_b$, *lb/ft*	R_s, *lb/ft*	F_s, *lb/ft*	$(F_s)_b$, *lb/ft*	R_s, *lb/ft*	F_s, *lb/ft*	$(F_s)_b$, *lb/ft*	R_s, *lb/ft*
12	2.5	259	518	385	415	829	577	415	829	818	415	829	962
12	3.0	259	518	405	415	829	607	415	829	860	415	829	1,011
12	4.0	259	518	444	415	829	666	415	829	944	415	829	1,110
12	6.0	259	518	523	415	829	785	415	829	1,112	415	829	1,308
12	8.0	259	518	603	415	829	904	415	829	1,280	415	829	1,506
12	10.0	259	518	682	415	829	1,023	415	829	1,449	415	829	1,704
14	2.5	300	601	450	481	961	675	481	961	956	481	961	1,125
14	3.0	300	601	473	481	961	709	481	961	1,005	481	961	1,182
14	4.0	300	601	519	481	961	778	481	961	1,103	481	961	1,297
14	6.0	300	601	611	481	961	916	481	961	1,298	481	961	1,527
14	8.0	300	601	702	481	961	1,054	481	961	1,493	481	961	1,756
14	10.0	300	601	794	481	961	1,191	481	961	1,688	481	961	1,986
16	2.5	342	683	516	547	1,093	775	547	1,093	1,097	547	1,093	1,291
16	3.0	342	683	542	547	1,093	814	547	1,093	1,153	547	1,093	1,356
16	4.0	342	683	595	547	1,093	892	547	1,093	1,264	547	1,093	1,487
16	6.0	342	683	699	547	1,093	1,049	547	1,093	1,485	547	1,093	1,748
16	8.0	342	683	803	547	1,093	1,205	547	1,093	1,707	547	1,093	2,009
16	10.0	342	683	908	547	1,093	1,362	547	1,093	1,929	547	1,093	2,270
18	2.5	383	766	584	613	1,225	876	613	1,225	1,241	613	1,225	1,459
18	3.0	383	766	613	613	1,225	920	613	1,225	1,303	613	1,225	1,533
18	4.0	383	766	672	613	1,225	1,007	613	1,225	1,427	613	1,225	1,679
18	6.0	383	766	789	613	1,225	1,183	613	1,225	1,676	613	1,225	1,971
18	8.0	383	766	906	613	1,225	1,358	613	1,225	1,924	613	1,225	2,264
18	10.0	383	766	1,023	613	1,225	1,534	613	1,225	2,173	613	1,225	2,556
20	2.5	424	848	652	679	1,357	978	679	1,357	1,386	679	1,357	1,631
20	3.0	424	848	685	679	1,357	1,027	679	1,357	1,455	679	1,357	1,712
20	4.0	424	848	750	679	1,357	1,124	679	1,357	1,593	679	1,357	1,874
20	6.0	424	848	879	679	1,357	1,319	679	1,357	1,868	679	1,357	2,198
20	8.0	424	848	1,009	679	1,357	1,513	679	1,357	2,144	679	1,357	2,522
20	10.0	424	848	1,138	679	1,357	1,707	679	1,357	2,419	679	1,357	2,846

(continued)

Table 8-7 Thrust restraint design for ductile-iron pipe for clay 2* soil type (*Continued*)

Soil Parameters:	A21.50—Laying Conditions:	2	3	4	5
φ = 0 degrees					
C_s = 300 lb/ft² (14.37 kN/m²)	f_φ =	0.00	0.00	0.00	0.00
γ = 90 lb/ft³ (14.14 kN/m³)	f_c =	0.50	0.80	0.80	0.80
	K_n =	0.40	0.60	0.85	1.00

		A21.50—Laying Conditions											
Nominal Pipe Size, *in.*	Depth, *ft*	2			3			4			5		
		F_s, *lb/ft*	$(F_s)_b$, *lb/ft*	R_s, *lb/ft*	F_s, *lb/ft*	$(F_s)_b$, *lb/ft*	R_s, *lb/ft*	F_s, *lb/ft*	$(F_s)_b$, *lb/ft*	R_s, *lb/ft*	F_s, *lb/ft*	$(F_s)_b$, *lb/ft*	R_s, *lb/ft*
24	2.5	507	1,013	793	811	1,621	1,189	811	1,621	1,684	811	1,621	1,982
24	3.0	507	1,013	831	811	1,621	1,247	811	1,621	1,767	811	1,621	2,079
24	4.0	507	1,013	909	811	1,621	1,363	811	1,621	1,931	811	1,621	2,272
24	6.0	507	1,013	1,064	811	1,621	1,595	811	1,621	2,260	811	1,621	2,659
24	8.0	507	1,013	1,218	811	1,621	1,828	811	1,621	2,589	811	1,621	3,046
24	10.0	507	1,013	1,373	811	1,621	2,060	811	1,621	2,918	811	1,621	3,433
30	2.5	628	1,257	1,008	1,005	2,011	1,512	1,005	2,011	2,142	1,005	2,011	2,520
30	3.0	628	1,257	1,056	1,005	2,011	1,584	1,005	2,011	2,244	1,005	2,011	2,640
30	4.0	628	1,257	1,152	1,005	2,011	1,728	1,005	2,011	2,448	1,005	2,011	2,880
30	6.0	628	1,257	1,344	1,005	2,011	2,016	1,005	2,011	2,856	1,005	2,011	3,360
30	8.0	628	1,257	1,536	1,005	2,011	2,304	1,005	2,011	3,264	1,005	2,011	3,840
30	10.0	628	1,257	1,728	1,005	2,011	2,592	1,005	2,011	3,672	1,005	2,011	4,320
36	2.5	752	1,504	1,237	1,203	2,406	1,855	1,203	2,406	2,628	1,203	2,406	3,092
36	3.0	752	1,504	1,294	1,203	2,406	1,941	1,203	2,406	2,750	1,203	2,406	3,235
36	4.0	752	1,504	1,409	1,203	2,406	2,113	1,203	2,406	2,994	1,203	2,406	3,522
36	6.0	752	1,504	1,639	1,203	2,406	2,458	1,203	2,406	3,482	1,203	2,406	4,097
36	8.0	752	1,504	1,869	1,203	2,406	2,803	1,203	2,406	3,971	1,203	2,406	4,671
36	10.0	752	1,504	2,098	1,203	2,406	3,148	1,203	2,406	4,459	1,203	2,406	5,246
42	2.5	874	1,748	1,471	1,398	2,796	2,207	1,398	2,796	3,126	1,398	2,796	3,678
42	3.0	874	1,748	1,538	1,398	2,796	2,307	1,398	2,796	3,268	1,398	2,796	3,845
42	4.0	874	1,748	1,672	1,398	2,796	2,507	1,398	2,796	3,552	1,398	2,796	4,179
42	6.0	874	1,748	1,939	1,398	2,796	2,908	1,398	2,796	4,119	1,398	2,796	4,846
42	8.0	874	1,748	2,206	1,398	2,796	3,308	1,398	2,796	4,687	1,398	2,796	5,514
42	10.0	874	1,748	2,473	1,398	2,796	3,709	1,398	2,796	5,254	1,398	2,796	6,181
48	2.5	997	1,995	1,720	1,596	3,192	2,579	1,596	3,192	3,654	1,596	3,192	4,299
48	3.0	997	1,995	1,796	1,596	3,192	2,694	1,596	3,192	3,816	1,596	3,192	4,489
48	4.0	997	1,995	1,948	1,596	3,192	2,922	1,596	3,192	4,140	1,596	3,192	4,870
48	6.0	997	1,995	2,253	1,596	3,192	3,379	1,596	3,192	4,788	1,596	3,192	5,632
48	8.0	997	1,995	2,558	1,596	3,192	3,837	1,596	3,192	5,435	1,596	3,192	6,394
48	10.0	997	1,995	2,863	1,596	3,192	4,294	1,596	3,192	6,083	1,596	3,192	7,156

(continued)

Table 8-7 Thrust restraint design for ductile-iron pipe for clay 2* soil type (*Continued*)

Soil Parameters:	A21.50—Laying Conditions:	2	3	4	5
φ = 0 degrees					
C_s = 300 lb/ft^2 (14.37 kN/m^2)	f_φ =	0.00	0.00	0.00	0.00
γ = 90 lb/ft^3 (14.14 kN/m^3)	f_c =	0.50	0.80	0.80	0.80
	K_n =	0.40	0.60	0.85	1.00

		A21.50—Laying Conditions											
Nominal Pipe Size, *in.*	Depth, *ft*	2			3			4			5		
		F_s, *lb/ft*	$(F_s)_b$, *lb/ft*	R_s, *lb/ft*	F_s, *lb/ft*	$(F_s)_b$, *lb/ft*	R_s, *lb/ft*	F_s, *lb/ft*	$(F_s)_b$, *lb/ft*	R_s, *lb/ft*	F_s, *lb/ft*	$(F_s)_b$, *lb/ft*	R_s, *lb/ft*
54	2.5	1,130	2,260	1,997	1,808	3,617	2,996	1,808	3,617	4,244	1,808	3,617	4,993
54	3.0	1,130	2,260	2,083	1,808	3,617	3,125	1,808	3,617	4,427	1,808	3,617	5,208
54	4.0	1,130	2,260	2,256	1,808	3,617	3,384	1,808	3,617	4,794	1,808	3,617	5,640
54	6.0	1,130	2,260	2,601	1,808	3,617	3,902	1,808	3,617	5,528	1,808	3,617	6,504
54	8.0	1,130	2,260	2,947	1,808	3,617	4,420	1,808	3,617	6,262	1,808	3,617	7,367
54	10.0	1,130	2,260	3,292	1,808	3,617	4,938	1,808	3,617	6,996	1,808	3,617	8,230
60	2.5	1,210	2,419	2,169	1,936	3,871	3,253	1,936	3,871	4,609	1,936	3,871	5,422
60	3.0	1,210	2,419	2,261	1,936	3,871	3,392	1,936	3,871	4,805	1,936	3,871	5,653
60	4.0	1,210	2,419	2,446	1,936	3,871	3,669	1,936	3,871	5,198	1,936	3,871	6,115
60	6.0	1,210	2,419	2,816	1,936	3,871	4,223	1,936	3,871	5,983	1,936	3,871	7,039
60	8.0	1,210	2,419	3,185	1,936	3,871	4,778	1,936	3,871	6,769	1,936	3,871	7,963
60	10.0	1,210	2,419	3,555	1,936	3,871	5,332	1,936	3,871	7,554	1,936	3,871	8,887
64	2.5	1,289	2,579	2,345	2,063	4,126	3,517	2,063	4,126	4,983	2,063	4,126	5,862
64	3.0	1,289	2,579	2,443	2,063	4,126	3,665	2,063	4,126	5,192	2,063	4,126	6,109
64	4.0	1,289	2,579	2,641	2,063	4,126	3,961	2,063	4,126	5,611	2,063	4,126	6,601
64	6.0	1,289	2,579	3,035	2,063	4,126	4,552	2,063	4,126	6,448	2,063	4,126	7,586
64	8.0	1,289	2,579	3,429	2,063	4,126	5,143	2,063	4,126	7,286	2,063	4,126	8,571
64	10.0	1,289	2,579	3,823	2,063	4,126	5,734	2,063	4,126	8,123	2,063	4,126	9,556

NOTE: To convert inches (in.) to millimeters (mm), multiply by 25.4; to convert feet (ft) to meters (m), multiply by 0.3048; to convert pounds per foot (lb/ft) to kilonewtons per meter (kN/m), multiply by 0.0146.
* See Tables 8-2 and 8-3 for soil description.

Table 8-8 Thrust restraint design for ductile-iron pipe for silt 2* soil type

Soil Parameters:		A21.50—Laying Conditions:			
		2	3	4	5
φ = 29 degrees	f_φ =	0.50	0.75	0.75	0.75
C_s = 0 lb/ft^2	f_c =	0.00	0.00	0.00	0.00
γ = 90 lb/ft^3 (14.14 kN/m^3)	K_n =	0.40	0.60	0.85	1.00
NOTE: $F_s = (F_s)_b$ in this soil type.					

Nominal Pipe Size, *in.*	Depth, *ft*	A21.50—Laying Conditions 2 F_s, *lb/ft*	2 R_s, *lb/ft*	3 F_s, *lb/ft*	3 R_s, *lb/ft*	4 F_s, *lb/ft*	4 R_s, *lb/ft*	5 F_s, *lb/ft*	5 R_s, *lb/ft*
3	2.5	42	91	65	137	65	194	65	228
3	3.0	50	108	77	163	77	230	77	271
3	4.0	65	143	100	214	100	303	100	357
3	6.0	96	211	148	317	148	449	148	528
3	8.0	127	280	195	419	195	594	195	699
3	10.0	157	348	243	522	243	740	243	870
4	2.5	51	112	79	168	79	238	79	280
4	3.0	61	133	93	199	93	282	93	332
4	4.0	79	174	122	261	122	370	122	436
4	6.0	116	257	180	386	180	547	180	643
4	8.0	154	340	237	510	237	723	237	851
4	10.0	191	423	294	635	294	900	294	1,058
6	2.5	75	166	116	249	116	353	116	416
6	3.0	88	196	136	294	136	417	136	490
6	4.0	115	256	178	384	178	544	178	639
6	6.0	169	375	260	563	260	797	260	938
6	8.0	222	494	343	742	343	1,051	343	1,236
6	10.0	276	614	425	921	425	1,304	425	1,534
8	2.5	100	225	155	338	155	478	155	563
8	3.0	118	264	182	396	182	562	182	661
8	4.0	153	342	236	514	236	728	236	856
8	6.0	223	499	344	748	344	1,060	344	1,247
8	8.0	293	655	452	983	452	1,393	452	1,639
8	10.0	363	812	561	1,218	561	1,725	561	2,030
10	2.5	125	284	193	426	193	604	193	711
10	3.0	147	332	226	498	226	706	226	831
10	4.0	190	428	292	642	292	910	292	1,071
10	6.0	276	620	425	930	425	1,318	425	1,551
10	8.0	362	812	558	1,218	558	1,726	558	2,030
10	10.0	448	1,004	691	1,506	691	2,134	691	2,510

(continued)

Table 8-8 Thrust restraint design for ductile-iron pipe for silt 2* soil type (*Continued*)

Soil Parameters:	A21.50—Laying Conditions:				
		2	3	4	5
φ = 29 degrees	f_φ =	0.50	0.75	0.75	0.75
C_s = 0 lb/ft^2	f_c =	0.00	0.00	0.00	0.00
γ = 90 lb/ft^3 (14.14 kN/m^3)	K_n =	0.40	0.60	0.85	1.00
NOTE: $F_s = (F_s)_b$ in this soil type.					

Nominal Pipe Size, *in.*	Depth, *ft*	A21.50—Laying Conditions 2 F_s, *lb/ft*	2 R_s, *lb/ft*	3 F_s, *lb/ft*	3 R_s, *lb/ft*	4 F_s, *lb/ft*	4 R_s, *lb/ft*	5 F_s, *lb/ft*	5 R_s, *lb/ft*
12	2.5	152	348	234	522	234	740	234	870
12	3.0	177	405	274	608	274	861	274	1,013
12	4.0	229	519	353	779	353	1,103	353	1,298
12	6.0	331	748	511	1,121	511	1,589	511	1,869
12	8.0	433	976	669	1,464	669	2,074	669	2,440
14	2.5	179	415	276	623	276	882	276	1,038
14	3.0	209	481	322	722	322	1,023	322	1,203
14	4.0	268	613	414	920	414	1,304	414	1,534
14	6.0	387	878	597	1,317	597	1,866	597	2,195
14	8.0	506	1,143	780	1,714	780	2,428	780	2,857
14	10.0	624	1,407	963	2,111	963	2,990	963	3,518
16	2.5	208	485	321	728	321	1,031	321	1,213
16	3.0	242	560	373	841	373	1,191	373	1,401
16	4.0	309	711	477	1,066	477	1,511	477	1,777
16	6.0	444	1,012	685	1,518	685	2,150	685	2,529
16	8.0	579	1,313	893	1,969	893	2,789	893	3,282
16	10.0	714	1,614	1,102	2,420	1,102	3,429	1,102	4,034
18	2.5	237	558	366	838	366	1,187	366	1,396
18	3.0	275	643	424	964	424	1,366	424	1,607
18	4.0	350	811	541	1,217	541	1,724	541	2,028
18	6.0	502	1,149	774	1,723	774	2,441	774	2,871
18	8.0	653	1,486	1,007	2,229	1,007	3,157	1,007	3,714
18	10.0	804	1,823	1,241	2,734	1,241	3,874	1,241	4,557
20	2.5	268	635	413	952	413	1,349	413	1,587
20	3.0	310	728	478	1,093	478	1,548	478	1,821
20	4.0	393	915	607	1,373	607	1,945	607	2,288
20	6.0	561	1,289	865	1,933	865	2,738	865	3,222
20	8.0	729	1,662	1,124	2,493	1,124	3,532	1,124	4,155
20	10.0	896	2,036	1,382	3,053	1,382	4,326	1,382	5,089

(continued)

Table 8-8 Thrust restraint design for ductile-iron pipe for silt 2* soil type (*Continued*)

Soil Parameters:	A21.50—Laying Conditions:				
		2	3	4	5
φ = 29 degrees	f_φ =	0.50	0.75	0.75	0.75
C_s = 0 lb/ft^2	f_c =	0.00	0.00	0.00	0.00
γ = 90 lb/ft^3 (14.14 kN/m^3)	K_n =	0.40	0.60	0.85	1.00
NOTE: $F_s = (F_s)_b$ in this soil type.					

Nominal Pipe Size, *in.*	Depth, *ft*	A21.50—Laying Conditions 2 F_s, *lb/ft*	2 R_s, *lb/ft*	3 F_s, *lb/ft*	3 R_s, *lb/ft*	4 F_s, *lb/ft*	4 R_s, *lb/ft*	5 F_s, *lb/ft*	5 R_s, *lb/ft*
24	2.5	329	797	508	1,196	508	1,695	508	1,994
24	3.0	379	909	585	1,364	585	1,932	585	2,273
24	4.0	479	1,132	739	1,698	739	2,406	739	2,830
24	6.0	679	1,578	1,048	2,367	1,048	3,354	1,048	3,946
24	8.0	880	2,024	1,357	3,037	1,357	4,302	1,357	5,061
24	10.0	1,080	2,471	1,666	3,706	1,666	5,250	1,666	6,176
30	2.5	427	1,061	659	1,591	659	2,254	659	2,651
30	3.0	489	1,199	755	1,798	755	2,548	755	2,997
30	4.0	613	1,476	946	2,213	946	3,136	946	3,689
30	6.0	862	2,029	1,329	3,043	1,329	4,312	1,329	5,072
30	8.0	1,110	2,582	1,712	3,873	1,712	5,487	1,712	6,456
30	10.0	1,358	3,136	2,095	4,704	2,095	6,663	2,095	7,839
36	2.5	536	1,356	827	2,034	827	2,882	827	3,391
36	3.0	610	1,522	941	2,283	941	3,234	941	3,805
36	4.0	759	1,853	1,171	2,780	1,171	3,938	1,171	4,633
36	6.0	1,056	2,515	1,629	3,773	1,629	5,345	1,629	6,288
36	8.0	1,353	3,178	2,087	4,766	2,087	6,753	2,087	7,944
36	10.0	1,650	3,840	2,546	5,760	2,546	8,160	2,546	9,600
42	2.5	651	1,675	1,004	2,513	1,004	3,560	1,004	4,188
42	3.0	737	1,868	1,137	2,801	1,137	3,969	1,137	4,669
42	4.0	910	2,252	1,404	3,379	1,404	4,786	1,404	5,631
42	6.0	1,255	3,022	1,936	4,533	1,936	6,422	1,936	7,555
42	8.0	1,600	3,791	2,469	5,687	2,469	8,057	2,469	9,479
42	10.0	1,946	4,561	3,001	6,841	3,001	9,692	3,001	11,402
48	2.5	777	2,028	1,198	3,042	1,198	4,309	1,198	5,069
48	3.0	875	2,247	1,350	3,371	1,350	4,776	1,350	5,618
48	4.0	1,072	2,687	1,654	4,030	1,654	5,709	1,654	6,716
48	6.0	1,467	3,565	2,262	5,348	2,262	7,576	2,262	8,913
48	8.0	1,861	4,444	2,871	6,665	2,871	9,442	2,871	11,109
48	10.0	2,255	5,322	3,479	7,983	3,479	11,309	3,479	13,305

(continued)

Table 8-8 Thrust restraint design for ductile-iron pipe for silt 2* soil type (*Continued*)

Soil Parameters:	A21.50—Laying Conditions:	2	3	4	5
φ = 29 degrees					
C_s = 0 lb/ft^2	f_φ =	0.50	0.75	0.75	0.75
γ = 90 lb/ft^3 (14.14 kN/m^3)	f_c =	0.00	0.00	0.00	0.00
NOTE: $F_s = (F_s)_b$ in this soil type.	K_n =	0.40	0.60	0.85	1.00

Nominal Pipe Size, *in.*	Depth, *ft*	A21.50—Laying Conditions 2 F_s, *lb/ft*	2 R_s, *lb/ft*	3 F_s, *lb/ft*	3 R_s, *lb/ft*	4 F_s, *lb/ft*	4 R_s, *lb/ft*	5 F_s, *lb/ft*	5 R_s, *lb/ft*
54	2.5	921	2,438	1,421	3,657	1,421	5,180	1,421	6,094
54	3.0	1,033	2,687	1,593	4,030	1,593	5,709	1,593	6,717
54	4.0	1,256	3,184	1,938	4,776	1,938	6,767	1,938	7,961
54	6.0	1,703	4,180	2,627	6,269	2,627	8,882	2,627	10,449
54	8.0	2,149	5,175	3,315	7,762	3,315	10,997	3,315	12,937
54	10.0	2,596	6,170	4,004	9,255	4,004	13,112	4,004	15,426
60	2.5	1,013	2,699	1,563	4,049	1,563	5,736	1,563	6,748
60	3.0	1,133	2,966	1,748	4,448	1,748	6,302	1,748	7,414
60	4.0	1,372	3,498	2,116	5,247	2,116	7,434	2,116	8,746
60	6.0	1,850	4,564	2,854	6,845	2,854	9,698	2,854	11,409
60	8.0	2,328	5,629	3,591	8,443	3,591	11,962	3,591	14,072
60	10.0	2,806	6,694	4,329	10,042	4,329	14,226	4,329	16,736
64	2.5	1,107	2,973	1,707	4,460	1,707	6,318	1,707	7,433
64	3.0	1,234	3,257	1,904	4,886	1,904	6,921	1,904	8,143
64	4.0	1,489	3,825	2,297	5,737	2,297	8,128	2,297	9,562
64	6.0	1,998	4,960	3,083	7,441	3,083	10,541	3,083	12,401
64	8.0	2,508	6,096	3,869	9,144	3,869	12,954	3,869	15,240
64	10.0	3,017	7,232	4,655	10,847	4,655	15,367	4,655	18,079

NOTE: To convert inches (in.) to millimeters (mm), multiply by 25.4; to convert feet (ft) to meters (m), multiply by 0.3048; to convert pounds per foot (lb/ft) to kilonewtons per meter (kN/m), multiply by 0.0146.
* See Tables 8-2 and 8-3 for soil description.

Table 8-9 Thrust restraint design for ductile-iron pipe for coh-gran* soil type

Soil Parameters:		A21.50—Laying Conditions:			
		2	3	4	5
φ = 20 degrees					
C_s = 200 lb/ft² (9.58 kN/m²)	f_φ =	0.40	0.65	0.65	0.65
γ = 90 lb/ft³ (14.14 kN/m³)	f_c =	0.40	0.40	0.40	0.40
	K_n =	0.40	0.60	0.85	1.00

Nominal Pipe Size, *in.*	Depth, *ft*	A21.50—Laying Conditions 2 F_s, *lb/ft*	2 $(F_s)_b$, *lb/ft*	2 R_s, *lb/ft*	3 F_s, *lb/ft*	3 $(F_s)_b$, *lb/ft*	3 R_s, *lb/ft*	4 F_s, *lb/ft*	4 $(F_s)_b$, *lb/ft*	4 R_s, *lb/ft*	5 F_s, *lb/ft*	5 $(F_s)_b$, *lb/ft*	5 R_s, *lb/ft*
3	2.5	64	106	140	79	120	210	79	120	297	79	120	350
3	3.0	68	110	152	86	127	228	86	127	323	86	127	380
3	4.0	77	118	176	100	141	264	100	141	375	100	141	441
3	6.0	94	135	225	127	168	337	127	168	478	127	168	562
3	8.0	110	152	273	154	196	410	154	196	581	154	196	683
3	10.0	127	168	322	182	223	483	182	223	684	182	223	804
4	2.5	78	128	171	96	146	256	96	146	363	96	146	427
4	3.0	83	133	185	104	155	278	104	155	394	104	155	463
4	4.0	93	144	215	121	171	322	121	171	456	121	171	537
4	6.0	114	164	273	154	204	410	154	204	581	154	204	684
4	8.0	134	184	332	187	238	498	187	238	706	187	238	831
4	10.0	154	204	391	221	271	586	221	271	831	221	271	977
6	2.5	113	185	249	139	211	374	139	211	529	139	211	623
6	3.0	120	193	270	151	223	405	151	223	574	151	223	675
6	4.0	135	207	312	175	247	469	175	247	664	175	247	781
6	6.0	164	236	397	223	295	595	223	295	843	223	295	992
6	8.0	193	265	481	271	343	722	271	343	1,023	271	343	1,203
6	10.0	222	294	566	318	391	849	318	391	1,202	318	391	1,414
8	2.5	149	244	332	184	279	497	184	279	705	184	279	829
8	3.0	159	254	359	200	295	539	200	295	764	200	295	898
8	4.0	178	273	415	231	326	622	231	326	881	231	326	1,037
8	6.0	216	311	525	294	389	788	294	389	1,117	294	389	1,314
8	8.0	254	349	636	357	451	954	357	451	1,352	357	451	1,591
8	10.0	292	387	747	419	514	1,120	419	514	1,587	419	514	1,867
10	2.5	184	300	413	228	344	619	228	344	877	228	344	1,031
10	3.0	196	312	447	247	363	670	247	363	949	247	363	1,116
10	4.0	219	335	514	285	402	772	285	402	1,093	285	402	1,286
10	6.0	266	382	650	362	479	975	362	479	1,382	362	479	1,626
10	8.0	313	429	786	439	555	1,179	439	555	1,671	439	555	1,965
10	10.0	360	476	922	516	632	1,383	516	632	1,959	516	632	2,305

(continued)

Table 8-9 Thrust restraint design for ductile-iron pipe for coh-gran* soil type (*Continued*)

Soil Parameters:

φ = 20 degrees

C_s = 200 lb/ft^2 (9.58 kN/m^2)

γ = 90 lb/ft^3 (14.14 kN/m^3)

A21.50—Laying Conditions:

	2	3	4	5
f_φ =	0.40	0.65	0.65	0.65
f_c =	0.40	0.40	0.40	0.40
K_n =	0.40	0.60	0.85	1.00

		A21.50—Laying Conditions											
Nominal		2			3			4			5		
Pipe Size, *in.*	Depth, *ft*	F_s, *lb/ft*	$(F_s)_b$, *lb/ft*	R_s, *lb/ft*	F_s, *lb/ft*	$(F_s)_b$, *lb/ft*	R_s, *lb/ft*	F_s, *lb/ft*	$(F_s)_b$, *lb/ft*	R_s, *lb/ft*	F_s, *lb/ft*	$(F_s)_b$, *lb/ft*	R_s, *lb/ft*
12	2.5	221	359	498	274	412	747	274	412	1,058	274	412	1,244
12	3.0	235	373	538	297	435	807	297	435	1,143	297	435	1,345
12	4.0	262	401	619	342	481	928	342	481	1,315	342	481	1,547
12	6.0	318	456	780	434	572	1,171	434	572	1,658	434	572	1,951
12	8.0	374	512	942	525	663	1,413	525	663	2,002	525	663	2,355
12	10.0	429	568	1,103	617	755	1,655	617	755	2,345	617	755	2,759
14	2.5	258	418	585	320	480	878	320	480	1,243	320	480	1,463
14	3.0	274	434	632	347	507	948	347	507	1,343	347	507	1,580
14	4.0	306	466	725	400	560	1,088	400	560	1,542	400	560	1,814
14	6.0	370	531	913	506	666	1,369	506	666	1,940	506	666	2,282
14	8.0	435	595	1,100	612	772	1,650	612	772	2,337	612	772	2,750
14	10.0	499	660	1,287	718	878	1,931	718	878	2,735	718	878	3,218
16	2.5	295	477	675	368	550	1,012	368	550	1,434	368	550	1,687
16	3.0	313	496	728	398	580	1,092	398	580	1,547	398	580	1,820
16	4.0	350	532	834	458	640	1,252	458	640	1,773	458	640	2,086
16	6.0	424	606	1,047	579	761	1,571	579	761	2,226	579	761	2,618
16	8.0	497	679	1,260	699	881	1,890	699	881	2,678	699	881	3,151
16	10.0	570	752	1,473	820	1,002	2,210	820	1,002	3,131	820	1,002	3,683
18	2.5	333	537	767	416	620	1,150	416	620	1,629	416	620	1,916
18	3.0	354	558	826	450	654	1,239	450	654	1,756	450	654	2,066
18	4.0	395	599	946	517	721	1,418	517	721	2,009	517	721	2,364
18	6.0	477	681	1,184	652	856	1,776	652	856	2,516	652	856	2,960
18	8.0	559	763	1,423	787	991	2,134	787	991	3,023	787	991	3,557
18	10.0	641	845	1,661	922	1,126	2,492	922	1,126	3,531	922	1,126	4,154
20	2.5	372	598	861	465	691	1,291	465	691	1,829	465	691	2,152
20	3.0	394	621	927	503	729	1,390	503	729	1,969	503	729	2,317
20	4.0	440	666	1,059	577	804	1,588	577	804	2,250	577	804	2,647
20	6.0	531	757	1,323	727	953	1,985	727	953	2,812	727	953	3,308
20	8.0	622	848	1,588	877	1,103	2,381	877	1,103	3,374	877	1,103	3,969
20	10.0	713	939	1,852	1,026	1,252	2,778	1,026	1,252	3,935	1,026	1,252	4,630

(continued)

Table 8-9 Thrust restraint design for ductile-iron pipe for coh-gran* soil type (*Continued*)

Soil Parameters:	A21.50—Laying Conditions:	2	3	4	5
φ = 20 degrees					
C_s = 200 lb/ft² (9.58 kN/m²)	f_φ =	0.40	0.65	0.65	0.65
γ = 90 lb/ft³ (14.14 kN/m³)	f_c =	0.40	0.40	0.40	0.40
	K_n =	0.40	0.60	0.85	1.00

		A21.50—Laying Conditions											
Nominal Pipe Size, *in.*	Depth, *ft*	2			3			4			5		
		F_s, *lb/ft*	$(F_s)_b$, *lb/ft*	R_s, *lb/ft*	F_s, *lb/ft*	$(F_s)_b$, *lb/ft*	R_s, *lb/ft*	F_s, *lb/ft*	$(F_s)_b$, *lb/ft*	R_s, *lb/ft*	F_s, *lb/ft*	$(F_s)_b$, *lb/ft*	R_s, *lb/ft*
24	2.5	449	719	1,056	564	834	1,583	564	834	2,243	564	834	2,639
24	3.0	476	746	1,135	609	879	1,702	609	879	2,411	609	879	2,836
24	4.0	531	801	1,292	698	968	1,939	698	968	2,746	698	968	3,231
24	6.0	639	910	1,608	877	1,147	2,412	877	1,147	3,417	877	1,147	4,020
24	8.0	748	1,018	1,924	1,055	1,326	2,886	1,055	1,326	4,088	1,055	1,326	4,810
24	10.0	857	1,127	2,240	1,234	1,504	3,359	1,234	1,504	4,759	1,234	1,504	5,599
30	2.5	567	902	1,360	716	1,052	2,040	716	1,052	2,890	716	1,052	3,400
30	3.0	601	936	1,458	772	1,107	2,187	772	1,107	3,098	772	1,107	3,645
30	4.0	668	1,004	1,654	883	1,218	2,480	883	1,218	3,514	883	1,218	4,134
30	6.0	803	1,138	2,045	1,104	1,439	3,068	1,104	1,439	4,346	1,104	1,439	5,113
30	8.0	938	1,273	2,437	1,326	1,661	3,655	1,326	1,661	5,178	1,326	1,661	6,092
30	10.0	1,073	1,408	2,828	1,548	1,883	4,243	1,548	1,883	6,010	1,548	1,883	7,071
36	2.5	692	1,093	1,689	879	1,281	2,534	879	1,281	3,589	879	1,281	4,223
36	3.0	733	1,134	1,806	946	1,347	2,710	946	1,347	3,838	946	1,347	4,516
36	4.0	813	1,215	2,041	1,078	1,480	3,061	1,078	1,480	4,336	1,078	1,480	5,102
36	6.0	975	1,376	2,509	1,344	1,745	3,764	1,344	1,745	5,332	1,344	1,745	6,273
36	8.0	1,136	1,537	2,978	1,609	2,010	4,467	1,609	2,010	6,328	1,609	2,010	7,445
36	10.0	1,298	1,699	3,447	1,874	2,275	5,170	1,874	2,275	7,324	1,874	2,275	8,617
42	2.5	820	1,286	2,033	1,047	1,513	3,049	1,047	1,513	4,320	1,047	1,513	5,082
42	3.0	867	1,333	2,169	1,124	1,590	3,254	1,124	1,590	4,609	1,124	1,590	5,423
42	4.0	960	1,426	2,441	1,278	1,744	3,662	1,278	1,744	5,188	1,278	1,744	6,103
42	6.0	1,148	1,614	2,986	1,586	2,052	4,479	1,586	2,052	6,345	1,586	2,052	7,465
42	8.0	1,336	1,802	3,531	1,895	2,361	5,296	1,895	2,361	7,502	1,895	2,361	8,826
42	10.0	1,523	1,989	4,075	2,203	2,669	6,113	2,203	2,669	8,660	2,203	2,669	10,188
48	2.5	954	1,486	2,402	1,226	1,757	3,604	1,226	1,757	5,105	1,226	1,757	6,006
48	3.0	1,008	1,540	2,558	1,313	1,845	3,837	1,313	1,845	5,435	1,313	1,845	6,394
48	4.0	1,115	1,647	2,869	1,489	2,021	4,303	1,489	2,021	6,096	1,489	2,021	7,172
48	6.0	1,329	1,861	3,490	1,841	2,373	5,235	1,841	2,373	7,417	1,841	2,373	8,726
48	8.0	1,543	2,075	4,112	2,193	2,725	6,168	2,193	2,725	8,738	2,193	2,725	10,280
48	10.0	1,757	2,289	4,734	2,545	3,077	7,100	2,545	3,077	10,059	2,545	3,077	11,834

(continued)

Table 8-9 Thrust restraint design for ductile-iron pipe for coh-gran* soil type (*Continued*)

Soil Parameters:	A21.50—Laying Conditions:	2	3	4	5
φ = 20 degrees					
C_s = 200 lb/ft² (9.58 kN/m²)	f_φ =	0.40	0.65	0.65	0.65
γ = 90 lb/ft³ (14.14 kN/m³)	f_c =	0.40	0.40	0.40	0.40
	K_n =	0.40	0.60	0.85	1.00

		A21.50—Laying Conditions											
Nominal Pipe Size, *in.*	Depth, *ft*	2			3			4			5		
		F_s, *lb/ft*	$(F_s)_b$, *lb/ft*	R_s, *lb/ft*	F_s, *lb/ft*	$(F_s)_b$, *lb/ft*	R_s, *lb/ft*	F_s, *lb/ft*	$(F_s)_b$, *lb/ft*	R_s, *lb/ft*	F_s, *lb/ft*	$(F_s)_b$, *lb/ft*	R_s, *lb/ft*
54	2.5	1,103	1,706	2,821	1,425	2,028	4,232	1,425	2,028	5,995	1,425	2,028	7,053
54	3.0	1,164	1,767	2,997	1,525	2,127	4,496	1,525	2,127	6,369	1,525	2,127	7,493
54	4.0	1,285	1,888	3,350	1,724	2,327	5,024	1,724	2,327	7,118	1,724	2,327	8,374
54	6.0	1,528	2,131	4,054	2,123	2,725	6,081	2,123	2,725	8,615	2,123	2,725	10,135
54	8.0	1,771	2,373	4,758	2,521	3,124	7,138	2,521	3,124	10,111	2,521	3,124	11,896
54	10.0	2,013	2,616	5,463	2,920	3,523	8,194	2,920	3,523	11,608	2,920	3,523	13,657
60	2.5	1,196	1,841	3,083	1,550	2,195	4,625	1,550	2,195	6,552	1,550	2,195	7,708
60	3.0	1,261	1,906	3,272	1,656	2,302	4,908	1,656	2,302	6,953	1,656	2,302	8,180
60	4.0	1,391	2,036	3,649	1,870	2,515	5,473	1,870	2,515	7,754	1,870	2,515	9,122
60	6.0	1,650	2,296	4,403	2,297	2,942	6,604	2,297	2,942	9,356	2,297	2,942	11,007
60	8.0	1,910	2,555	5,157	2,723	3,368	7,735	2,723	3,368	10,958	2,723	3,368	12,892
60	10.0	2,170	2,815	5,911	3,150	3,795	8,866	3,150	3,795	12,560	3,150	3,795	14,777
64	2.5	1,289	1,977	3,355	1,676	2,363	5,032	1,676	2,363	7,128	1,676	2,363	8,386
64	3.0	1,358	2,046	3,555	1,789	2,477	5,333	1,789	2,477	7,555	1,789	2,477	8,889
64	4.0	1,497	2,185	3,957	2,017	2,705	5,936	2,017	2,705	8,409	2,017	2,705	9,893
64	6.0	1,774	2,461	4,761	2,472	3,159	7,141	2,472	3,159	10,117	2,472	3,159	11,902
64	8.0	2,051	2,738	5,565	2,927	3,614	8,347	2,927	3,614	11,825	2,927	3,614	13,911
64	10.0	2,327	3,015	6,368	3,381	4,069	9,552	3,381	4,069	13,532	3,381	4,069	15,921

NOTE: To convert inches (in.) to millimeters (mm), multiply by 25.4; to convert feet (ft) to meters (m), multiply by 0.3048; to convert pounds per foot (lb/ft) to kilonewtons per meter (kN/m), multiply by 0.0146.
* See Tables 8-2 and 8-3 for soil description.

Table 8-10 Thrust restraint design for ductile-iron pipe for sand silt* soil type

Soil Parameters:	A21.50—Laying Conditions:			
	2	3	4	5
φ = 30 degrees	f_φ = 0.50	0.75	0.75	0.75
C_s = 0 lb/ft^2	f_c = 0.00	0.00	0.00	0.00
γ = 90 lb/ft^3 (14.14 kN/m^3)	K_n = 0.40	0.60	0.85	1.00
NOTE: $F_s = (F_s)_b$ in this soil type.				

Nominal Pipe Size, *in.*	Depth, *ft*	A21.50—Laying Conditions 2 F_s, *lb/ft*	2 R_s, *lb/ft*	3 F_s, *lb/ft*	3 R_s, *lb/ft*	4 F_s, *lb/ft*	4 R_s, *lb/ft*	5 F_s, *lb/ft*	5 R_s, *lb/ft*
3	2.5	44	95	67	142	67	202	67	237
3	3.0	51	113	80	169	80	240	80	282
3	4.0	67	148	104	223	104	315	104	371
3	6.0	99	220	153	330	153	467	153	549
3	8.0	131	291	203	437	203	618	203	728
3	10.0	163	362	252	543	252	770	252	906
4	2.5	53	117	82	175	82	248	82	292
4	3.0	63	138	97	207	97	294	97	346
4	4.0	82	181	127	272	127	386	127	454
4	6.0	121	268	186	402	186	569	186	670
4	8.0	159	354	246	531	246	753	246	886
4	10.0	198	441	306	661	306	936	306	1,102
6	2.5	78	173	120	260	120	368	120	433
6	3.0	92	204	141	306	141	434	141	510
6	4.0	119	266	184	399	184	566	184	666
6	6.0	175	390	270	586	270	830	270	976
6	8.0	230	515	356	772	356	1,094	356	1,287
6	10.0	286	639	442	958	442	1,358	442	1,597
8	2.5	104	234	160	352	160	498	160	586
8	3.0	122	275	189	413	189	585	189	688
8	4.0	158	357	245	535	245	758	245	891
8	6.0	231	519	357	779	357	1,104	357	1,299
8	8.0	304	682	470	1,023	470	1,450	470	1,706
8	10.0	377	845	582	1,268	582	1,796	582	2,113
10	2.5	129	296	200	444	200	629	200	740
10	3.0	152	346	235	519	235	735	235	865
10	4.0	196	446	304	669	304	947	304	1,115
10	6.0	286	646	442	968	442	1,372	442	1,614
10	8.0	375	845	579	1,268	579	1,796	579	2,114
10	10.0	464	1,045	717	1,568	717	2,221	717	2,613

(continued)

Table 8-10 Thrust restraint design for ductile-iron pipe for sand silt* soil type (*Continued*)

Soil Parameters:	A21.50—Laying Conditions:	2	3	4	5
φ = 30 degrees					
C_s = 0 lb/ft^2	f_φ =	0.50	0.75	0.75	0.75
γ = 90 lb/ft^3 (14.14 kN/m^3)	f_c =	0.00	0.00	0.00	0.00
NOTE: $F_s = (F_s)_b$ in this soil type.	K_n =	0.40	0.60	0.85	1.00

Nominal Pipe Size, *in.*	Depth, *ft*	A21.50—Laying Conditions 2 F_s, *lb/ft*	2 R_s, *lb/ft*	3 F_s, *lb/ft*	3 R_s, *lb/ft*	4 F_s, *lb/ft*	4 R_s, *lb/ft*	5 F_s, *lb/ft*	5 R_s, *lb/ft*
12	2.5	157	362	243	544	243	770	243	906
12	3.0	184	422	284	633	284	896	284	1,054
12	4.0	237	541	366	811	366	1,149	366	1,351
12	6.0	343	778	530	1,167	530	1,654	530	1,945
12	8.0	449	1,016	694	1,524	694	2,158	694	2,539
12	10.0	555	1,253	858	1,880	858	2,663	858	3,133
14	2.5	186	432	287	648	287	918	287	1,080
14	3.0	216	501	334	751	334	1,064	334	1,252
14	4.0	278	639	430	958	430	1,357	430	1,596
14	6.0	401	914	620	1,371	620	1,942	620	2,285
14	8.0	524	1,189	810	1,784	810	2,527	810	2,973
14	10.0	647	1,465	1,000	2,197	1,000	3,113	1,000	3,662
16	2.5	215	505	333	758	333	1,073	333	1,263
16	3.0	250	583	387	875	387	1,240	387	1,458
16	4.0	320	740	495	1,110	495	1,572	495	1,850
16	6.0	460	1,053	711	1,580	711	2,238	711	2,633
16	8.0	600	1,366	927	2,050	927	2,903	927	3,416
16	10.0	740	1,680	1,144	2,519	1,144	3,569	1,144	4,199
18	2.5	246	581	380	872	380	1,235	380	1,453
18	3.0	285	669	440	1,004	440	1,422	440	1,673
18	4.0	363	845	561	1,267	561	1,795	561	2,111
18	6.0	520	1,196	804	1,793	804	2,541	804	2,989
18	8.0	677	1,547	1,046	2,320	1,046	3,287	1,046	3,866
18	10.0	833	1,898	1,288	2,846	1,288	4,032	1,288	4,744
20	2.5	277	661	429	991	429	1,405	429	1,652
20	3.0	321	758	496	1,137	496	1,611	496	1,895
20	4.0	408	953	630	1,429	630	2,024	630	2,381
20	6.0	581	1,341	898	2,012	898	2,850	898	3,353
20	8.0	755	1,730	1,167	2,595	1,167	3,677	1,167	4,325
20	10.0	928	2,119	1,435	3,178	1,435	4,503	1,435	5,297

(continued)

Table 8-10 Thrust restraint design for ductile-iron pipe for sand silt* soil type (*Continued*)

Soil Parameters:	A21.50—Laying Conditions:			
		2	3	4	5
φ = 30 degrees					
C_s = 0 lb/ft^2	f_φ =	0.50	0.75	0.75	0.75
γ = 90 lb/ft^3 (14.14 kN/m^3)	f_c =	0.00	0.00	0.00	0.00
NOTE: $F_s = (F_s)_b$ in this soil type.	K_n =	0.40	0.60	0.85	1.00

Nominal Pipe Size, *in.*	Depth, *ft*	2 F_s, *lb/ft*	2 R_s, *lb/ft*	3 F_s, *lb/ft*	3 R_s, *lb/ft*	4 F_s, *lb/ft*	4 R_s, *lb/ft*	5 F_s, *lb/ft*	5 R_s, *lb/ft*
24	2.5	341	830	527	1,245	527	1,764	527	2,075
24	3.0	393	946	607	1,419	607	2,011	607	2,366
24	4.0	497	1,178	768	1,768	768	2,504	768	2,946
24	6.0	704	1,643	1,088	2,464	1,088	3,491	1,088	4,107
24	8.0	911	2,107	1,409	3,161	1,409	4,478	1,409	5,268
24	10.0	1,119	2,572	1,729	3,857	1,729	5,465	1,729	6,429
30	2.5	443	1,104	684	1,656	684	2,346	684	2,760
30	3.0	507	1,248	784	1,872	784	2,652	784	3,120
30	4.0	636	1,536	983	2,304	983	3,264	983	3,840
30	6.0	893	2,112	1,380	3,168	1,380	4,488	1,380	5,280
30	8.0	1,150	2,688	1,778	4,032	1,778	5,712	1,778	6,720
30	10.0	1,407	3,264	2,175	4,896	2,175	6,936	2,175	8,160
36	2.5	555	1,412	858	2,118	858	3,000	858	3,530
36	3.0	632	1,584	977	2,376	977	3,366	977	3,960
36	4.0	786	1,929	1,215	2,893	1,215	4,099	1,215	4,822
36	6.0	1,094	2,618	1,691	3,927	1,691	5,564	1,691	6,546
36	8.0	1,402	3,308	2,167	4,962	2,167	7,029	2,167	8,269
36	10.0	1,710	3,997	2,643	5,996	2,643	8,494	2,643	9,993
42	2.5	674	1,744	1,042	2,616	1,042	3,706	1,042	4,360
42	3.0	764	1,944	1,181	2,916	1,181	4,131	1,181	4,860
42	4.0	943	2,345	1,457	3,517	1,457	4,982	1,457	5,861
42	6.0	1,300	3,146	2,010	4,718	2,010	6,684	2,010	7,864
42	8.0	1,658	3,947	2,563	5,920	2,563	8,387	2,563	9,866
42	10.0	2,016	4,748	3,116	7,121	3,116	10,089	3,116	11,869
48	2.5	805	2,111	1,244	3,166	1,244	4,485	1,244	5,277
48	3.0	907	2,339	1,402	3,509	1,402	4,971	1,402	5,848
48	4.0	1,111	2,797	1,718	4,195	1,718	5,943	1,718	6,991
48	6.0	1,520	3,711	2,349	5,566	2,349	7,886	2,349	9,277
48	8.0	1,928	4,625	2,980	6,938	2,980	9,829	2,980	11,563
48	10.0	2,336	5,540	3,612	8,310	3,612	11,772	3,612	13,849

(continued)

Table 8-10 Thrust restraint design for ductile-iron pipe for sand silt* soil type (*Continued*)

Soil Parameters:	A21.50—Laying Conditions:				
φ = 30 degrees		2	3	4	5
C_s = 0 lb/ft^2	f_φ =	0.50	0.75	0.75	0.75
γ = 90 lb/ft^3 (14.14 kN/m^3)	f_c =	0.00	0.00	0.00	0.00
NOTE: $F_s = (F_s)_b$ in this soil type.	K_n =	0.40	0.60	0.85	1.00

Nominal Pipe Size, *in.*	Depth, *ft*	A21.50—Laying Conditions 2 F_s, *lb/ft*	2 R_s, *lb/ft*	3 F_s, *lb/ft*	3 R_s, *lb/ft*	4 F_s, *lb/ft*	4 R_s, *lb/ft*	5 F_s, *lb/ft*	5 R_s, *lb/ft*
54	2.5	954	2,538	1,475	3,806	1,475	5,392	1,475	6,344
54	3.0	1,070	2,797	1,654	4,195	1,654	5,943	1,654	6,991
54	4.0	1,301	3,315	2,012	4,972	2,012	7,044	2,012	8,286
54	6.0	1,764	4,351	2,727	6,526	2,727	9,245	2,727	10,877
54	8.0	2,227	5,387	3,442	8,080	3,442	11,447	3,442	13,467
54	10.0	2,689	6,423	4,157	9,634	4,157	13,649	4,157	16,057
60	2.5	1,050	2,810	1,623	4,214	1,623	5,971	1,623	7,024
60	3.0	1,174	3,087	1,814	4,630	1,814	6,560	1,814	7,717
60	4.0	1,421	3,641	2,197	5,462	2,197	7,738	2,197	9,103
60	6.0	1,917	4,750	2,963	7,126	2,963	10,095	2,963	11,876
60	8.0	2,412	5,859	3,728	8,789	3,728	12,451	3,728	14,648
60	10.0	2,907	6,968	4,494	10,452	4,494	14,808	4,494	17,421
64	2.5	1,147	3,095	1,773	4,642	1,773	6,576	1,773	7,737
64	3.0	1,279	3,390	1,977	5,085	1,977	7,204	1,977	8,476
64	4.0	1,543	3,981	2,385	5,972	2,385	8,460	2,385	9,953
64	6.0	2,071	5,163	3,201	7,745	3,201	10,972	3,201	12,908
64	8.0	2,598	6,345	4,017	9,518	4,017	13,484	4,017	15,864
64	10.0	3,126	7,528	4,833	11,291	4,833	15,996	4,833	18,819

NOTE: To convert inches (in.) to millimeters (mm), multiply by 25.4; to convert feet (ft) to meters (m), multiply by 0.3048; to convert pounds per foot (lb/ft) to kilonewtons per meter (kN/m), multiply by 0.0146.
* See Tables 8-2 and 8-3 for soil description.

Table 8-11 Thrust restraint design for ductile-iron pipe for good sand* soil type

Soil Parameters:	A21.50—Laying Conditions:			
	2	3	4	5
φ = 36 degrees	f_φ = 0.75	0.80	0.80	0.80
C_s = 0 lb/ft^2	f_c = 0.00	0.00	0.00	0.00
γ = 100 lb/ft^3 (15.71 kN/m^3)	K_n = 0.40	0.60	0.85	1.00
NOTE: $F_s = (F_s)_b$ in this soil type.				

Nominal Pipe Size, *in.*	Depth, *ft*	A21.50—Laying Conditions 2 F_s, *lb/ft*	2 R_s, *lb/ft*	3 F_s, *lb/ft*	3 R_s, *lb/ft*	4 F_s, *lb/ft*	4 R_s, *lb/ft*	5 F_s, *lb/ft*	5 R_s, *lb/ft*
3	2.5	91	136	136	203	98	288	98	339
3	3.0	108	161	161	241	117	342	117	402
3	4.0	142	212	212	318	153	450	153	529
3	6.0	209	313	313	470	225	666	225	784
3	8.0	276	415	415	623	298	882	298	1,038
3	10.0	343	517	517	775	371	1,098	371	1,292
4	2.5	111	166	166	250	120	354	120	416
4	3.0	131	197	197	296	142	419	142	493
4	4.0	172	259	259	388	186	550	186	647
4	6.0	254	382	382	573	274	812	274	955
4	8.0	335	505	505	758	362	1,074	362	1,263
4	10.0	417	629	629	943	450	1,336	450	1,572
6	2.5	162	247	247	370	175	525	175	617
6	3.0	192	291	291	437	207	619	207	728
6	4.0	250	380	380	570	270	807	270	950
6	6.0	367	557	557	836	396	1,184	396	1,393
6	8.0	485	734	734	1,101	523	1,560	523	1,836
6	10.0	602	911	911	1,367	649	1,937	649	2,278
8	2.5	217	334	334	501	234	710	234	836
8	3.0	255	392	392	589	275	834	275	981
8	4.0	332	509	509	763	358	1,081	358	1,272
8	6.0	486	741	741	1,111	524	1,575	524	1,852
8	8.0	639	973	973	1,460	690	2,068	690	2,433
8	10.0	793	1,206	1,206	1,809	856	2,562	856	3,014
10	2.5	270	422	422	633	291	897	291	1,056
10	3.0	317	493	493	740	342	1,049	342	1,234
10	4.0	411	636	636	954	444	1,351	444	1,590
10	6.0	600	921	921	1,382	647	1,957	647	2,303
10	8.0	788	1,206	1,206	1,809	850	2,563	850	3,015
10	10.0	977	1,491	1,491	2,237	1,054	3,169	1,054	3,728
12	2.5	327	517	353	775	353	1,098	353	1,292
12	3.0	383	602	413	902	413	1,279	413	1,504
12	4.0	495	771	534	1,157	534	1,639	534	1,928
12	6.0	719	1,110	776	1,665	776	2,359	776	2,775
12	8.0	944	1,449	1,018	2,174	1,018	3,079	1,018	3,623
12	10.0	1,168	1,788	1,260	2,682	1,260	3,800	1,260	4,470

(continued)

Table 8-11 Thrust restraint design for ductile-iron pipe for good sand* soil type (*Continued*)

Soil Parameters:	A21.50—Laying Conditions:				
		2	3	4	5
φ = 36 degrees	f_φ =	0.75	0.80	0.80	0.80
C_s = 0 lb/ft^2	f_c =	0.00	0.00	0.00	0.00
γ = 100 lb/ft^3 (15.71 kN/m^3)	K_n =	0.40	0.60	0.85	1.00
NOTE: $F_s = (F_s)_b$ in this soil type.					

Nominal Pipe Size, *in.*	Depth, *ft*	A21.50—Laying Conditions 2 F_s, *lb/ft*	2 R_s, *lb/ft*	3 F_s, *lb/ft*	3 R_s, *lb/ft*	4 F_s, *lb/ft*	4 R_s, *lb/ft*	5 F_s, *lb/ft*	5 R_s, *lb/ft*
14	2.5	385	616	416	925	416	1,310	416	1,541
14	3.0	450	715	486	1,072	486	1,518	486	1,786
14	4.0	580	911	626	1,367	626	1,936	626	2,278
14	6.0	840	1,304	907	1,956	907	2,771	907	3,260
14	8.0	1,100	1,697	1,187	2,545	1,187	3,606	1,187	4,242
14	10.0	1,360	2,090	1,467	3,135	1,467	4,441	1,467	5,224
16	2.5	446	720	482	1,081	482	1,531	482	1,801
16	3.0	520	832	561	1,248	561	1,768	561	2,080
16	4.0	668	1,056	721	1,583	721	2,243	721	2,639
16	6.0	964	1,502	1,040	2,254	1,040	3,193	1,040	3,756
16	8.0	1,259	1,949	1,358	2,924	1,358	4,142	1,358	4,873
16	10.0	1,555	2,396	1,677	3,594	1,677	5,092	1,677	5,990
18	2.5	508	829	548	1,244	548	1,762	548	2,073
18	3.0	591	955	638	1,432	638	2,028	638	2,386
18	4.0	757	1,205	816	1,807	816	2,560	816	3,012
18	6.0	1,088	1,706	1,174	2,558	1,174	3,624	1,174	4,264
18	8.0	1,419	2,206	1,531	3,310	1,531	4,689	1,531	5,516
18	10.0	1,750	2,707	1,888	4,061	1,888	5,753	1,888	6,768
20	2.5	573	943	618	1,414	618	2,004	618	2,357
20	3.0	665	1,082	717	1,622	717	2,298	717	2,704
20	4.0	848	1,359	915	2,038	915	2,888	915	3,397
20	6.0	1,215	1,914	1,311	2,870	1,311	4,066	1,311	4,784
20	8.0	1,582	2,468	1,707	3,702	1,707	5,245	1,707	6,171
20	10.0	1,949	3,023	2,103	4,534	2,103	6,424	2,103	7,557
24	2.5	703	1,184	759	1,776	759	2,517	759	2,961
24	3.0	813	1,350	877	2,025	877	2,868	877	3,375
24	4.0	1,032	1,681	1,113	2,522	1,113	3,572	1,113	4,203
24	6.0	1,470	2,344	1,586	3,515	1,586	4,980	1,586	5,859
24	8.0	1,908	3,006	2,059	4,509	2,059	6,388	2,059	7,515
24	10.0	2,346	3,669	2,532	5,503	2,532	7,796	2,532	9,172
30	2.5	910	1,575	981	2,362	981	3,347	981	3,937
30	3.0	1,046	1,780	1,128	2,671	1,128	3,783	1,128	4,451
30	4.0	1,317	2,191	1,421	3,287	1,421	4,656	1,421	5,478
30	6.0	1,861	3,013	2,008	4,519	2,008	6,403	2,008	7,532
30	8.0	2,404	3,835	2,594	5,752	2,594	8,149	2,594	9,587

(continued)

Table 8-11 Thrust restraint design for ductile-iron pipe for good sand* soil type (*Continued*)

Soil Parameters:	A21.50—Laying Conditions:	2	3	4	5
φ = 36 degrees		2	3	4	5
C_s = 0 lb/ft^2	f_φ =	0.75	0.80	0.80	0.80
γ = 100 lb/ft^3 (15.71 kN/m^3)	f_c =	0.00	0.00	0.00	0.00
NOTE: $F_s = (F_s)_b$ in this soil type.	K_n =	0.40	0.60	0.85	1.00

Nominal Pipe Size, *in.*	Depth, *ft*	A21.50—Laying Conditions 2		3		4		5	
		F_s, *lb/ft*	R_s, *lb/ft*	F_s, *lb/ft*	R_s, *lb/ft*	F_s, *lb/ft*	R_s, *lb/ft*	F_s, *lb/ft*	R_s, *lb/ft*
30	10.0	2,948	4,656	3,181	6,985	3,181	9,895	3,181	11,641
36	2.5	1,137	2,014	1,227	3,021	1,227	4,280	1,227	5,035
36	3.0	1,300	2,260	1,402	3,390	1,402	4,803	1,402	5,650
36	4.0	1,625	2,752	1,753	4,128	1,753	5,847	1,753	6,879
36	6.0	2,276	3,735	2,455	5,603	2,455	7,937	2,455	9,338
36	8.0	2,926	4,719	3,157	7,078	3,157	10,027	3,157	11,797
36	10.0	3,577	5,702	3,859	8,553	3,859	12,117	3,859	14,256
42	2.5	1,377	2,488	1,486	3,732	1,486	5,287	1,486	6,219
42	3.0	1,566	2,773	1,689	4,160	1,689	5,894	1,689	6,934
42	4.0	1,944	3,345	2,097	5,017	2,097	7,108	2,097	8,362
42	6.0	2,699	4,488	2,913	6,731	2,913	9,536	2,913	11,219
42	8.0	3,455	5,630	3,728	8,445	3,728	11,964	3,728	14,076
42	10.0	4,211	6,773	4,544	10,159	4,544	14,393	4,544	16,932
48	2.5	1,638	3,011	1,768	4,517	1,768	6,399	1,768	7,528
48	3.0	1,854	3,337	2,001	5,006	2,001	7,092	2,001	8,343
48	4.0	2,286	3,990	2,466	5,984	2,466	8,478	2,466	9,974
48	6.0	3,148	5,294	3,397	7,941	3,397	11,250	3,397	13,235
48	8.0	4,011	6,599	4,328	9,898	4,328	14,022	4,328	16,496
48	10.0	4,874	7,903	5,259	11,855	5,259	16,794	5,259	19,758
54	2.5	1,937	3,620	2,090	5,430	2,090	7,693	2,090	9,050
54	3.0	2,181	3,990	2,353	5,984	2,353	8,478	2,353	9,974
54	4.0	2,670	4,729	2,881	7,093	2,881	10,048	2,881	11,822
54	6.0	3,648	6,207	3,936	9,310	3,936	13,189	3,936	15,517
54	8.0	4,625	7,685	4,990	11,527	4,990	16,330	4,990	19,212
54	10.0	5,603	9,163	6,045	13,744	6,045	19,471	6,045	22,907
60	2.5	2,127	4,008	2,295	6,012	2,295	8,518	2,295	10,021
60	3.0	2,389	4,404	2,578	6,606	2,578	9,358	2,578	11,009
60	4.0	2,912	5,195	3,142	7,792	3,142	11,039	3,142	12,987
60	6.0	3,959	6,777	4,271	10,165	4,271	14,401	4,271	16,942
60	8.0	5,005	8,359	5,400	12,538	5,400	17,763	5,400	20,897
60	10.0	6,051	9,941	6,529	14,912	6,529	21,125	6,529	24,853
64	2.5	2,320	4,415	2,503	6,623	2,503	9,382	2,503	11,038
64	3.0	2,599	4,837	2,804	7,255	2,804	10,278	2,804	12,092

(continued)

Table 8-11 Thrust restraint design for ductile-iron pipe for good sand* soil type (*Continued*)

Soil Parameters:	A21.50—Laying Conditions:			
	2	3	4	5
φ = 36 degrees				
C_s = 0 lb/ft^2	f_φ = 0.75	0.80	0.80	0.80
γ = 100 lb/ft^3 (15.71 kN/m^3)	f_c = 0.00	0.00	0.00	0.00
NOTE: $F_s = (F_s)_b$ in this soil type.	K_n = 0.40	0.60	0.85	1.00

Nominal Pipe Size, *in.*	Depth, *ft*	A21.50—Laying Conditions 2 F_s, *lb/ft*	2 R_s, *lb/ft*	3 F_s, *lb/ft*	3 R_s, *lb/ft*	4 F_s, *lb/ft*	4 R_s, *lb/ft*	5 F_s, *lb/ft*	5 R_s, *lb/ft*
64	4.0	3,157	5,680	3,406	8,520	3,406	12,070	3,406	14,199
64	6.0	4,272	7,366	4,609	11,049	4,609	15,653	4,609	18,415
64	8.0	5,387	9,052	5,813	13,579	5,813	19,236	5,813	22,631
64	10.0	6,503	10,739	7,016	16,108	7,016	22,820	7,016	26,847

NOTE: To convert inches (in.) to millimeters (mm), multiply by 25.4; to convert feet (ft) to meters (m), multiply by 0.3048; to convert pounds per foot (lb/ft) to kilonewtons per meter (kN/m), multiply by 0.0146.
* See Tables 8-2 and 8-3 for soil description.

8.17 RESTRAINED JOINT DESIGN TABLES FOR HORIZONTAL BENDS

Tables 8-12 through 8-18 are provided for the convenience of the designer of restrained joint systems. The tables are based on seven general soil types as described in Table 8-2. The suggested values of soil parameters for the general soil types are conservative within the range of actual values that may be measured for a given soil. In providing the suggested values in Table 8-2 and the following design tables, AWWA does not assume responsibility that the suggested values accurately reflect the actual values of the soil parameters in a particular installation.

8.17.1 Use of the Design Tables

As indicated, Tables 8-12 through 8-18 are based on 90° bends and 100-psi (689-kPa) internal pressure. For other design pressures, the required restrained length is equal to the tabulated length multiplied by $\frac{P}{100}$ where P is the design pressure. For other bend angles, the required restrained length is equal to the tabulated length multiplied by tan (θ/2).

Bend Angle, θ	$\tan\left(\frac{\theta}{2}\right)$
90	1.000
45	0.414
22½	0.199
11¼	0.098

Note that in actual installations, the restrained length will be in multiples of the length of one piece of pipe (normally 18 or 20 ft [5.5 or 6.1 m]). Thus, for example, if the required restrained length is 30 ft (9.1 m), two full restrained lengths of pipe will result in 36 or 40 ft (11.0 or 12.2 m) of restraint, providing an additional safety factor. Where short connecting or closure pieces are needed to position fittings, they should be placed outside the restrained section for maximum economy when possible. If there is polyethylene encasement, multiply friction factor, F_s, $(F_s)_b$, by 0.7.

Table 8-12 Restrained joint lengths for horizontal bends for clay 1* soil type

Soil Parameters:	A21.50—Laying Conditions:	2	3	4	5
φ = 0 degrees					
C_s = 300 lb/ft^2 (14.37 kN/m^2)	f_φ =	0.00	0.00	0.00	0.00
γ = 90 lb/ft^3 (14.14 kN/m^3)	f_c =	0.50	0.80	0.80	0.80
S_f = 1.5	K_n =	0.20	0.40	0.60	0.85

Values listed here are based on a 90° horizontal bend; pressure class listed in Table 8-4; and an internal pressure of 100 psi (689 kPa).

For other pressures, multiply by ratio to 100 psi (689 kPa).

For other horizontal bends, multiply by the following coefficients: 45°–0.414; 22½° –0.199; 11¼° –0.098. Use alternate method for other type fitting or dead end.

Values in parentheses are for pipe encased in polyethylene.

Nominal Pipe Size, *in.*	Depth, *ft*	A21.50—Laying Conditions 2 Restrained Length, *ft*		3 Restrained Length, *ft*		4 Restrained Length, *ft*		5 Restrained Length, *ft*	
3	2.5	18	(22)	10	(13)	9	(11)	8	(9)
3	3.0	17	(22)	10	(13)	9	(11)	7	(9)
3	4.0	17	(21)	10	(12)	8	(10)	7	(8)
3	6.0	16	(20)	9	(11)	8	(9)	6	(7)
3	8.0	15	(19)	9	(11)	7	(8)	6	(7)
3	10.0	14	(18)	8	(10)	7	(8)	5	(6)
4	2.5	21	(27)	12	(16)	11	(13)	9	(11)
4	3.0	21	(27)	12	(15)	11	(13)	9	(11)
4	4.0	20	(26)	12	(15)	10	(12)	9	(10)
4	6.0	19	(24)	11	(14)	9	(11)	8	(9)
4	8.0	18	(23)	11	(13)	9	(10)	7	(8)
4	10.0	18	(21)	10	(12)	8	(9)	7	(7)
6	2.5	30	(39)	18	(22)	15	(19)	13	(16)
6	3.0	30	(38)	18	(22)	15	(18)	13	(15)
6	4.0	29	(37)	17	(21)	14	(17)	12	(14)
6	6.0	28	(35)	16	(20)	13	(16)	11	(13)
6	8.0	26	(33)	15	(18)	12	(15)	10	(12)
6	10.0	25	(31)	14	(17)	12	(14)	10	(11)
8	2.5	40	(51)	23	(29)	20	(25)	17	(20)
8	3.0	39	(50)	23	(29)	20	(24)	17	(20)
8	4.0	38	(48)	22	(28)	19	(23)	16	(19)
8	6.0	36	(45)	21	(26)	18	(21)	15	(17)
8	8.0	34	(43)	20	(24)	16	(19)	13	(15)
8	10.0	33	(40)	19	(22)	15	(18)	12	(14)
10	2.5	49	(62)	29	(36)	25	(30)	21	(25)

(continued)

Table 8-12 Restrained joint lengths for horizontal bends for clay 1* soil type (*Continued*)

Soil Parameters:	A21.50—Laying Conditions:				
		2	3	4	5
φ = 0 degrees	f_φ =	0.00	0.00	0.00	0.00
C_s = 300 lb/ft² (14.37 kN/m²)	f_c =	0.50	0.80	0.80	0.80
γ = 90 lb/ft³ (14.14 kN/m³)	K_n =	0.20	0.40	0.60	0.85
S_f = 1.5					

Values listed here are based on a 90° horizontal bend; pressure class listed in Table 8-4; and an internal pressure of 100 psi (689 kPa).

For other pressures, multiply by ratio to 100 psi (689 kPa).

For other horizontal bends, multiply by the following coefficients: 45°–0.414; 22½° –0.199; 11¼° –0.098. Use alternate method for other type fitting or dead end.

Values in parentheses are for pipe encased in polyethylene.

		A21.50—Laying Conditions							
		2		3		4		5	
Nominal Pipe Size, *in.*	Depth, *ft*	Restrained Length, *ft*		Restrained Length, *ft*		Restrained Length, *ft*		Restrained Length, *ft*	
10	3.0	48	(61)	28	(35)	24	(29)	21	(24)
10	4.0	47	(59)	27	(34)	23	(28)	20	(23)
10	6.0	44	(55)	26	(31)	21	(25)	18	(20)
10	8.0	42	(52)	24	(29)	20	(23)	16	(19)
10	10.0	40	(49)	23	(27)	19	(22)	15	(17)
12	2.5	58	(74)	34	(43)	29	(35)	25	(29)
12	3.0	57	(73)	33	(42)	29	(35)	24	(29)
12	4.0	55	(70)	32	(40)	27	(33)	23	(27)
12	6.0	53	(66)	30	(37)	25	(30)	21	(24)
12	8.0	50	(62)	29	(35)	24	(28)	19	(22)
12	10.0	48	(58)	27	(33)	22	(26)	18	(20)
14	2.5	67	(85)	39	(49)	34	(41)	29	(34)
14	3.0	66	(84)	38	(48)	33	(40)	28	(33)
14	4.0	64	(81)	37	(46)	32	(38)	27	(31)
14	6.0	61	(76)	35	(43)	29	(35)	24	(28)
14	8.0	58	(71)	33	(40)	27	(32)	22	(25)
14	10.0	55	(67)	31	(38)	26	(30)	21	(23)
16	2.5	76	(97)	44	(56)	38	(46)	33	(38)
16	3.0	75	(95)	44	(55)	37	(45)	32	(37)
16	4.0	73	(92)	42	(52)	36	(43)	30	(35)
16	6.0	69	(86)	40	(49)	33	(39)	28	(32)
16	8.0	66	(81)	38	(45)	31	(36)	25	(29)
16	10.0	63	(77)	36	(43)	29	(34)	24	(26)
18	2.5	85	(108)	50	(62)	43	(52)	36	(43)

(continued)

Table 8-12 Restrained joint lengths for horizontal bends for clay 1* soil type (*Continued*)

Soil Parameters:		A21.50—Laying Conditions:			
		2	3	4	5
φ = 0 degrees	f_φ =	0.00	0.00	0.00	0.00
C_s = 300 lb/ft^2 (14.37 kN/m^2)	f_c =	0.50	0.80	0.80	0.80
γ = 90 lb/ft^3 (14.14 kN/m^3)	K_n =	0.20	0.40	0.60	0.85
S_f = 1.5					

Values listed here are based on a 90° horizontal bend; pressure class listed in Table 8-4; and an internal pressure of 100 psi (689 kPa).

For other pressures, multiply by ratio to 100 psi (689 kPa).

For other horizontal bends, multiply by the following coefficients: 45°–0.414; 22½° –0.199; 11¼° –0.098. Use alternate method for other type fitting or dead end.

Values in parentheses are for pipe encased in polyethylene.

Nominal Pipe Size, *in.*	Depth, *ft*	A21.50—Laying Conditions 2 Restrained Length, *ft*		3 Restrained Length, *ft*		4 Restrained Length, *ft*		5 Restrained Length, *ft*	
18	3.0	84	(106)	49	(61)	42	(50)	35	(41)
18	4.0	81	(103)	47	(59)	40	(48)	34	(39)
18	6.0	77	(96)	44	(54)	37	(44)	31	(35)
18	8.0	74	(91)	42	(51)	35	(40)	28	(32)
18	10.0	70	(86)	40	(48)	32	(37)	26	(30)
20	2.5	94	(120)	55	(69)	47	(57)	40	(47)
20	3.0	92	(117)	54	(67)	46	(56)	39	(46)
20	4.0	90	(114)	52	(65)	44	(53)	37	(43)
20	6.0	85	(106)	49	(60)	41	(48)	34	(39)
20	8.0	81	(100)	46	(56)	38	(45)	31	(36)
20	10.0	78	(95)	44	(53)	36	(41)	29	(33)
24	2.5	111	(142)	65	(81)	56	(67)	47	(56)
24	3.0	110	(139)	64	(80)	55	(66)	46	(54)
24	4.0	107	(135)	62	(77)	53	(63)	44	(51)
24	6.0	102	(126)	58	(71)	49	(57)	40	(46)
24	8.0	97	(119)	55	(67)	45	(53)	37	(42)
24	10.0	92	(112)	52	(63)	43	(49)	35	(39)
30	2.5	137	(174)	80	(100)	68	(83)	58	(68)
30	3.0	135	(171)	79	(98)	67	(81)	57	(66)
30	4.0	132	(166)	76	(94)	65	(77)	54	(63)
30	6.0	125	(155)	72	(88)	60	(70)	50	(57)
30	8.0	119	(146)	68	(82)	56	(65)	46	(52)
30	10.0	114	(138)	65	(77)	52	(60)	42	(48)
36	2.5	163	(207)	95	(118)	81	(98)	69	(80)
36	3.0	161	(203)	93	(116)	79	(95)	67	(78)

(continued)

Table 8-12 Restrained joint lengths for horizontal bends for clay 1* soil type (*Continued*)

Soil Parameters:	A21.50—Laying Conditions:				
		2	3	4	5
φ = 0 degrees	f_φ =	0.00	0.00	0.00	0.00
C_s = 300 lb/ft^2 (14.37 kN/m^2)	f_c =	0.50	0.80	0.80	0.80
γ = 90 lb/ft^3 (14.14 kN/m^3)	K_n =	0.20	0.40	0.60	0.85
S_f = 1.5					

Values listed here are based on a 90° horizontal bend; pressure class listed in Table 8-4; and an internal pressure of 100 psi (689 kPa).

For other pressures, multiply by ratio to 100 psi (689 kPa).

For other horizontal bends, multiply by the following coefficients: 45°–0.414; 22½° –0.199; 11¼° –0.098. Use alternate method for other type fitting or dead end.

Values in parentheses are for pipe encased in polyethylene.

Nominal Pipe Size, *in.*	Depth, *ft*	A21.50—Laying Conditions 2 Restrained Length, *ft*		3 Restrained Length, *ft*		4 Restrained Length, *ft*		5 Restrained Length, *ft*	
36	4.0	156	(197)	91	(112)	76	(91)	64	(74)
36	6.0	149	(185)	85	(104)	71	(83)	59	(67)
36	8.0	142	(174)	81	(97)	66	(77)	54	(61)
36	10.0	135	(164)	77	(91)	62	(72)	50	(56)
42	2.5	188	(238)	109	(136)	93	(112)	79	(92)
42	3.0	185	(234)	108	(133)	91	(109)	77	(89)
42	4.0	181	(227)	104	(129)	88	(105)	74	(85)
42	6.0	172	(213)	99	(120)	82	(96)	67	(77)
42	8.0	164	(201)	93	(112)	76	(89)	62	(70)
42	10.0	156	(190)	89	(105)	72	(82)	58	(65)
48	2.5	213	(269)	124	(154)	105	(126)	89	(103)
48	3.0	210	(265)	122	(151)	103	(123)	87	(100)
48	4.0	205	(257)	118	(145)	99	(118)	83	(95)
48	6.0	195	(241)	112	(136)	93	(108)	76	(87)
48	8.0	186	(227)	106	(127)	87	(100)	70	(79)
48	10.0	177	(215)	100	(119)	81	(93)	66	(73)
54	2.5	240	(302)	139	(172)	118	(141)	99	(115)
54	3.0	236	(298)	137	(169)	116	(138)	97	(112)
54	4.0	230	(288)	133	(163)	112	(132)	93	(107)
54	6.0	219	(271)	126	(152)	104	(121)	85	(97)
54	8.0	209	(255)	119	(142)	97	(112)	79	(89)
54	10.0	200	(242)	113	(134)	91	(105)	74	(82)
60	2.5	255	(322)	148	(183)	126	(150)	105	(122)
60	3.0	252	(317)	146	(180)	123	(147)	103	(119)
60	4.0	246	(307)	142	(173)	119	(140)	99	(113)

(continued)

Table 8-12 Restrained joint lengths for horizontal bends for clay 1* soil type (*Continued*)

Soil Parameters:	A21.50—Laying Conditions:	2	3	4	5
φ = 0 degrees					
C_s = 300 lb/ft^2 (14.37 kN/m^2)	f_φ =	0.00	0.00	0.00	0.00
γ = 90 lb/ft^3 (14.14 kN/m^3)	f_c =	0.50	0.80	0.80	0.80
S_f = 1.5	K_n =	0.20	0.40	0.60	0.85

Values listed here are based on a 90° horizontal bend; pressure class listed in Table 8-4; and an internal pressure of 100 psi (689 kPa).

For other pressures, multiply by ratio to 100 psi (689 kPa).

For other horizontal bends, multiply by the following coefficients: 45°–0.414; 22½° –0.199; 11¼° –0.098. Use alternate method for other type fitting or dead end.

Values in parentheses are for pipe encased in polyethylene.

Nominal Pipe Size, *in.*	Depth, *ft*	A21.50—Laying Conditions 2: Restrained Length, *ft*		3: Restrained Length, *ft*		4: Restrained Length, *ft*		5: Restrained Length, *ft*	
60	6.0	234	(288)	134	(162)	110	(129)	91	(103)
60	8.0	223	(272)	127	(152)	103	(119)	84	(94)
60	10.0	213	(258)	120	(143)	97	(111)	78	(87)
64	2.5	271	(341)	157	(194)	133	(159)	112	(129)
64	3.0	267	(336)	155	(191)	130	(155)	109	(126)
64	4.0	261	(325)	150	(184)	126	(148)	104	(120)
64	6.0	248	(306)	142	(172)	117	(137)	96	(109)
64	8.0	237	(289)	135	(161)	110	(127)	89	(100)
64	10.0	226	(273)	128	(151)	103	(118)	83	(92)

NOTE: To convert inches (in.) to millimeters (mm), multiply by 25.4; to convert feet (ft) to meters (m), multiply by 0.3048.
* See Tables 8-2 and 8-3 for soil description.

Table 8-13 Restrained joint lengths for horizontal bends for silt 1* soil type

Soil Parameters:	A21.50—Laying Conditions:				
		2	3	4	5
φ = 29 degrees					
C_s = 0 lb/ft^2	f_φ =	0.50	0.75	0.75	0.75
γ = 90 lb/ft^3 (14.14 kN/m^3)	f_c =	0.00	0.00	0.00	0.00
S_f = 1.5	K_n =	0.20	0.40	0.60	0.85

Values listed here are based on a 90° horizontal bend; pressure class listed in Table 8-4; and an internal pressure of 100 psi (689 kPa).

For other pressures, multiply by ratio to 100 psi (689 kPa).

For other horizontal bends, multiply by the following coefficients: 45°–0.414; 22½° –0.199; 11¼° –0.098. Use alternate method for other type fitting or dead end.

Values in parentheses are for pipe encased in polyethylene.

		A21.50—Laying Conditions							
		2		3		4		5	
Nominal Pipe Size, *in.*	Depth, *ft*	Restrained Length, *ft*		Restrained Length, *ft*		Restrained Length, *ft*		Restrained Length, *ft*	
3	2.5	28	(35)	17	(20)	14	(16)	11	(13)
3	3.0	24	(30)	14	(17)	12	(14)	10	(11)
3	4.0	18	(23)	11	(13)	9	(10)	7	(8)
3	6.0	12	(15)	7	(9)	6	(7)	5	(6)
3	8.0	9	(12)	6	(7)	5	(5)	4	(4)
3	10.0	8	(9)	4	(5)	4	(4)	3	(3)
4	2.5	34	(43)	20	(24)	17	(19)	14	(16)
4	3.0	29	(36)	17	(21)	14	(16)	12	(13)
4	4.0	22	(27)	13	(16)	11	(13)	9	(10)
4	6.0	15	(19)	9	(11)	7	(9)	6	(7)
4	8.0	11	(14)	7	(8)	6	(6)	5	(5)
4	10.0	9	(11)	5	(6)	4	(5)	4	(4)
6	2.5	48	(60)	28	(34)	23	(27)	19	(22)
6	3.0	41	(51)	24	(29)	20	(23)	16	(18)
6	4.0	31	(39)	18	(22)	15	(18)	12	(14)
6	6.0	21	(26)	13	(15)	10	(12)	9	(10)
6	8.0	16	(20)	10	(12)	8	(9)	6	(7)
6	10.0	13	(16)	8	(9)	6	(7)	5	(6)
8	2.5	62	(76)	36	(44)	30	(35)	25	(28)
8	3.0	52	(65)	31	(37)	25	(30)	21	(24)
8	4.0	40	(50)	24	(29)	20	(23)	16	(18)
8	6.0	28	(34)	16	(20)	13	(16)	11	(13)
8	8.0	21	(26)	12	(15)	10	(12)	8	(10)
8	10.0	17	(21)	10	(12)	8	(10)	7	(8)
10	2.5	74	(92)	43	(52)	36	(42)	29	(33)

(continued)

Table 8-13 Restrained joint lengths for horizontal bends for silt 1* soil type (*Continued*)

Soil Parameters:		A21.50—Laying Conditions:			
φ = 29 degrees		2	3	4	5
C_s = 0 lb/ft^2	f_φ =	0.50	0.75	0.75	0.75
γ = 90 lb/ft^3 (14.14 kN/m^3)	f_c =	0.00	0.00	0.00	0.00
S_f = 1.5	K_n =	0.20	0.40	0.60	0.85

Values listed here are based on a 90° horizontal bend; pressure class listed in Table 8-4; and an internal pressure of 100 psi (689 kPa).

For other pressures, multiply by ratio to 100 psi (689 kPa).

For other horizontal bends, multiply by the following coefficients: 45°–0.414; 22½° –0.199; 11¼° –0.098. Use alternate method for other type fitting or dead end.

Values in parentheses are for pipe encased in polyethylene.

Nominal Pipe Size, *in.*	Depth, *ft*	2 Restrained Length, *ft*		3 Restrained Length, *ft*		4 Restrained Length, *ft*		5 Restrained Length, *ft*	
10	3.0	63	(78)	37	(45)	31	(36)	25	(28)
10	4.0	49	(61)	29	(35)	24	(28)	19	(22)
10	6.0	34	(42)	20	(24)	16	(19)	13	(15)
10	8.0	26	(32)	15	(18)	12	(15)	10	(12)
10	10.0	21	(26)	12	(15)	10	(12)	8	(9)
12	2.5	86	(106)	50	(61)	41	(48)	34	(38)
12	3.0	74	(91)	43	(52)	36	(41)	29	(33)
12	4.0	57	(71)	34	(41)	28	(32)	23	(26)
12	6.0	40	(49)	23	(28)	19	(22)	16	(18)
12	8.0	30	(38)	18	(21)	15	(17)	12	(14)
12	10.0	25	(30)	14	(17)	12	(14)	10	(11)
14	2.5	97	(120)	57	(69)	47	(55)	38	(43)
14	3.0	84	(103)	49	(59)	40	(47)	33	(37)
14	4.0	65	(81)	38	(46)	32	(37)	26	(29)
14	6.0	45	(56)	27	(32)	22	(26)	18	(20)
14	8.0	35	(43)	20	(25)	17	(20)	14	(16)
14	10.0	28	(35)	17	(20)	14	(16)	11	(13)
16	2.5	108	(134)	63	(76)	52	(61)	43	(48)
16	3.0	93	(115)	55	(66)	45	(52)	37	(42)
16	4.0	73	(91)	43	(52)	35	(41)	29	(33)
16	6.0	51	(63)	30	(36)	25	(29)	20	(23)
16	8.0	39	(49)	23	(28)	19	(22)	16	(18)
16	10.0	32	(39)	19	(23)	15	(18)	13	(14)
18	2.5	119	(147)	69	(84)	57	(66)	47	(53)
18	3.0	103	(127)	60	(72)	49	(58)	40	(46)

(continued)

Table 8-13 Restrained joint lengths for horizontal bends for silt 1* soil type (*Continued*)

Soil Parameters:	A21.50—Laying Conditions:				
		2	3	4	5
φ = 29 degrees					
C_s = 0 lb/ft^2	f_φ =	0.50	0.75	0.75	0.75
γ = 90 lb/ft^3 (14.14 kN/m^3)	f_c =	0.00	0.00	0.00	0.00
S_f = 1.5	K_n =	0.20	0.40	0.60	0.85

Values listed here are based on a 90° horizontal bend; pressure class listed in Table 8-4; and an internal pressure of 100 psi (689 kPa).

For other pressures, multiply by ratio to 100 psi (689 kPa).

For other horizontal bends, multiply by the following coefficients: 45°–0.414; 22½° –0.199; 11¼° –0.098. Use alternate method for other type fitting or dead end.

Values in parentheses are for pipe encased in polyethylene.

		A21.50—Laying Conditions							
		2		3		4		5	
Nominal Pipe Size, *in.*	Depth, *ft*	Restrained Length, *ft*		Restrained Length, *ft*		Restrained Length, *ft*		Restrained Length, *ft*	
18	4.0	81	(100)	47	(57)	39	(45)	32	(36)
18	6.0	57	(70)	33	(40)	27	(32)	22	(25)
18	8.0	44	(54)	26	(31)	21	(25)	17	(20)
18	10.0	36	(44)	21	(25)	17	(20)	14	(16)
20	2.5	129	(159)	75	(91)	62	(72)	51	(57)
20	3.0	112	(138)	65	(79)	54	(62)	44	(50)
20	4.0	88	(109)	52	(62)	43	(49)	35	(39)
20	6.0	62	(77)	36	(44)	30	(35)	25	(28)
20	8.0	48	(59)	28	(34)	23	(27)	19	(22)
20	10.0	39	(48)	23	(28)	19	(22)	16	(18)
24	2.5	148	(182)	87	(104)	71	(82)	58	(65)
24	3.0	129	(159)	75	(91)	62	(72)	51	(57)
24	4.0	103	(127)	60	(72)	49	(57)	40	(46)
24	6.0	73	(90)	43	(51)	35	(41)	29	(33)
24	8.0	57	(70)	33	(40)	27	(32)	22	(25)
24	10.0	46	(57)	27	(33)	22	(26)	18	(21)
30	2.5	174	(214)	101	(122)	83	(96)	68	(76)
30	3.0	153	(188)	89	(107)	73	(85)	59	(67)
30	4.0	123	(151)	72	(86)	59	(68)	48	(54)
30	6.0	88	(109)	51	(62)	42	(49)	35	(39)
30	8.0	69	(85)	40	(48)	33	(38)	27	(31)
30	10.0	56	(70)	33	(40)	27	(32)	22	(25)
36	2.5	198	(242)	115	(137)	94	(108)	76	(86)
36	3.0	174	(214)	102	(122)	83	(96)	68	(76)
36	4.0	141	(174)	82	(99)	67	(78)	55	(62)

(continued)

Table 8-13 Restrained joint lengths for horizontal bends for silt 1* soil type (*Continued*)

Soil Parameters:		A21.50—Laying Conditions:			
		2	3	4	5
φ = 29 degrees	f_φ =	0.50	0.75	0.75	0.75
C_s = 0 lb/ft^2	f_c =	0.00	0.00	0.00	0.00
γ = 90 lb/ft^3 (14.14 kN/m^3)	K_n =	0.20	0.40	0.60	0.85
S_f = 1.5					

Values listed here are based on a 90° horizontal bend; pressure class listed in Table 8-4; and an internal pressure of 100 psi (689 kPa).

For other pressures, multiply by ratio to 100 psi (689 kPa).

For other horizontal bends, multiply by the following coefficients: 45°–0.414; 22½° –0.199; 11¼° –0.098. Use alternate method for other type fitting or dead end.

Values in parentheses are for pipe encased in polyethylene.

Nominal Pipe Size, *in.*	Depth, *ft*	A21.50—Laying Conditions 2: Restrained Length, *ft*		3: Restrained Length, *ft*		4: Restrained Length, *ft*		5: Restrained Length, *ft*	
36	6.0	103	(126)	60	(72)	49	(57)	40	(45)
36	8.0	80	(99)	47	(57)	39	(45)	32	(36)
36	10.0	66	(92)	39	(47)	32	(37)	26	(29)
42	2.5	218	(267)	127	(151)	103	(119)	84	(94)
42	3.0	194	(237)	113	(135)	92	(106)	75	(84)
42	4.0	158	(194)	92	(111)	75	(87)	61	(69)
42	6.0	116	(143)	68	(81)	56	(64)	45	(51)
42	8.0	92	(113)	53	(64)	44	(51)	36	(41)
42	10.0	76	(93)	44	(53)	36	(42)	30	(34)
48	2.5	237	(289)	137	(164)	112	(129)	91	(102)
48	3.0	212	(259)	123	(147)	100	(116)	81	(91)
48	4.0	174	(214)	101	(122)	83	(96)	67	(76)
48	6.0	129	(159)	75	(90)	62	(71)	50	(57)
48	8.0	102	(126)	60	(72)	49	(57)	40	(45)
48	10.0	85	(105)	50	(60)	41	(47)	33	(38)
54	2.5	255	(311)	148	(176)	120	(138)	97	(109)
54	3.0	229	(280)	133	(159)	108	(125)	88	(98)
54	4.0	190	(233)	111	(132)	90	(104)	73	(82)
54	6.0	142	(175)	83	(99)	68	(78)	55	(62)
54	8.0	113	(139)	66	(80)	54	(63)	44	(50)
54	10.0	94	(116)	55	(66)	45	(53)	37	(42)
60	2.5	265	(323)	154	(183)	125	(143)	101	(113)
60	3.0	239	(291)	138	(165)	113	(130)	91	(102)
60	4.0	199	(244)	116	(138)	94	(109)	77	(86)
60	6.0	150	(184)	87	(104)	71	(83)	58	(65)

(continued)

Table 8-13 Restrained joint lengths for horizontal bends for silt 1* soil type (*Continued*)

Soil Parameters:	A21.50—Laying Conditions:				
		2	3	4	5
φ = 29 degrees	f_φ =	0.50	0.75	0.75	0.75
C_s = 0 lb/ft^2	f_c =	0.00	0.00	0.00	0.00
γ = 90 lb/ft^3 (14.14 kN/m^3)	K_n =	0.20	0.40	0.60	0.85
S_f = 1.5					

Values listed here are based on a 90° horizontal bend; pressure class listed in Table 8-4; and an internal pressure of 100 psi (689 kPa).

For other pressures, multiply by ratio to 100 psi (689 kPa).

For other horizontal bends, multiply by the following coefficients: 45°–0.414; 22½° –0.199; 11¼° –0.098. Use alternate method for other type fitting or dead end.

Values in parentheses are for pipe encased in polyethylene.

Nominal Pipe Size, *in.*	Depth, *ft*	A21.50—Laying Conditions 2: Restrained Length, *ft*		3: Restrained Length, *ft*		4: Restrained Length, *ft*		5: Restrained Length, *ft*	
60	8.0	120	(147)	70	(84)	57	(66)	47	(53)
60	10.0	100	(123)	58	(70)	48	(56)	39	(44)
64	2.5	275	(335)	159	(189)	129	(148)	104	(117)
64	3.0	248	(303)	144	(172)	117	(135)	95	(106)
64	4.0	208	(254)	121	(144)	98	(113)	80	(90)
64	6.0	157	(193)	91	(110)	75	(86)	61	(68)
64	8.0	126	(155)	73	(88)	60	(70)	49	(55)
64	10.0	105	(130)	61	(74)	50	(59)	41	(46)

NOTE: To convert inches (in.) to millimeters (mm), multiply by 25.4; to convert feet (ft) to meters (m), multiply by 0.3048.
* See Tables 8-2 and 8-3 for soil description.

Table 8-14 Restrained joint lengths for horizontal bends for clay 2* soil type

Soil Parameters:	A21.50—Laying Conditions:			
φ = 0 degrees	2	3	4	5
C_s = 300 lb/ft^2 (14.37 kN/m^2)	f_φ = 0.00	0.00	0.00	0.00
γ = 90 lb/ft^3 (14.14 kN/m^3)	f_c = 0.50	0.80	0.80	0.80
S_f = 1.5	K_n = 0.40	0.60	0.85	1.00

Values listed here are based on a 90° horizontal bend; pressure class listed in Table 8-4; and an internal pressure of 100 psi (689 kPa).

For other pressures, multiply by ratio to 100 psi (689 kPa).

For other horizontal bends, multiply by the following coefficients: 45°–0.414; 22½° –0.199; 11¼° –0.098. Use alternate method for other type fitting or dead end.

Values in parentheses are for pipe encased in polyethylene.

		A21.50—Laying Conditions							
		2		3		4		5	
Nominal Pipe Size, *in.*	Depth, *ft*	Restrained Length, *ft*		Restrained Length, *ft*		Restrained Length, *ft*		Restrained Length, *ft*	
3	2.5	14	(17)	9	(11)	8	(9)	7	(8)
3	3.0	14	(16)	9	(11)	7	(9)	7	(8)
3	4.0	13	(16)	8	(10)	7	(8)	6	(7)
3	6.0	12	(14)	8	(9)	6	(7)	6	(7)
3	8.0	11	(13)	7	(8)	6	(7)	5	(6)
3	10.0	10	(12)	7	(8)	5	(6)	5	(5)
4	2.5	17	(20)	11	(13)	9	(11)	8	(10)
4	3.0	16	(20)	11	(13)	9	(11)	8	(10)
4	4.0	16	(19)	10	(12)	9	(10)	8	(9)
4	6.0	15	(17)	9	(11)	8	(9)	7	(8)
4	8.0	13	(16)	9	(10)	7	(8)	6	(7)
4	10.0	13	(14)	8	(9)	7	(7)	6	(7)
6	2.5	24	(29)	15	(19)	13	(16)	12	(14)
6	3.0	24	(28)	15	(18)	13	(15)	12	(14)
6	4.0	23	(27)	14	(17)	12	(14)	11	(13)
6	6.0	21	(25)	13	(16)	11	(13)	10	(12)
6	8.0	19	(22)	12	(15)	10	(12)	9	(10)
6	10.0	18	(21)	12	(14)	10	(11)	9	(9)
8	2.5	31	(38)	20	(25)	17	(20)	16	(18)
8	3.0	31	(37)	20	(24)	17	(20)	15	(18)
8	4.0	29	(35)	19	(23)	16	(19)	15	(17)
8	6.0	27	(32)	18	(21)	15	(17)	13	(15)
8	8.0	25	(29)	16	(19)	13	(15)	12	(14)
8	10.0	24	(27)	15	(18)	12	(14)	11	(12)

(continued)

Table 8-14 Restrained joint lengths for horizontal bends for clay 2* soil type (*Continued*)

Soil Parameters:	A21.50—Laying Conditions:			
	2	3	4	5
φ = 0 degrees				
C_s = 300 lb/ft^2 (14.37 kN/m^2)	f_φ = 0.00	0.00	0.00	0.00
γ = 90 lb/ft^3 (14.14 kN/m^3)	f_c = 0.50	0.80	0.80	0.80
S_f = 1.5	K_n = 0.40	0.60	0.85	1.00

Values listed here are based on a 90° horizontal bend; pressure class listed in Table 8-4; and an internal pressure of 100 psi (689 kPa).

For other pressures, multiply by ratio to 100 psi (689 kPa).

For other horizontal bends, multiply by the following coefficients: 45°–0.414; 22½° –0.199; 11¼° –0.098. Use alternate method for other type fitting or dead end.

Values in parentheses are for pipe encased in polyethylene.

Nominal Pipe Size, *in.*	Depth, *ft*	A21.50—Laying Conditions 2 Restrained Length, *ft*		3 Restrained Length, *ft*		4 Restrained Length, *ft*		5 Restrained Length, *ft*	
10	2.5	38	(46)	25	(30)	21	(25)	19	(23)
10	3.0	38	(45)	24	(29)	21	(24)	19	(22)
10	4.0	36	(43)	23	(28)	20	(23)	18	(21)
10	6.0	33	(39)	21	(25)	18	(20)	16	(18)
10	8.0	31	(36)	20	(23)	16	(19)	15	(17)
10	10.0	29	(33)	19	(22)	15	(17)	14	(15)
12	2.5	45	(55)	29	(35)	25	(29)	23	(27)
12	3.0	44	(53)	29	(35)	24	(29)	22	(26)
12	4.0	43	(51)	27	(33)	23	(27)	21	(24)
12	6.0	39	(46)	25	(30)	21	(24)	19	(22)
12	8.0	37	(43)	24	(28)	19	(22)	18	(20)
12	10.0	34	(39)	22	(26)	18	(20)	16	(18)
14	2.5	52	(63)	34	(41)	29	(34)	26	(31)
14	3.0	51	(62)	33	(40)	28	(33)	26	(30)
14	4.0	49	(59)	32	(38)	27	(31)	24	(28)
14	6.0	46	(53)	29	(35)	24	(28)	22	(25)
14	8.0	42	(49)	27	(32)	22	(25)	20	(23)
14	10.0	40	(45)	26	(30)	21	(23)	19	(21)
16	2.5	59	(72)	38	(46)	33	(38)	30	(35)
16	3.0	58	(70)	37	(45)	32	(37)	29	(34)
16	4.0	56	(66)	36	(43)	30	(35)	28	(32)
16	6.0	52	(61)	33	(39)	28	(32)	25	(28)
16	8.0	48	(56)	31	(36)	25	(29)	23	(26)
16	10.0	45	(51)	29	(34)	24	(26)	21	(24)

(continued)

Table 8-14 Restrained joint lengths for horizontal bends for clay 2* soil type (*Continued*)

Soil Parameters:	A21.50—Laying Conditions:				
		2	3	4	5
φ = 0 degrees	f_φ =	0.00	0.00	0.00	0.00
C_s = 300 lb/ft^2 (14.37 kN/m^2)	f_c =	0.50	0.80	0.80	0.80
γ = 90 lb/ft^3 (14.14 kN/m^3)	K_n =	0.40	0.60	0.85	1.00
S_f = 1.5					

Values listed here are based on a 90° horizontal bend; pressure class listed in Table 8-4; and an internal pressure of 100 psi (689 kPa).

For other pressures, multiply by ratio to 100 psi (689 kPa).

For other horizontal bends, multiply by the following coefficients: 45°–0.414; 22½° –0.199; 11¼° –0.098. Use alternate method for other type fitting or dead end.

Values in parentheses are for pipe encased in polyethylene.

Nominal Pipe Size, *in.*	Depth, *ft*	A21.50—Laying Conditions 2 Restrained Length, *ft*		3 Restrained Length, *ft*		4 Restrained Length, *ft*		5 Restrained Length, *ft*	
18	2.5	66	(80)	43	(52)	36	(43)	33	(39)
18	3.0	65	(78)	42	(50)	35	(41)	32	(37)
18	4.0	62	(74)	40	(48)	34	(39)	31	(35)
18	6.0	58	(68)	37	(44)	31	(35)	28	(32)
18	8.0	54	(62)	35	(40)	28	(32)	26	(29)
18	10.0	50	(57)	32	(37)	26	(30)	24	(26)
20	2.5	73	(88)	47	(57)	40	(47)	37	(43)
20	3.0	72	(86)	46	(56)	39	(46)	36	(41)
20	4.0	69	(82)	44	(53)	37	(43)	34	(39)
20	6.0	64	(75)	41	(48)	34	(39)	31	(35)
20	8.0	59	(69)	38	(45)	31	(36)	28	(32)
20	10.0	55	(63)	36	(41)	29	(33)	26	(29)
24	2.5	87	(104)	56	(67)	47	(56)	44	(50)
24	3.0	85	(102)	55	(66)	46	(54)	42	(49)
24	4.0	82	(97)	53	(63)	44	(51)	40	(46)
24	6.0	76	(88)	49	(57)	40	(46)	37	(41)
24	8.0	70	(81)	45	(53)	37	(42)	34	(38)
24	10.0	66	(75)	43	(49)	35	(39)	31	(34)
30	2.5	107	(128)	68	(83)	58	(68)	53	(61)
30	3.0	104	(125)	67	(81)	57	(66)	52	(60)
30	4.0	100	(119)	65	(77)	54	(63)	49	(56)
30	6.0	93	(109)	60	(70)	50	(57)	45	(51)
30	8.0	86	(100)	56	(65)	46	(52)	41	(46)
30	10.0	81	(93)	52	(60)	42	(48)	38	(42)

(continued)

Table 8-14 Restrained joint lengths for horizontal bends for clay 2* soil type (*Continued*)

Soil Parameters:	A21.50—Laying Conditions:	2	3	4	5
φ = 0 degrees					
C_s = 300 lb/ft^2 (14.37 kN/m^2)	f_φ =	0.00	0.00	0.00	0.00
γ = 90 lb/ft^3 (14.14 kN/m^3)	f_c =	0.50	0.80	0.80	0.80
S_f = 1.5	K_n =	0.40	0.60	0.85	1.00

Values listed here are based on a 90° horizontal bend; pressure class listed in Table 8-4; and an internal pressure of 100 psi (689 kPa).

For other pressures, multiply by ratio to 100 psi (689 kPa).

For other horizontal bends, multiply by the following coefficients: 45°–0.414; 22½° –0.199; 11¼° –0.098. Use alternate method for other type fitting or dead end.

Values in parentheses are for pipe encased in polyethylene.

Nominal Pipe Size, *in.*	Depth, *ft*	A21.50—Laying Conditions 2 Restrained Length, *ft*		3 Restrained Length, *ft*		4 Restrained Length, *ft*		5 Restrained Length, *ft*	
36	2.5	126	(151)	81	(98)	69	(80)	63	(72)
36	3.0	124	(147)	79	(85)	67	(78)	61	(70)
36	4.0	119	(140)	76	(91)	64	(74)	58	(66)
36	6.0	110	(128)	71	(83)	59	(67)	53	(60)
36	8.0	102	(118)	66	(77)	54	(61)	49	(54)
36	10.0	96	(110)	62	(72)	50	(56)	45	(50)
42	2.5	145	(173)	93	(112)	79	(92)	72	(83)
42	3.0	142	(169)	91	(109)	77	(89)	70	(80)
42	4.0	136	(161)	88	(105)	74	(85)	67	(76)
42	6.0	127	(148)	82	(96)	67	(77)	61	(69)
42	8.0	118	(136)	76	(89)	62	(70)	56	(62)
42	10.0	111	(126)	72	(82)	58	(65)	52	(57)
48	2.5	164	(195)	105	(126)	89	(103)	81	(93)
48	3.0	160	(190)	103	(123)	87	(100)	79	(90)
48	4.0	154	(182)	99	(118)	83	(95)	75	(86)
48	6.0	143	(167)	93	(108)	76	(87)	69	(77)
48	8.0	134	(154)	87	(100)	70	(79)	63	(70)
48	10.0	125	(143)	81	(93)	66	(73)	59	(65)
54	2.5	183	(218)	118	(141)	99	(115)	91	(104)
54	3.0	180	(213)	116	(138)	97	(112)	88	(101)
54	4.0	173	(203)	112	(132)	93	(107)	84	(96)
54	6.0	161	(187)	104	(121)	85	(97)	77	(86)
54	8.0	150	(172)	97	(112)	79	(89)	71	(79)
54	10.0	141	(160)	91	(105)	74	(82)	66	(73)

(continued)

Table 8-14 Restrained joint lengths for horizontal bends for clay 2* soil type (*Continued*)

Soil Parameters:	A21.50—Laying Conditions:				
		2	3	4	5
φ = 0 degrees					
C_s = 300 lb/ft^2 (14.37 kN/m^2)	f_φ =	0.00	0.00	0.00	0.00
γ = 90 lb/ft^3 (14.14 kN/m^3)	f_c =	0.50	0.80	0.80	0.80
S_f = 1.5	K_n =	0.40	0.60	0.85	1.00

Values listed here are based on a 90° horizontal bend; pressure class listed in Table 8-4; and an internal pressure of 100 psi (689 kPa).

For other pressures, multiply by ratio to 100 psi (689 kPa).

For other horizontal bends, multiply by the following coefficients: 45°–0.414; 22½° –0.199; 11¼° –0.098. Use alternate method for other type fitting or dead end.

Values in parentheses are for pipe encased in polyethylene.

		A21.50—Laying Conditions							
		2		3		4		5	
Nominal Pipe Size, *in.*	Depth, *ft*	Restrained Length, *ft*		Restrained Length, *ft*		Restrained Length, *ft*		Restrained Length, *ft*	
60	2.5	195	(232)	126	(150)	105	(122)	96	(110)
60	3.0	191	(226)	123	(147)	103	(119)	94	(107)
60	4.0	184	(216)	119	(140)	99	(113)	90	(101)
60	6.0	171	(198)	110	(129)	91	(103)	82	(92)
60	8.0	160	(183)	103	(119)	84	(94)	76	(84)
60	10.0	150	(170)	97	(111)	78	(87)	70	(77)
64	2.5	206	(245)	133	(159)	112	(129)	102	(116)
64	3.0	202	(239)	130	(155)	109	(126)	99	(113)
64	4.0	195	(229)	126	(148)	104	(120)	95	(107)
64	6.0	181	(210)	117	(137)	96	(109)	87	(97)
64	8.0	169	(194)	110	(127)	89	(100)	80	(89)
64	10.0	159	(181)	103	(118)	83	(92)	74	(82)

Note: To convert inches (in.) to millimeters (mm), multiply by 25.4; to convert feet (ft) to meters (m), multiply by 0.3048.
* See Tables 8-2 and 8-3 for soil description.

Table 8-15 Restrained joint lengths for horizontal bends for silt 2* soil type

Soil Parameters:	A21.50—Laying Conditions:			
	2	3	4	5
φ = 29 degrees	f_φ = 0.50	0.75	0.75	0.75
C_s = 0 lb/ft^2	f_c = 0.00	0.00	0.00	0.00
γ = 90 lb/ft^3 (14.14 kN/m^3)	K_n = 0.40	0.60	0.85	1.00
S_f = 1.5				

Values listed here are based on a 90° horizontal bend; pressure class listed in Table 8-4; and an internal pressure of 100 psi (689 kPa).

For other pressures, multiply by ratio to 100 psi (689 kPa).

For other horizontal bends, multiply by the following coefficients: 45°–0.414; 22½° –0.199; 11¼° –0.098. Use alternate method for other type fitting or dead end.

Values in parentheses are for pipe encased in polyethylene.

		A21.50—Laying Conditions							
		2		3		4		5	
Nominal Pipe Size, *in.*	Depth, *ft*	Restrained Length, *ft*		Restrained Length, *ft*		Restrained Length, *ft*		Restrained Length, *ft*	
3	2.5	21	(25)	14	(16)	11	(13)	10	(12)
3	3.0	18	(21)	12	(14)	10	(11)	9	(10)
3	4.0	14	(16)	9	(10)	7	(8)	7	(7)
3	6.0	9	(11)	6	(7)	5	(6)	4	(5)
3	8.0	7	(8)	5	(5)	4	(4)	3	(4)
3	10.0	6	(7)	4	(4)	3	(3)	3	(3)
4	2.5	25	(30)	17	(19)	14	(16)	12	(14)
4	3.0	21	(25)	14	(16)	12	(13)	10	(12)
4	4.0	16	(19)	11	(13)	9	(10)	8	(9)
4	6.0	11	(13)	7	(9)	6	(7)	5	(6)
4	8.0	8	(10)	6	(6)	5	(5)	4	(5)
4	10.0	7	(8)	4	(5)	4	(4)	3	(4)
6	2.5	35	(41)	23	(27)	19	(22)	17	(19)
6	3.0	30	(35)	20	(23)	16	(18)	15	(16)
6	4.0	23	(27)	15	(18)	12	(14)	11	(13)
6	6.0	16	(18)	10	(12)	9	(10)	8	(9)
6	8.0	12	(14)	8	(9)	6	(7)	6	(7)
6	10.0	10	(11)	6	(7)	5	(6)	5	(5)
8	2.5	45	(53)	30	(35)	25	(28)	22	(25)
8	3.0	39	(45)	25	(30)	21	(24)	19	(21)
8	4.0	30	(35)	20	(23)	16	(18)	15	(16)
8	6.0	20	(24)	13	(16)	11	(13)	10	(11)
8	8.0	16	(18)	10	(12)	8	(10)	8	(8)
8	10.0	13	(15)	8	(10)	7	(8)	6	(7)
10	2.5	54	(63)	36	(42)	29	(33)	26	(30)

(continued)

Table 8-15 Restrained joint lengths for horizontal bends for silt 2* soil type (*Continued*)

Soil Parameters:	A21.50—Laying Conditions:	2	3	4	5
φ = 29 degrees					
C_s = 0 lb/ft^2	f_φ =	0.50	0.75	0.75	0.75
γ = 90 lb/ft^3 (14.14 kN/m^3)	f_c =	0.00	0.00	0.00	0.00
S_f = 1.5	K_n =	0.40	0.60	0.85	1.00

Values listed here are based on a 90° horizontal bend; pressure class listed in Table 8-4; and an internal pressure of 100 psi (689 kPa).

For other pressures, multiply by ratio to 100 psi (689 kPa).

For other horizontal bends, multiply by the following coefficients: 45°–0.414; 22½° –0.199; 11¼° –0.098. Use alternate method for other type fitting or dead end.

Values in parentheses are for pipe encased in polyethylene.

Nominal Pipe Size, *in.*	Depth, *ft*	A21.50—Laying Conditions 2 Restrained Length, *ft*		3 Restrained Length, *ft*		4 Restrained Length, *ft*		5 Restrained Length, *ft*	
10	3.0	46	(54)	31	(36)	25	(28)	23	(25)
10	4.0	36	(42)	24	(28)	19	(22)	18	(20)
10	6.0	25	(29)	16	(19)	13	(15)	12	(14)
10	8.0	19	(22)	12	(15)	10	(12)	9	(10)
10	10.0	15	(18)	10	(12)	8	(9)	7	(8)
12	2.5	63	(73)	41	(48)	34	(38)	31	(34)
12	3.0	54	(63)	36	(41)	29	(33)	26	(29)
12	4.0	42	(49)	28	(32)	23	(26)	20	(23)
12	6.0	29	(34)	19	(22)	16	(18)	14	(16)
12	8.0	22	(26)	15	(17)	12	(14)	11	(12)
12	10.0	18	(21)	12	(14)	10	(11)	9	(10)
14	2.5	71	(83)	47	(55)	38	(43)	35	(39)
14	3.0	61	(71)	40	(47)	33	(37)	30	(33)
14	4.0	48	(56)	32	(37)	26	(29)	23	(26)
14	6.0	33	(39)	22	(26)	18	(20)	16	(18)
14	8.0	26	(30)	17	(20)	14	(16)	12	(14)
14	10.0	21	(24)	14	(16)	11	(13)	10	(11)
16	2.5	79	(92)	52	(61)	43	(48)	38	(43)
16	3.0	68	(79)	45	(52)	37	(42)	33	(37)
16	4.0	54	(62)	35	(41)	29	(33)	26	(29)
16	6.0	38	(44)	25	(29)	20	(23)	18	(20)
16	8.0	29	(34)	19	(22)	16	(18)	14	(16)
16	10.0	23	(27)	15	(18)	13	(14)	11	(13)
18	2.5	87	(101)	57	(66)	47	(53)	42	(47)

(continued)

Table 8-15 Restrained joint lengths for horizontal bends for silt 2* soil type (*Continued*)

Soil Parameters:	A21.50—Laying Conditions:				
		2	3	4	5
φ = 29 degrees	f_φ =	0.50	0.75	0.75	0.75
C_s = 0 lb/ft²	f_c =	0.00	0.00	0.00	0.00
γ = 90 lb/ft³ (14.14 kN/m³)	K_n =	0.40	0.60	0.85	1.00
S_f = 1.5					

Values listed here are based on a 90° horizontal bend; pressure class listed in Table 8-4; and an internal pressure of 100 psi (689 kPa).

For other pressures, multiply by ratio to 100 psi (689 kPa).

For other horizontal bends, multiply by the following coefficients: 45°–0.414; 22½° –0.199; 11¼° –0.098. Use alternate method for other type fitting or dead end.

Values in parentheses are for pipe encased in polyethylene.

Nominal Pipe Size, *in.*	Depth, *ft*	A21.50—Laying Conditions 2 Restrained Length, *ft*		3 Restrained Length, *ft*		4 Restrained Length, *ft*		5 Restrained Length, *ft*	
18	3.0	75	(87)	49	(58)	40	(46)	36	(41)
18	4.0	59	(69)	39	(45)	32	(36)	29	(32)
18	6.0	42	(48)	27	(32)	22	(25)	20	(23)
18	8.0	32	(37)	21	(25)	17	(20)	16	(17)
18	10.0	26	(30)	17	(20)	14	(16)	13	(14)
20	2.5	94	(109)	62	(72)	51	(57)	46	(51)
20	3.0	82	(95)	54	(62)	44	(50)	40	(44)
20	4.0	65	(75)	43	(49)	35	(39)	31	(35)
20	6.0	46	(53)	30	(35)	25	(28)	22	(25)
20	8.0	35	(41)	23	(27)	19	(22)	17	(19)
20	10.0	29	(33)	19	(22)	16	(18)	14	(16)
24	2.5	108	(125)	71	(82)	58	(65)	52	(58)
24	3.0	94	(109)	62	(72)	51	(57)	46	(51)
24	4.0	75	(87)	49	(57)	40	(46)	36	(41)
24	6.0	53	(62)	35	(41)	29	(33)	26	(29)
24	8.0	41	(48)	27	(32)	22	(25)	20	(23)
24	10.0	34	(39)	22	(26)	18	(21)	16	(18)
30	2.5	126	(145)	83	(96)	68	(76)	61	(68)
30	3.0	111	(128)	73	(85)	59	(67)	54	(60)
30	4.0	89	(103)	59	(68)	48	(54)	43	(48)
30	6.0	64	(75)	42	(49)	35	(39)	31	(35)
30	8.0	50	(58)	33	(38)	27	(31)	24	(27)
30	10.0	41	(48)	27	(32)	22	(25)	20	(22)
36	2.5	142	(164)	94	(108)	76	(86)	69	(76)

(continued)

Table 8-15 Restrained joint lengths for horizontal bends for silt 2* soil type (*Continued*)

Soil Parameters:	A21.50—Laying Conditions:				
		2	3	4	5
φ = 29 degrees	f_φ =	0.50	0.75	0.75	0.75
C_s = 0 lb/ft^2	f_c =	0.00	0.00	0.00	0.00
γ = 90 lb/ft^3 (14.14 kN/m^3)	K_n =	0.40	0.60	0.85	1.00
S_f = 1.5					

Values listed here are based on a 90° horizontal bend; pressure class listed in Table 8-4; and an internal pressure of 100 psi (689 kPa).

For other pressures, multiply by ratio to 100 psi (689 kPa).

For other horizontal bends, multiply by the following coefficients: 45°–0.414; 22½° –0.199; 11¼° –0.098. Use alternate method for other type fitting or dead end.

Values in parentheses are for pipe encased in polyethylene.

Nominal Pipe Size, *in.*	Depth, *ft*	A21.50—Laying Conditions 2 Restrained Length, *ft*		3 Restrained Length, *ft*		4 Restrained Length, *ft*		5 Restrained Length, *ft*	
36	3.0	126	(145)	83	(96)	68	(76)	61	(67)
36	4.0	103	(119)	67	(78)	55	(62)	50	(55)
36	6.0	75	(87)	49	(57)	40	(45)	36	(40)
36	8.0	59	(68)	39	(45)	32	(36)	29	(32)
36	10.0	48	(56)	32	(37)	26	(29)	24	(26)
42	2.5	157	(180)	103	(119)	84	(94)	75	(83)
42	3.0	140	(161)	92	(106)	75	(84)	67	(75)
42	4.0	115	(132)	75	(87)	61	(69)	55	(61)
42	6.0	84	(98)	56	(64)	45	(51)	41	(45)
42	8.0	67	(77)	44	(51)	36	(41)	32	(36)
42	10.0	55	(64)	36	(42)	30	(34)	27	(30)
48	2.5	170	(195)	112	(129)	91	(102)	81	(90)
48	3.0	152	(175)	100	(116)	81	(91)	73	(81)
48	4.0	126	(145)	83	(96)	67	(76)	61	(67)
48	6.0	94	(108)	62	(71)	50	(57)	45	(50)
48	8.0	74	(86)	49	(57)	40	(45)	36	(40)
48	10.0	62	(72)	41	(47)	33	(38)	30	(33)
54	2.5	182	(209)	120	(138)	97	(109)	87	(97)
54	3.0	164	(189)	108	(125)	88	(98)	79	(87)
54	4.0	137	(158)	90	(104)	73	(82)	66	(73)
54	6.0	103	(119)	68	(78)	55	(62)	50	(55)
54	8.0	82	(95)	54	(63)	44	(50)	40	(44)
54	10.0	69	(80)	45	(53)	37	(42)	33	(37)
60	2.5	189	(217)	125	(143)	101	(113)	91	(100)

(continued)

Table 8-15 Restrained joint lengths for horizontal bends for silt 2* soil type (*Continued*)

Soil Parameters:	A21.50—Laying Conditions:				
		2	3	4	5
φ = 29 degrees	f_φ =	0.50	0.75	0.75	0.75
C_s = 0 lb/ft^2	f_c =	0.00	0.00	0.00	0.00
γ = 90 lb/ft^3 (14.14 kN/m^3)	K_n =	0.40	0.60	0.85	1.00
S_f = 1.5					

Values listed here are based on a 90° horizontal bend; pressure class listed in Table 8-4; and an internal pressure of 100 psi (689 kPa).

For other pressures, multiply by ratio to 100 psi (689 kPa).

For other horizontal bends, multiply by the following coefficients: 45°–0.414; 22½° –0.199; 11¼° –0.098. Use alternate method for other type fitting or dead end.

Values in parentheses are for pipe encased in polyethylene.

		A21.50—Laying Conditions							
		2		3		4		5	
Nominal Pipe Size, *in.*	Depth, *ft*	Restrained Length, *ft*		Restrained Length, *ft*		Restrained Length, *ft*		Restrained Length, *ft*	
60	3.0	171	(196)	113	(130)	91	(102)	82	(91)
60	4.0	143	(165)	94	(109)	77	(86)	69	(76)
60	6.0	108	(125)	71	(83)	58	(65)	52	(58)
60	8.0	87	(101)	57	(66)	47	(53)	42	(47)
60	10.0	73	(84)	48	(56)	39	(44)	35	(39)
64	2.5	196	(225)	129	(148)	104	(117)	94	(103)
64	3.0	177	(204)	117	(135)	95	(106)	85	(94)
64	4.0	149	(172)	98	(113)	80	(90)	72	(80)
64	6.0	113	(131)	75	(86)	61	(68)	55	(61)
64	8.0	91	(106)	60	(70)	49	(55)	44	(49)
64	10.0	77	(89)	50	(59)	41	(46)	37	(41)

NOTE: To convert inches (in.) to millimeters (mm), multiply by 25.4; to convert feet (ft) to meters (m), multiply by 0.3048.
* See Tables 8-2 and 8-3 for soil description.

Table 8-16 Restrained joint lengths for horizontal bends for coh-gran* soil type

Soil Parameters:	A21.50—Laying Conditions:				
		2	3	4	5
φ = 20 degrees					
C_s = 200 lb/ft² (9.58 kN/m²)	f_φ =	0.40	0.65	0.65	0.65
γ = 90 lb/ft³ (14.14 kN/m³)	f_c =	0.40	0.40	0.40	0.40
S_f = 1.5	K_n =	0.40	0.60	0.85	1.00

Values listed here are based on a 90° horizontal bend; pressure class listed in Table 8-4; and an internal pressure of 100 psi (689 kPa).

For other pressures, multiply by ratio to 100 psi (689 kPa).

For other horizontal bends, multiply by the following coefficients: 45°–0.414; 22½° –0.199; 11¼° –0.098. Use alternate method for other type fitting or dead end.

Values in parentheses are for pipe encased in polyethylene.

Nominal Pipe Size, *in.*	Depth, *ft*	A21.50—Laying Conditions 2 Restrained Length, *ft*		3 Restrained Length, *ft*		4 Restrained Length, *ft*		5 Restrained Length, *ft*	
3	2.5	14	(16)	10	(12)	8	(9)	7	(8)
3	3.0	13	(15)	9	(11)	7	(8)	7	(7)
3	4.0	11	(13)	8	(9)	6	(7)	6	(6)
3	6.0	9	(10)	6	(7)	5	(6)	5	(5)
3	8.0	7	(9)	5	(6)	4	(5)	4	(4)
3	10.0	6	(7)	4	(5)	4	(4)	3	(3)
4	2.5	17	(19)	12	(14)	10	(11)	9	(10)
4	3.0	15	(18)	11	(13)	9	(10)	8	(9)
4	4.0	14	(16)	10	(11)	8	(9)	7	(8)
4	6.0	11	(13)	8	(9)	6	(7)	5	(6)
4	8.0	9	(10)	6	(7)	5	(6)	5	(5)
4	10.0	8	(9)	5	(6)	4	(5)	4	(4)
6	2.5	24	(28)	17	(20)	14	(15)	12	(14)
6	3.0	22	(26)	16	(18)	13	(14)	11	(13)
6	4.0	19	(22)	14	(16)	11	(12)	10	(11)
6	6.0	15	(18)	11	(12)	9	(10)	8	(9)
6	8.0	13	(15)	9	(10)	7	(8)	6	(7)
6	10.0	11	(13)	8	(9)	6	(7)	5	(6)
8	2.5	31	(36)	22	(26)	18	(20)	16	(18)
8	3.0	29	(33)	21	(24)	17	(18)	15	(16)
8	4.0	25	(29)	18	(20)	14	(16)	13	(14)
8	6.0	20	(23)	14	(16)	11	(13)	10	(11)
8	8.0	17	(19)	12	(13)	9	(10)	8	(9)
8	10.0	14	(17)	10	(11)	8	(9)	7	(8)
10	2.5	37	(43)	27	(31)	22	(24)	20	(21)

(continued)

Table 8-16 Restrained joint lengths for horizontal bends for coh-gran* soil type (*Continued*)

Soil Parameters:	A21.50—Laying Conditions:				
		2	3	4	5
φ = 20 degrees					
C_s = 200 lb/ft² (9.58 kN/m²)	f_φ =	0.40	0.65	0.65	0.65
γ = 90 lb/ft³ (14.14 kN/m³)	f_c =	0.40	0.40	0.40	0.40
S_f = 1.5	K_n =	0.40	0.60	0.85	1.00

Values listed here are based on a 90° horizontal bend; pressure class listed in Table 8-4; and an internal pressure of 100 psi (689 kPa).

For other pressures, multiply by ratio to 100 psi (689 kPa).

For other horizontal bends, multiply by the following coefficients: 45°–0.414; 22½° –0.199; 11¼° –0.098. Use alternate method for other type fitting or dead end.

Values in parentheses are for pipe encased in polyethylene.

		A21.50—Laying Conditions							
		2		3		4		5	
Nominal Pipe Size, *in.*	Depth, *ft*	Restrained Length, *ft*		Restrained Length, *ft*		Restrained Length, *ft*		Restrained Length, *ft*	
10	3.0	35	(40)	25	(29)	20	(22)	18	(20)
10	4.0	30	(35)	22	(25)	17	(19)	16	(17)
10	6.0	25	(28)	17	(20)	14	(15)	12	(14)
10	8.0	21	(24)	14	(16)	11	(13)	10	(11)
10	10.0	18	(20)	12	(14)	10	(11)	9	(10)
12	2.5	44	(51)	32	(36)	26	(28)	23	(25)
12	3.0	41	(47)	29	(34)	24	(26)	21	(23)
12	4.0	36	(42)	25	(29)	21	(23)	18	(20)
12	6.0	29	(33)	20	(23)	16	(18)	15	(16)
12	8.0	24	(28)	17	(19)	13	(15)	12	(13)
12	10.0	21	(24)	14	(16)	11	(13)	10	(11)
14	2.5	50	(58)	36	(42)	29	(33)	26	(29)
14	3.0	47	(54)	34	(38)	27	(30)	24	(27)
14	4.0	41	(48)	29	(33)	24	(26)	21	(23)
14	6.0	33	(39)	23	(27)	19	(21)	17	(18)
14	8.0	28	(32)	19	(22)	15	(17)	14	(15)
14	10.0	24	(28)	16	(19)	13	(15)	12	(13)
16	2.5	56	(66)	41	(47)	33	(37)	29	(32)
16	3.0	53	(61)	38	(43)	30	(34)	27	(30)
16	4.0	46	(54)	33	(38)	27	(30)	24	(26)
16	6.0	38	(43)	26	(30)	21	(23)	19	(21)
16	8.0	32	(36)	22	(25)	18	(20)	16	(17)
16	10.0	27	(31)	19	(21)	15	(17)	13	(15)
18	2.5	63	(73)	45	(52)	36	(41)	33	(36)

(continued)

Table 8-16 Restrained joint lengths for horizontal bends for coh-gran* soil type (*Continued*)

Soil Parameters:	A21.50—Laying Conditions:			
	2	3	4	5
φ = 20 degrees				
C_s = 200 lb/ft^2 (9.58 kN/m^2)	f_φ = 0.40	0.65	0.65	0.65
γ = 90 lb/ft^3 (14.14 kN/m^3)	f_c = 0.40	0.40	0.40	0.40
S_f = 1.5	K_n = 0.40	0.60	0.85	1.00

Values listed here are based on a 90° horizontal bend; pressure class listed in Table 8-4; and an internal pressure of 100 psi (689 kPa).

For other pressures, multiply by ratio to 100 psi (689 kPa).

For other horizontal bends, multiply by the following coefficients: 45°–0.414; 22½° –0.199; 11¼° –0.098. Use alternate method for other type fitting or dead end.

Values in parentheses are for pipe encased in polyethylene.

		A21.50—Laying Conditions							
		2		3		4		5	
Nominal Pipe Size, *in.*	Depth, *ft*	Restrained Length, *ft*		Restrained Length, *ft*		Restrained Length, *ft*		Restrained Length, *ft*	
18	3.0	58	(68)	42	(48)	34	(38)	30	(33)
18	4.0	52	(60)	37	(42)	29	(33)	26	(29)
18	6.0	42	(48)	29	(33)	23	(26)	21	(23)
18	8.0	35	(41)	24	(28)	19	(22)	17	(19)
18	10.0	30	(35)	21	(24)	17	(19)	15	(16)
20	2.5	69	(80)	49	(57)	40	(44)	36	(39)
20	3.0	64	(74)	46	(53)	37	(41)	33	(36)
20	4.0	57	(66)	40	(46)	32	(36)	29	(32)
20	6.0	46	(53)	32	(37)	26	(29)	23	(25)
20	8.0	39	(45)	27	(30)	21	(24)	19	(21)
20	10.0	34	(39)	23	(26)	18	(20)	16	(18)
24	2.5	80	(93)	58	(66)	47	(52)	42	(46)
24	3.0	75	(87)	54	(61)	43	(48)	39	(43)
24	4.0	67	(77)	47	(54)	38	(42)	34	(37)
24	6.0	54	(63)	38	(43)	30	(34)	27	(30)
24	8.0	46	(53)	31	(36)	25	(28)	23	(25)
24	10.0	40	(46)	27	(31)	22	(24)	19	(21)
30	2.5	97	(112)	69	(79)	56	(62)	50	(55)
30	3.0	91	(105)	65	(74)	52	(58)	47	(51)
30	4.0	81	(93)	57	(65)	46	(51)	41	(45)
30	6.0	66	(76)	46	(52)	37	(41)	33	(36)
30	8.0	56	(64)	38	(44)	31	(34)	28	(30)
30	10.0	48	(56)	33	(38)	26	(30)	24	(26)
36	2.5	112	(130)	81	(92)	65	(72)	58	(63)

(continued)

Table 8-16 Restrained joint lengths for horizontal bends for coh-gran* soil type (*Continued*)

Soil Parameters:	A21.50—Laying Conditions:				
		2	3	4	5
φ = 20 degrees					
C_s = 200 lb/ft² (9.58 kN/m²)	f_φ =	0.40	0.65	0.65	0.65
γ = 90 lb/ft³ (14.14 kN/m³)	f_c =	0.40	0.40	0.40	0.40
S_f = 1.5	K_n =	0.40	0.60	0.85	1.00

Values listed here are based on a 90° horizontal bend; pressure class listed in Table 8-4; and an internal pressure of 100 psi (689 kPa).

For other pressures, multiply by ratio to 100 psi (689 kPa).

For other horizontal bends, multiply by the following coefficients: 45°–0.414; 22½° –0.199; 11¼° –0.098. Use alternate method for other type fitting or dead end.

Values in parentheses are for pipe encased in polyethylene.

Nominal Pipe Size, *in.*	Depth, *ft*	A21.50—Laying Conditions 2 Restrained Length, *ft*		3 Restrained Length, *ft*		4 Restrained Length, *ft*		5 Restrained Length, *ft*	
36	3.0	106	(122)	75	(86)	60	(67)	54	(59)
36	4.0	94	(109)	66	(76)	53	(59)	48	(52)
36	6.0	78	(89)	54	(61)	43	(48)	39	(42)
36	8.0	66	(76)	45	(51)	36	(40)	32	(36)
36	10.0	57	(66)	39	(44)	31	(35)	28	(31)
42	2.5	127	(147)	91	(103)	73	(81)	65	(71)
42	3.0	120	(138)	85	(97)	68	(75)	61	(67)
42	4.0	107	(123)	75	(86)	60	(67)	54	(59)
42	6.0	88	(102)	61	(70)	49	(54)	44	(48)
42	8.0	75	(86)	51	(59)	41	(46)	37	(41)
42	10.0	66	(75)	44	(51)	36	(40)	32	(35)
48	2.5	141	(163)	100	(114)	80	(89)	72	(79)
48	3.0	133	(153)	94	(107)	75	(84)	67	(74)
48	4.0	119	(137)	84	(95)	67	(74)	60	(66)
48	6.0	99	(114)	68	(78)	55	(61)	49	(54)
48	8.0	84	(97)	58	(66)	46	(51)	41	(46)
48	10.0	74	(85)	50	(57)	40	(45)	36	(39)
54	2.5	155	(179)	110	(125)	88	(98)	79	(86)
54	3.0	147	(169)	103	(118)	83	(92)	74	(81)
54	4.0	132	(152)	92	(105)	74	(82)	66	(72)
54	6.0	110	(126)	76	(86)	61	(67)	54	(60)
54	8.0	94	(108)	64	(73)	52	(57)	46	(51)
54	10.0	82	(94)	56	(64)	45	(50)	40	(44)
60	2.5	163	(188)	116	(132)	93	(103)	83	(91)

(continued)

Table 8-16 Restrained joint lengths for horizontal bends for coh-gran* soil type (*Continued*)

Soil Parameters:	A21.50—Laying Conditions:			
	2	3	4	5
φ = 20 degrees				
C_s = 200 lb/ft^2 (9.58 kN/m^2)	f_φ = 0.40	0.65	0.65	0.65
γ = 90 lb/ft^3 (14.14 kN/m^3)	f_c = 0.40	0.40	0.40	0.40
S_f = 1.5	K_n = 0.40	0.60	0.85	1.00

Values listed here are based on a 90° horizontal bend; pressure class listed in Table 8-4; and an internal pressure of 100 psi (689 kPa).

For other pressures, multiply by ratio to 100 psi (689 kPa).

For other horizontal bends, multiply by the following coefficients: 45°–0.414; 22½° –0.199; 11¼° –0.098. Use alternate method for other type fitting or dead end.

Values in parentheses are for pipe encased in polyethylene.

Nominal Pipe Size, *in.*	Depth, *ft*	A21.50—Laying Conditions 2 Restrained Length, *ft*		3 Restrained Length, *ft*		4 Restrained Length, *ft*		5 Restrained Length, *ft*	
60	3.0	154	(178)	109	(124)	87	(96)	78	(85)
60	4.0	139	(160)	97	(111)	78	(86)	70	(76)
60	6.0	116	(133)	80	(91)	64	(71)	57	(63)
60	8.0	100	(114)	68	(77)	55	(61)	49	(54)
60	10.0	87	(100)	59	(67)	47	(53)	42	(47)
64	2.5	171	(197)	121	(138)	97	(107)	87	(95)
64	3.0	162	(186)	114	(130)	91	(101)	82	(89)
64	4.0	146	(168)	102	(116)	82	(90)	73	(80)
64	6.0	122	(140)	84	(96)	67	(75)	60	(66)
64	8.0	105	(120)	72	(82)	57	(64)	51	(56)
64	10.0	92	(106)	62	(71)	50	(56)	45	(49)

NOTE: To convert inches (in.) to millimeters (mm), multiply by 25.4; to convert feet (ft) to meters (m), multiply by 0.3048.
* See Tables 8-2 and 8-3 for soil description.

Table 8-17 Restrained joint lengths for horizontal bends for sand silt* soil type

Soil Parameters:	A21.50—Laying Conditions:			
	2	3	4	5
φ = 30 degrees				
C_s = 0 lb/ft^2	f_φ = 0.50	0.75	0.75	0.75
γ = 90 lb/ft^3 (14.14 kN/m^3)	f_c = 0.00	0.00	0.00	0.00
S_f = 1.5	K_n = 0.40	0.60	0.85	1.00

Values listed here are based on a 90° horizontal bend; pressure class listed in Table 8-4; and an internal pressure of 100 psi (689 kPa).

For other pressures, multiply by ratio to 100 psi (689 kPa).

For other horizontal bends, multiply by the following coefficients: 45°–0.414; 22½° –0.199; 11¼° –0.098. Use alternate method for other type fitting or dead end.

Values in parentheses are for pipe encased in polyethylene.

Nominal Pipe Size, *in.*	Depth, *ft*	A21.50—Laying Conditions 2: Restrained Length, *ft*		3: Restrained Length, *ft*		4: Restrained Length, *ft*		5: Restrained Length, *ft*	
3	2.5	20	(24)	13	(16)	11	(12)	10	(11)
3	3.0	17	(20)	11	(13)	9	(11)	8	(9)
3	4.0	13	(15)	9	(10)	7	(8)	6	(7)
3	6.0	9	(10)	6	(7)	5	(5)	4	(5)
3	8.0	7	(8)	4	(5)	4	(4)	3	(4)
3	10.0	5	(6)	4	(4)	3	(3)	3	(3)
4	2.5	24	(28)	16	(19)	13	(15)	12	(13)
4	3.0	21	(24)	14	(16)	11	(13)	10	(11)
4	4.0	16	(18)	10	(12)	8	(10)	8	(9)
4	6.0	11	(12)	7	(8)	6	(7)	5	(6)
4	8.0	8	(9)	5	(6)	4	(5)	4	(4)
4	10.0	6	(8)	4	(5)	4	(4)	3	(4)
6	2.5	34	(40)	22	(26)	18	(21)	17	(19)
6	3.0	29	(34)	19	(22)	16	(18)	14	(16)
6	4.0	22	(26)	15	(17)	12	(14)	11	(12)
6	6.0	15	(18)	10	(12)	8	(9)	7	(8)
6	8.0	12	(13)	8	(9)	6	(7)	6	(6)
6	10.0	9	(11)	6	(7)	5	(6)	5	(5)
8	2.5	44	(51)	29	(33)	24	(27)	21	(24)
8	3.0	37	(43)	24	(29)	20	(23)	18	(20)
8	4.0	29	(33)	19	(22)	15	(18)	14	(16)
8	6.0	20	(23)	13	(15)	11	(12)	10	(11)
8	8.0	15	(17)	10	(11)	8	(9)	7	(8)
8	10.0	12	(14)	8	(9)	7	(7)	6	(7)
10	2.5	52	(61)	34	(40)	28	(32)	25	(28)

(continued)

Table 8-17 Restrained joint lengths for horizontal bends for sand silt* soil type (*Continued*)

Soil Parameters:	A21.50—Laying Conditions:			
φ = 30 degrees	2	3	4	5
C_s = 0 lb/ft^2	f_φ = 0.50	0.75	0.75	0.75
γ = 90 lb/ft^3 (14.14 kN/m^3)	f_c = 0.00	0.00	0.00	0.00
S_f = 1.5	K_n = 0.40	0.60	0.85	1.00

Values listed here are based on a 90° horizontal bend; pressure class listed in Table 8-4; and an internal pressure of 100 psi (689 kPa).

For other pressures, multiply by ratio to 100 psi (689 kPa).

For other horizontal bends, multiply by the following coefficients: 45°–0.414; 22½° –0.199; 11¼° –0.098. Use alternate method for other type fitting or dead end.

Values in parentheses are for pipe encased in polyethylene.

Nominal Pipe Size, *in.*	Depth, *ft*	A21.50—Laying Conditions 2 Restrained Length, *ft*		3 Restrained Length, *ft*		4 Restrained Length, *ft*		5 Restrained Length, *ft*	
10	3.0	45	(52)	29	(34)	24	(27)	22	(24)
10	4.0	35	(40)	23	(27)	19	(21)	17	(19)
10	6.0	24	(28)	16	(18)	13	(15)	12	(13)
10	8.0	18	(21)	12	(14)	10	(11)	9	(10)
10	10.0	15	(17)	10	(11)	8	(9)	7	(8)
12	2.5	61	(70)	40	(46)	33	(37)	29	(33)
12	3.0	52	(60)	34	(40)	28	(32)	25	(28)
12	4.0	40	(47)	27	(31)	22	(25)	20	(22)
12	6.0	28	(33)	18	(22)	15	(17)	14	(15)
12	8.0	21	(25)	14	(16)	12	(13)	10	(12)
12	10.0	17	(20)	11	(13)	9	(11)	8	(9)
14	2.5	69	(80)	45	(53)	37	(42)	33	(37)
14	3.0	59	(69)	39	(45)	32	(36)	29	(32)
14	4.0	46	(54)	30	(35)	25	(28)	22	(25)
14	6.0	32	(37)	21	(25)	17	(20)	16	(17)
14	8.0	25	(29)	16	(19)	13	(15)	12	(13)
14	10.0	20	(23)	13	(15)	11	(12)	10	(11)
16	2.5	76	(88)	50	(58)	41	(46)	37	(41)
16	3.0	66	(76)	43	(50)	35	(40)	32	(36)
16	4.0	52	(60)	34	(40)	28	(31)	25	(28)
16	6.0	36	(42)	24	(28)	19	(22)	18	(20)
16	8.0	28	(32)	18	(21)	15	(17)	14	(15)
16	10.0	23	(26)	15	(17)	12	(14)	11	(12)
18	2.5	84	(97)	55	(64)	45	(51)	40	(45)

(continued)

Table 8-17 Restrained joint lengths for horizontal bends for sand silt* soil type (*Continued*)

Soil Parameters:	A21.50—Laying Conditions:	2	3	4	5
φ = 30 degrees					
C_s = 0 lb/ft^2	f_φ =	0.50	0.75	0.75	0.75
γ = 90 lb/ft^3 (14.14 kN/m^3)	f_c =	0.00	0.00	0.00	0.00
S_f = 1.5	K_n =	0.40	0.60	0.85	1.00

Values listed here are based on a 90° horizontal bend; pressure class listed in Table 8-4; and an internal pressure of 100 psi (689 kPa).

For other pressures, multiply by ratio to 100 psi (689 kPa).

For other horizontal bends, multiply by the following coefficients: 45°–0.414; 22½° –0.199; 11¼° –0.098. Use alternate method for other type fitting or dead end.

Values in parentheses are for pipe encased in polyethylene.

Nominal Pipe Size, *in.*	Depth, *ft*	A21.50—Laying Conditions 2 Restrained Length, *ft*		3 Restrained Length, *ft*		4 Restrained Length, *ft*		5 Restrained Length, *ft*	
18	3.0	72	(84)	48	(55)	39	(44)	35	(39)
18	4.0	57	(66)	37	(44)	31	(35)	28	(31)
18	6.0	40	(47)	26	(31)	22	(24)	19	(22)
18	8.0	31	(36)	20	(24)	17	(19)	15	(17)
18	10.0	25	(29)	17	(19)	14	(15)	12	(14)
20	2.5	90	(105)	59	(69)	49	(55)	44	(49)
20	3.0	79	(91)	52	(60)	42	(48)	38	(42)
20	4.0	62	(72)	41	(48)	33	(38)	30	(34)
20	6.0	44	(51)	29	(34)	24	(27)	21	(24)
20	8.0	34	(39)	22	(26)	18	(21)	17	(18)
20	10.0	28	(32)	18	(21)	15	(17)	13	(15)
24	2.5	104	(120)	68	(79)	56	(63)	50	(56)
24	3.0	91	(105)	60	(69)	49	(55)	44	(49)
24	4.0	72	(84)	47	(55)	39	(44)	35	(39)
24	6.0	51	(60)	34	(39)	28	(31)	25	(28)
24	8.0	40	(46)	26	(31)	21	(24)	19	(22)
24	10.0	33	(38)	21	(25)	18	(20)	16	(18)
30	2.5	121	(140)	80	(92)	65	(73)	58	(65)
30	3.0	107	(123)	70	(81)	57	(64)	51	(57)
30	4.0	86	(99)	57	(66)	46	(52)	42	(46)
30	6.0	62	(72)	41	(47)	33	(38)	30	(33)
30	8.0	48	(56)	32	(37)	26	(29)	23	(26)
30	10.0	40	(46)	26	(30)	21	(24)	19	(22)
36	2.5	137	(158)	90	(104)	73	(82)	66	(73)

(continued)

Table 8-17 Restrained joint lengths for horizontal bends for sand silt* soil type (*Continued*)

Soil Parameters:	A21.50—Laying Conditions:			
	2	3	4	5
φ = 30 degrees				
C_s = 0 lb/ft^2	f_φ = 0.50	0.75	0.75	0.75
γ = 90 lb/ft^3 (14.14 kN/m^3)	f_c = 0.00	0.00	0.00	0.00
S_f = 1.5	K_n = 0.40	0.60	0.85	1.00

Values listed here are based on a 90° horizontal bend; pressure class listed in Table 8-4; and an internal pressure of 100 psi (689 kPa).

For other pressures, multiply by ratio to 100 psi (689 kPa).

For other horizontal bends, multiply by the following coefficients: 45°–0.414; 22½° –0.199; 11¼° –0.098. Use alternate method for other type fitting or dead end.

Values in parentheses are for pipe encased in polyethylene.

Nominal Pipe Size, *in.*	Depth, *ft*	A21.50—Laying Conditions 2 Restrained Length, *ft*		3 Restrained Length, *ft*		4 Restrained Length, *ft*		5 Restrained Length, *ft*	
36	3.0	121	(140)	80	(92)	65	(73)	58	(65)
36	4.0	99	(114)	65	(75)	53	(60)	48	(53)
36	6.0	72	(83)	47	(55)	39	(44)	35	(39)
36	8.0	57	(66)	37	(43)	30	(34)	27	(31)
36	10.0	47	(54)	31	(36)	25	(28)	23	(25)
42	2.5	151	(174)	99	(114)	81	(90)	72	(80)
42	3.0	134	(155)	88	(102)	72	(81)	65	(72)
42	4.0	110	(127)	73	(84)	59	(66)	53	(59)
42	6.0	81	(94)	53	(62)	44	(49)	39	(44)
42	8.0	64	(74)	42	(49)	35	(39)	31	(35)
42	10.0	53	(62)	35	(41)	29	(32)	26	(29)
48	2.5	163	(188)	108	(124)	87	(98)	78	(87)
48	3.0	146	(168)	96	(111)	78	(88)	70	(78)
48	4.0	121	(140)	80	(92)	65	(73)	58	(65)
48	6.0	90	(104)	59	(69)	48	(54)	44	(48)
48	8.0	72	(83)	47	(55)	39	(43)	35	(39)
48	10.0	60	(69)	39	(45)	32	(36)	29	(32)
54	2.5	176	(202)	116	(133)	94	(105)	84	(93)
54	3.0	158	(182)	104	(120)	84	(95)	76	(84)
54	4.0	132	(152)	87	(100)	71	(79)	63	(70)
54	6.0	99	(114)	65	(75)	53	(60)	48	(53)
54	8.0	79	(92)	52	(61)	43	(48)	38	(43)
54	10.0	66	(77)	43	(51)	36	(40)	32	(36)
60	2.5	182	(209)	120	(138)	97	(109)	87	(96)

(continued)

Table 8-17 Restrained joint lengths for horizontal bends for sand silt* soil type (*continued*)

Soil Parameters:	A21.50—Laying Conditions:	2	3	4	5
φ = 30 degrees		2	3	4	5
C_s = 0 lb/ft^2	f_φ =	0.50	0.75	0.75	0.75
γ = 90 lb/ft^3 (14.14 kN/m^3)	f_c =	0.00	0.00	0.00	0.00
S_f = 1.5	K_n =	0.40	0.60	0.85	1.00

Values listed here are based on a 90° horizontal bend; pressure class listed in Table 8-4; and an internal pressure of 100 psi (689 kPa).

For other pressures, multiply by ratio to 100 psi (689 kPa).

For other horizontal bends, multiply by the following coefficients: 45°–0.414; 22½° –0.199; 11¼° –0.098. Use alternate method for other type fitting or dead end.

Values in parentheses are for pipe encased in polyethylene.

Nominal Pipe Size, *in.*	Depth, *ft*	A21.50—Laying Conditions: 2 Restrained Length, *ft*		3 Restrained Length, *ft*		4 Restrained Length, *ft*		5 Restrained Length, *ft*	
60	3.0	165	(189)	108	(125)	88	(98)	79	(87)
60	4.0	138	(159)	91	(105)	74	(83)	66	(73)
60	6.0	104	(120)	69	(79)	56	(63)	50	(56)
60	8.0	84	(97)	55	(64)	45	(51)	40	(45)
60	10.0	70	(81)	46	(53)	38	(42)	34	(38)
64	2.5	189	(216)	124	(143)	100	(112)	90	(99)
64	3.0	171	(196)	112	(129)	91	(102)	82	(90)
64	4.0	144	(165)	95	(109)	77	(86)	69	(76)
64	6.0	109	(126)	72	(83)	58	(66)	53	(58)
64	8.0	88	(102)	58	(67)	47	(53)	43	(47)
64	10.0	74	(85)	48	(56)	40	(45)	36	(40)

NOTE: To convert inches (in.) to millimeters (mm), multiply by 25.4; to convert feet (ft) to meters (m), multiply by 0.3048.
* See Tables 8-2 and 8-3 for soil description.

Table 8-18 Restrained joint lengths for horizontal bends for good sand* soil type

Soil Parameters:		A21.50—Laying Conditions:			
		2	3	4	5
φ = 36 degrees	f_φ =	0.75	0.80	0.80	0.80
C_s = 0 lb/ft^2	f_c =	0.00	0.00	0.00	0.00
γ = 100 lb/ft^3 (15.71 kN/m^3)					
S_f = 1.5	K_n =	0.40	0.60	0.85	1.00

Values listed here are based on a 90° horizontal bend; pressure class listed in Table 8-4; and an internal pressure of 100 psi (689 kPa).

For other pressures, multiply by ratio to 100 psi (689 kPa).

For other horizontal bends, multiply by the following coefficients: 45°–0.414; 22½° –0.199; 11¼° –0.098. Use alternate method for other type fitting or dead end.

Values in parentheses are for pipe encased in polyethylene.

		A21.50—Laying Conditions							
		2		3		4		5	
Nominal Pipe Size, *in.*	Depth, *ft*	Restrained Length, *ft*		Restrained Length, *ft*		Restrained Length, *ft*		Restrained Length, *ft*	
3	2.5	12	(14)	9	(11)	8	(9)	7	(8)
3	3.0	10	(12)	8	(9)	6	(7)	6	(7)
3	4.0	7	(9)	6	(7)	5	(6)	4	(5)
3	6.0	5	(6)	4	(5)	3	(4)	3	(3)
3	8.0	4	(5)	3	(4)	2	(3)	2	(3)
3	10.0	3	(4)	2	(3)	2	(2)	2	(2)
4	2.5	14	(17)	11	(13)	9	(10)	8	(9)
4	3.0	12	(14)	9	(11)	8	(9)	7	(8)
4	4.0	9	(11)	7	(8)	6	(7)	5	(6)
4	6.0	6	(7)	5	(6)	4	(5)	4	(4)
4	8.0	5	(6)	4	(4)	3	(3)	3	(3)
4	10.0	4	(4)	3	(3)	2	(3)	2	(2)
6	2.5	20	(24)	16	(18)	13	(15)	12	(13)
6	3.0	17	(20)	13	(15)	11	(12)	10	(11)
6	4.0	13	(15)	10	(12)	8	(9)	8	(8)
6	6.0	9	(10)	7	(8)	6	(6)	5	(6)
6	8.0	7	(8)	5	(6)	4	(5)	4	(4)
6	10.0	5	(6)	4	(5)	3	(4)	3	(4)
8	2.5	25	(30)	20	(23)	16	(19)	15	(17)
8	3.0	21	(26)	17	(20)	14	(16)	13	(14)
8	4.0	16	(20)	13	(15)	11	(12)	10	(11)
8	6.0	11	(14)	9	(10)	7	(8)	7	(7)
8	8.0	9	(10)	7	(8)	6	(6)	5	(6)
8	10.0	7	(8)	5	(6)	5	(5)	4	(5)
10	2.5	30	(36)	24	(28)	20	(22)	18	(20)
10	3.0	26	(31)	20	(24)	17	(19)	15	(17)

(continued)

Table 8-18 Restrained joint lengths for horizontal bends for good sand* soil type (*Continued*)

Soil Parameters:	A21.50—Laying Conditions:	2	3	4	5
φ = 36 degrees					
C_s = 0 lb/ft^2	f_φ =	0.75	0.80	0.80	0.80
γ = 100 lb/ft^3 (15.71 kN/m^3)	f_c =	0.00	0.00	0.00	0.00
S_f = 1.5	K_n =	0.40	0.60	0.85	1.00

Values listed here are based on a 90° horizontal bend; pressure class listed in Table 8-4; and an internal pressure of 100 psi (689 kPa).

For other pressures, multiply by ratio to 100 psi (689 kPa).

For other horizontal bends, multiply by the following coefficients: 45°–0.414; 22½° –0.199; 11¼° –0.098. Use alternate method for other type fitting or dead end.

Values in parentheses are for pipe encased in polyethylene.

Nominal Pipe Size, *in.*	Depth, *ft*	A21.50—Laying Conditions 2 Restrained Length, *ft*		3 Restrained Length, *ft*		4 Restrained Length, *ft*		5 Restrained Length, *ft*	
10	4.0	20	(24)	16	(18)	13	(15)	12	(13)
10	6.0	14	(16)	11	(13)	9	(10)	8	(9)
10	8.0	10	(13)	8	(10)	7	(8)	6	(7)
10	10.0	8	(10)	7	(8)	6	(6)	5	(6)
12	2.5	35	(42)	28	(32)	23	(26)	21	(23)
12	3.0	30	(36)	24	(28)	19	(22)	18	(20)
12	4.0	23	(28)	18	(22)	15	(17)	14	(15)
12	6.0	16	(19)	13	(15)	10	(12)	9	(11)
12	8.0	12	(15)	10	(11)	8	(9)	7	(8)
12	10.0	10	(12)	8	(9)	6	(7)	6	(7)
14	2.5	40	(48)	31	(37)	26	(29)	23	(26)
14	3.0	34	(41)	27	(31)	22	(25)	20	(22)
14	4.0	27	(32)	21	(25)	17	(20)	16	(17)
14	6.0	18	(22)	15	(17)	12	(14)	11	(12)
14	8.0	14	(17)	11	(13)	9	(10)	8	(9)
14	10.0	11	(14)	9	(11)	7	(8)	7	(8)
16	2.5	44	(53)	35	(41)	29	(32)	26	(29)
16	3.0	38	(46)	30	(35)	25	(28)	22	(25)
16	4.0	30	(36)	24	(28)	19	(22)	17	(20)
16	6.0	21	(25)	16	(19)	14	(15)	12	(14)
16	8.0	16	(19)	13	(15)	10	(12)	9	(11)
16	10.0	13	(16)	10	(12)	8	(10)	8	(9)
18	2.5	49	(58)	38	(45)	31	(35)	28	(32)
18	3.0	42	(50)	33	(39)	27	(31)	24	(27)
18	4.0	33	(40)	26	(30)	21	(24)	19	(22)
18	6.0	23	(28)	18	(21)	15	(17)	14	(15)

(continued)

Table 8-18 Restrained joint lengths for horizontal bends for good sand* soil type (*Continued*)

Soil Parameters:		A21.50—Laying Conditions:			
		2	3	4	5
$\varphi = 36$ degrees					
$C_s = 0$ lb/ft^2	f_φ =	0.75	0.80	0.80	0.80
$\gamma = 100$ lb/ft^3 (15.71 kN/m^3)	f_c =	0.00	0.00	0.00	0.00
$S_f = 1.5$	K_n =	0.40	0.60	0.85	1.00

Values listed here are based on a 90° horizontal bend; pressure class listed in Table 8-4; and an internal pressure of 100 psi (689 kPa).

For other pressures, multiply by ratio to 100 psi (689 kPa).

For other horizontal bends, multiply by the following coefficients: 45°–0.414; 22½° –0.199; 11¼° –0.098. Use alternate method for other type fitting or dead end.

Values in parentheses are for pipe encased in polyethylene.

		A21.50—Laying Conditions							
Nominal Pipe Size, *in.*	Depth, *ft*	2		3		4		5	
		Restrained Length, *ft*		Restrained Length, *ft*		Restrained Length, *ft*		Restrained Length, *ft*	
18	8.0	18	(21)	14	(16)	12	(13)	10	(12)
18	10.0	14	(17)	11	(13)	9	(11)	8	(10)
20	2.5	53	(63)	41	(48)	34	(38)	31	(34)
20	3.0	46	(55)	36	(42)	29	(33)	27	(30)
20	4.0	36	(43)	28	(33)	23	(26)	21	(23)
20	6.0	25	(30)	20	(23)	16	(19)	15	(17)
20	8.0	20	(23)	15	(18)	13	(14)	11	(13)
20	10.0	16	(19)	13	(15)	10	(12)	9	(10)
24	2.5	61	(72)	48	(55)	39	(44)	35	(39)
24	3.0	53	(63)	42	(48)	34	(38)	31	(34)
24	4.0	42	(50)	33	(38)	27	(31)	24	(27)
24	6.0	30	(36)	23	(27)	19	(22)	17	(19)
24	8.0	23	(28)	18	(21)	15	(17)	13	(15)
24	10.0	19	(23)	15	(17)	12	(14)	11	(12)
30	2.5	71	(85)	56	(65)	45	(51)	41	(45)
30	3.0	62	(74)	49	(57)	40	(45)	36	(40)
30	4.0	50	(60)	39	(46)	32	(36)	29	(32)
30	6.0	36	(43)	28	(33)	23	(26)	21	(23)
30	8.0	28	(34)	22	(26)	18	(20)	16	(18)
30	10.0	23	(27)	18	(21)	15	(17)	13	(15)
36	2.5	81	(96)	63	(73)	51	(58)	46	(51)
36	3.0	71	(85)	56	(65)	45	(51)	41	(45)
36	4.0	58	(69)	45	(53)	37	(42)	33	(37)
36	6.0	42	(50)	33	(38)	27	(30)	24	(27)
36	8.0	33	(39)	26	(30)	21	(24)	19	(21)
36	10.0	27	(32)	21	(25)	17	(20)	16	(18)

(continued)

Table 8-18 Restrained joint lengths for horizontal bends for good sand* soil type (*Continued*)

Soil Parameters:	A21.50—Laying Conditions:	2	3	4	5
φ = 36 degrees					
C_s = 0 lb/ft^2	f_φ =	0.75	0.80	0.80	0.80
γ = 100 lb/ft^3 (15.71 kN/m^3)	f_c =	0.00	0.00	0.00	0.00
S_f = 1.5	K_n =	0.40	0.60	0.85	1.00

Values listed here are based on a 90° horizontal bend; pressure class listed in Table 8-4; and an internal pressure of 100 psi (689 kPa).

For other pressures, multiply by ratio to 100 psi (689 kPa).

For other horizontal bends, multiply by the following coefficients: 45°–0.414; 22½° –0.199; 11¼° –0.098. Use alternate method for other type fitting or dead end.

Values in parentheses are for pipe encased in polyethylene.

Nominal Pipe Size, *in.*	Depth, *ft*	A21.50—Laying Conditions 2: Restrained Length, *ft*		3: Restrained Length, *ft*		4: Restrained Length, *ft*		5: Restrained Length, *ft*	
42	2.5	89	(106)	70	(80)	57	(63)	51	(56)
42	3.0	79	(94)	62	(72)	50	(56)	45	(50)
42	4.0	65	(77)	51	(59)	41	(46)	37	(41)
42	6.0	47	(56)	37	(43)	30	(34)	27	(31)
42	8.0	37	(45)	29	(34)	24	(27)	22	(24)
42	10.0	31	(37)	24	(28)	20	(22)	18	(20)
48	2.5	97	(115)	76	(87)	61	(69)	55	(61)
48	3.0	86	(102)	68	(78)	55	(61)	49	(55)
48	4.0	71	(85)	56	(64)	45	(51)	41	(45)
48	6.0	52	(63)	41	(48)	34	(38)	30	(34)
48	8.0	42	(50)	33	(38)	27	(30)	24	(27)
48	10.0	34	(41)	27	(32)	22	(25)	20	(22)
54	2.5	104	(123)	81	(93)	66	(74)	59	(65)
54	3.0	93	(111)	73	(84)	59	(66)	53	(59)
54	4.0	78	(92)	61	(70)	49	(55)	44	(49)
54	6.0	58	(69)	45	(53)	37	(42)	33	(37)
54	8.0	46	(55)	36	(42)	30	(33)	27	(30)
54	10.0	38	(46)	30	(35)	25	(28)	22	(25)
60	2.5	108	(128)	84	(97)	68	(76)	61	(68)
60	3.0	97	(115)	76	(88)	62	(69)	55	(61)
60	4.0	81	(96)	64	(73)	52	(58)	46	(51)
60	6.0	61	(73)	48	(55)	39	(44)	35	(39)
60	8.0	49	(58)	38	(44)	31	(35)	28	(31)
60	10.0	41	(49)	32	(37)	26	(30)	24	(26)
64	2.5	112	(133)	87	(100)	71	(79)	63	(70)
64	3.0	101	(120)	79	(91)	64	(72)	57	(63)

(continued)

Table 8-18 Restrained joint lengths for horizontal bends for good sand* soil type (*Continued*)

Soil Parameters:		A21.50—Laying Conditions:			
		2	3	4	5
φ = 36 degrees					
C_s = 0 lb/ft^2	f_φ =	0.75	0.80	0.80	0.80
γ = 100 lb/ft^3 (15.71 kN/m^3)	f_c =	0.00	0.00	0.00	0.00
S_f = 1.5	K_n =	0.40	0.60	0.85	1.00

Values listed here are based on a 90° horizontal bend; pressure class listed in Table 8-4; and an internal pressure of 100 psi (689 kPa).

For other pressures, multiply by ratio to 100 psi (689 kPa).

For other horizontal bends, multiply by the following coefficients: 45°–0.414; 22½° –0.199; 11¼° –0.098. Use alternate method for other type fitting or dead end.

Values in parentheses are for pipe encased in polyethylene.

Nominal Pipe Size, *in.*	Depth, *ft*	A21.50—Laying Conditions 2 Restrained Length, *ft*		3 Restrained Length, *ft*		4 Restrained Length, *ft*		5 Restrained Length, *ft*	
64	4.0	85	(101)	66	(76)	54	(60)	48	(54)
64	6.0	64	(76)	50	(58)	41	(46)	37	(41)
64	8.0	51	(61)	40	(47)	33	(37)	30	(33)
64	10.0	43	(51)	34	(39)	28	(31)	25	(28)

NOTE: To convert inches (in.) to millimeters (mm), multiply by 25.4; to convert feet (ft) to meters (m), multiply by 0.3048.
* See Tables 8-2 and 8-3 for soil description.

REFERENCES

ASTM. 1993. Standard Practice for Classification of Soils for Engineering Purposes (Unified Soil Classification System). ASTM D2487. West Conshohocken, PA: ASTM.

Carlsen, R.J. 1975. "Thrust Restraint for Underground Piping Systems." *Cast Iron Pipe News*. Oakbrook, IL: Cast Iron Pipe Research Association.

Conner, R.C. 1988. "Thrust Restraint of Buried Ductile Iron Pipe." Proc. Pipeline Infrastructure Conf., New York.

Ductile Iron Pipe Research Association. 2016 (7th ed.). *Thrust Restraint Design for Ductile Iron Pipe*, 7th edition. Birmingham, AL: DIPRA.

Potyondy, J.G. 1961. "Skin Friction Between Various Soils and Construction Materials." *Geotechnique*, 11:4:339–353.

Chapter 9

Interior Linings and Hydraulics

9.1 CEMENT–MORTAR-LINED DUCTILE-IRON PIPE

9.1.1 Historical Development

The first cast-iron water mains were not coated or lined but were installed in the same condition in which they came after they were removed from the casting flasks and cleaned. After extensive use, it became evident that the interior of the pipe could be affected by certain types of water. The use of bituminous coatings was proposed, and most of the cast-iron pipe sold for water supply service after approximately 1860 was provided with a hot-dip bituminous lining and coating.

In systems where the water was relatively hard and slightly alkaline, the bituminous linings were generally satisfactory. In systems with soft or acidic waters, however, problems were frequently encountered, such as the water becoming red or rusty and a gradual reduction of the flow rate through the pipe. In these cases, it was established that corrosive water caused tuberculation. The need for a better pipe lining to combat tuberculation led to experiments and research with cement mortar as a lining material. Evidence from the use of cement–mortar lining indicates that it prevents tuberculation while improving flow characteristics.

The first cement–mortar-lined cast-iron pipe was installed in the water distribution system of Charleston, S.C., in 1922. This pipe was lined by means of a projectile drawn through the pipe. In 2019, after more than 97 years of service, friction flow tests showed that this original pipe retained a Hazen–Williams coefficient (*C* factor) of 140.

Since 1922, many improvements have been made in the production of cement–mortar-lined pipe, and the projectile method of cement–mortar lining has progressed to a centrifugal lining process. Today, cement–mortar-lined pipes are centrifugally lined at the

Figure 9-1 Cement–mortar-lined ductile-iron pipe

factory to ensure the best possible quality control is maintained and a uniform thickness of mortar is distributed throughout the entire length of the pipe. These smooth linings result in high flow coefficients. Ductile-iron pipe installed in water systems today is normally provided with a cement–mortar lining (Figure 9-1), unless otherwise specified by the purchaser. For installed unlined cast-iron pipe, in situ cleaning and lining may be economically feasible to restore hydraulic capacity.

9.1.2 Seal Coat

When cement–mortar linings were first introduced into systems with soft water, some customers complained about increased hardness of water during the first few months after installation. The use of a thin paint coating on the freshly placed cement–mortar lining sealed the surface of the lining, and the complaints stopped. It was also found that this seal coat helped cure the lining by minimizing loss of moisture during hydration, resulting in controlled cure of the mortar. The use of such seal coats is described in ANSI/AWWA C104/A21.4, Cement–Mortar Lining for Ductile-Iron Pipe and Fittings.

Cement–mortar linings may be provided with or without a seal coat. The purchaser of cement–mortar-lined pipe or fittings for use with water that is corrosive to cement, such as very soft water, should determine by using the appropriate test that a lining without a seal coat will not impart objectional hardness or alkalinity to the water. The procedure outlined in Section 5.2.2.2 of ANSI/AWWA C104/A21.4, modified by the substitution of the water with which the pipe is to be used, is recommended. The concentration of leachates increases with the aggressiveness of the water and its residual time in the pipe and is indirectly proportional to the diameter of the pipe.

Examination of numerous cement linings following service in various types of water has shown that high flow characteristics have been maintained by both seal-coated and non–seal coated linings.

9.1.3 Lining Processes

The centrifugal process and the projection method of applying cement–mortar linings are both used in modern practice. Excellent quality control of the cement mortar and lining operation can be maintained by using either of these methods. The linings produced are smooth, uniform, and meet the requirements of ANSI/AWWA Standard C104/A21.4. This standard stipulates that the thickness of the linings for pipe and fittings shall not be less than 1/16 in. (1.6 mm) for 3-in. to 12-in. pipe; 3/32 in. (2.4 mm) for 14-in. to 24-in. pipe; and 1/8 in. (3.2 mm) for 30-in. to 64-in. pipe. Double thickness linings, which are twice the minimum thicknesses set forth in the standard, can be provided if specified by the purchaser.

The centrifugal cement–mortar lining process consists of distributing the mortar evenly throughout the length of the pipe by means of a lance that moves while the pipe is spinning. After introduction of the mortar at a relatively low speed, the pipe is spun at a higher rate and vibrated to produce a dense lining that adheres to the pipe wall.

The projection method of cement–mortar lining consists of spraying or slinging mortar on the pipe wall with a rapidly revolving head that has been inserted through the stationary pipe at its centerline. The pipe is then spun and vibrated to smooth and compact the lining.

The linings produced by each of these methods are dense, smooth, and offer very little frictional resistance to the flow of water. Fittings are lined by the projection method or by hand application.

To provide for proper curing of cement linings in pipe by preventing a too rapid loss of moisture from the mortar, the lining can be (a) seal coated immediately, (b) stored in a moist atmosphere for a period of time, or (c) processed through an elevated temperature "curing tunnel" to accelerate the cure. The cement lining has sufficient adherence to the wall of the pipe that the pipe may be cut and tapped without concern for damage to the lining.

9.1.4 Properties of Cement Lining

The protective properties of cement–mortar lining result from two properties of cement. The first is the chemically alkaline reaction of the cement, and the second is the gradual reduction in the amount of water in contact with the iron. When a cement–mortar-lined pipe is filled with water, water permeates the pores of the lining, freeing a considerable amount of calcium hydrate. The calcium hydrate reacts with the calcium bicarbonate in the water to precipitate calcium carbonate. This tends to clog the pores of the mortar and prevent further passage of water. The first water in contact with iron through the lining dissolves some of the iron, but free lime tends to precipitate the iron as iron hydroxide, which also closes the pores of the cement. Sulfates are also precipitated as calcium sulfate. Through these reactions, the lining provides a physical as well as a chemical barrier to the corrosive water.

9.1.5 Autogenous Healing

Occasionally, cracks and looseness of cement–mortar linings in pipe and fittings have been detected before installation. These can occur because of shrinkage of linings, temperature variations, and improper handling. In some instances, there have been concerns that the lining would not provide the protection for which it was intended, or that it would be dislodged by the flow of the water. However, as reported by Wagner (1974), tests have been conducted that show lining fissures, developed while the pipe is in storage, will heal when put into contact with water, either flowing or nonflowing.

Cracks in cement–mortar linings are generally of two types. One is a surface crazing that occurs as a checkerboard or cobweb pattern of hairline cracks. ANSI/AWWA C104/A21.4 permits this type of crack without limitation. The other type of cracking is circumferential or longitudinal. If circumferential, it may extend completely around the pipe. Loosening of the lining may sometimes accompany this type of crack. Such cracks and looseness, with limitations on the length of longitudinal cracks, are also permitted by ANSI/AWWA C104/A21.4 as long as the lining is intact.

When a cement–mortar-lined pipe is placed in service and filled with water, some water is absorbed by the lining. Water is absorbed not only into the pores and voids in the mortar but also into the capillary channels of the calcium silicate gel. The ultimate result of this water absorption is that the lining swells almost to its initial volume. Thus the lining is restored to intimate contact with the pipe wall, and the cracks in the lining close. This swelling process is relatively slow, and it may take up to several weeks for the lining to be completely restored to its initial volume.

Not only do the cracks close and the lining becomes tight after a period of exposure to water, but finally the cracks actually knit back together. This occurs by a process known as *autogenous healing*, a phenomenon long recognized by the cement industry that occurs as a result of the formation of the calcium carbonate and the continuing hydration of the cement grains in the lining. Any cracks that might remain slightly open because of inadequate swelling are subsequently closed by the formation of calcium carbonate.

9.1.6 Repair of Defective or Damaged Areas of Linings

Defective or damaged areas of linings may be patched by cutting out the defective or damaged lining to the metal so that the edges of the lining not removed are perpendicular to the pipe wall or slightly undercut. A stiff mortar is prepared containing not less than one part of cement to two parts of sand by volume. The cutout area and the adjoining lining are thoroughly wetted. The mortar is then applied to the cutout area and troweled smooth with the adjoining lining. After any surface water has evaporated but while the patch is still moist, it should be cured in such a manner as to produce a properly hydrated mortar lining that is hard and durable.

9.2 FLOW CHARACTERISTICS OF DUCTILE-IRON PIPE

9.2.1 Head Loss

The flow of water through cement–mortar-lined ductile-iron pipe is usually computed by the widely used Hazen–Williams formula:

$$Q = 0.006756\, CD^{2.63} H^{0.54}$$

Where

Q = discharge, gal/min (gpm)
C = Hazen–Williams flow coefficient
D = actual inside diameter (ID), in.
H = head loss, ft/1,000 ft

In metric units,

$$Q = 0.0066815\, CD^{2.63} H^{0.54}$$

Where

Q = discharge, m^3/s
C = Hazen–Williams flow coefficient
D = actual ID, m
H = head loss, m/1,000 m

The flow coefficient C in the Hazen–Williams formula (sometimes known as the friction or roughness coefficient) is inversely related to the roughness of the interior pipe surface and used as an approximate indicator of the interior condition of the pipe. Tests

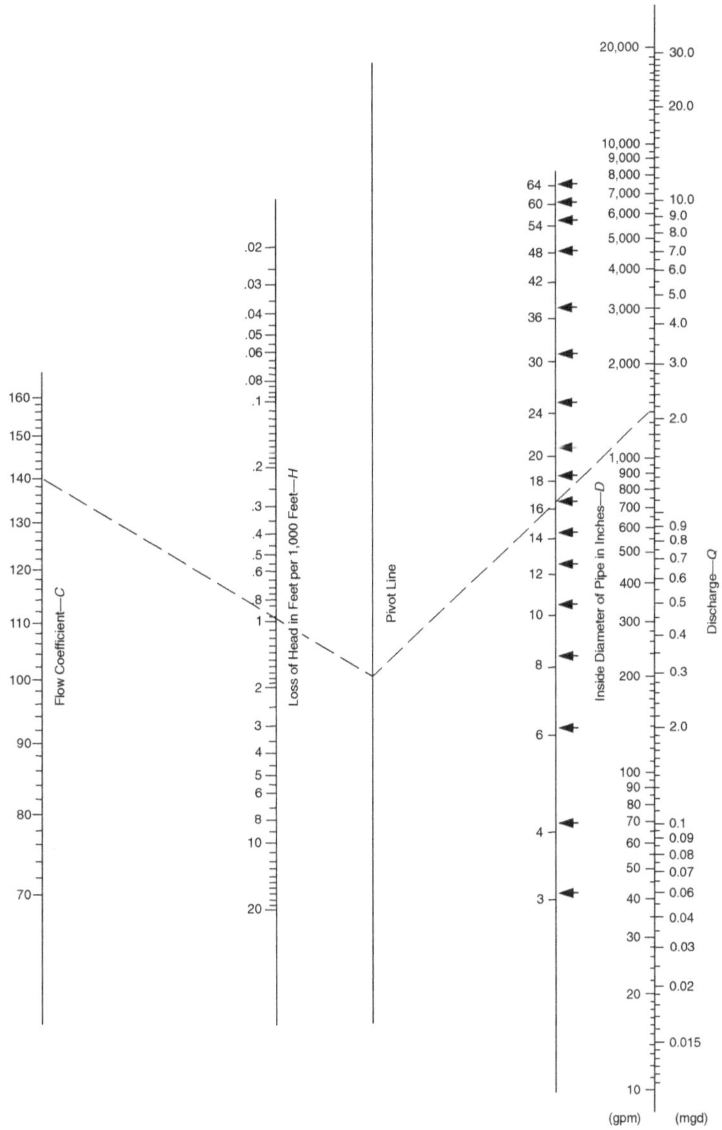

Note: Based on the Hazen–Williams formula: $Q = 0.006756CD^{2.63}H^{0.54}$ For cement–mortar-lined ductile-iron pipe, C = 140. (⟵) Shown are actual inside diameters of cement–mortar-lined ductile-iron pipe of the lowest available pressure classes.

Figure 9-2 Nomograph for pipe size, head loss, and discharge for ductile-iron pipe

conducted by the Ductile Iron Pipe Research Association suggest that a *C*-value of 140 is reasonable for cement-lined ductile-iron pipe.

Ductile-iron pipe is manufactured with the same outside diameters previously used for gray-iron pipe. Because the outside diameter is constant for any given nominal pipe size, ductile-iron pipe has larger than nominal IDs in the most commonly used pressure classes. This results in additional flow or less head loss with ductile-iron pipe compared to pipe with nominal or smaller IDs (see Section 9.3).

Figure 9-2 is a nomograph based on the Hazen–Williams formula, relating the flow coefficient, head loss, internal pipe diameter, and discharge rate. If any three of these factors are known, the fourth may be determined from the nomograph by locating a point on the pivot line that point lies on a common line with two of the known factors. Once the pivot point is established, the unknown factor will lie on a straight line between the pivot point and the third known factor. Actual IDs of the lowest pressure class available cement–mortar-lined ductile-iron pipe are indicated by arrows on the ID line.

Example 9-1. Maximum Delivery

Find the maximum delivery of an 8-in. (203-mm), pressure class 350, cement–mortar-lined ductile-iron pipe, 15,000 ft (4,572 m) in length under 150 ft (46 m) of head.

$$\text{The available head per 1,000 ft is } \frac{150}{15} = 10 \text{ ft/1,000 ft}$$

Using the nomograph, the result is determined to be 900 gpm or 1.30 million gal per day (mgd).

Example 9-2. Determination of Diameter

Find the diameter of pipe necessary to deliver 3 mgd (0.131 m^3/s) through a pipeline 25,000 ft (7,620 m) long under 150 ft (46 m) of head.

$$\text{The available head per 1,000 ft is } \frac{150}{25} = 6 \text{ ft/1,000 ft}$$

Using Figure 9-2, the result is determined to be 12-in. ductile-iron pipe.

Example 9-3. Friction Loss

Find the loss of head through a 10-in. (254-mm) pipeline 4,000-ft (1,219-m) long, delivering 1.4 mgd (0.061 m^3/s) using a *C* factor of 140.

Using Figure 9-2, a rate of head loss of 4 ft/1,000 ft is obtained. This is equivalent to a total head loss of 16 ft for the pipeline. (If water is delivered at a point 100 ft above the pump, total head against the pump is 100 ft [static] plus 16 ft [friction] or a total of 116 ft.)

Example 9-4. Determination of Rate of Flow From Measured Pressure Reduction

Attach two accurate pressure gauges to the pipeline in question, measure the horizontal and vertical distances between the two gauges, and read the pressures indicated. For example, assume a 12-in. (305-mm) pipe with the pressure gauges attached 500 ft (152 m) apart horizontally and 1.8 ft (0.55 m) vertically. Also assume the two gauges show a pressure difference of 2 psi (13.8 kPa) (4.6 ft of head).

The actual head loss will be either 4.6 ft + 1.8 ft = 6.4 ft or 4.6 ft – 1.8 ft = 2.8 ft/500 ft, equivalent to 12.8 ft/1,000 ft or 5.6 ft/1,000 ft, depending on elevation of the downstream gauge. Assuming that the downstream gauge is at the higher elevation, head loss due to friction is 5.6 ft/1,000 ft. Using the nomograph (Figure 9-2), the result of 1,900 gpm or 2.74 million gpd is obtained.

9.3 PUMPING COST

In the most commonly specified pressure classes of cement–mortar-lined ductile-iron pipe, the IDs are larger than the nominal pipe size. One way to realize the savings available through the use of ductile-iron pipe is to consider pumping cost. The energy required to pump water through pipe is proportional to the quantity of water being pumped, as well as to the hydraulic gradient or head loss as defined by the Hazen–Williams equation.

Pumping cost is directly proportional to the energy required, so pumping cost can be expressed by the following equation:

$$\text{PC} = 1.65\, H_L Q \left(\frac{a}{E}\right) \qquad \text{(Eq 9-1)}$$

Where

- PC = pumping cost, \$/year, based on 24-h/d pump operation/1,000 ft
- H_L = hydraulic gradient or head loss, ft/1,000 ft
- Q = flow, gpm
- a = unit cost of electricity, \$/kW × h
- E = total efficiency of pump system (pump, motor, transmission), %/100

In metric units,

$$\mathrm{PC} = 1.041 \times 10^{-4}\, H_L Q \left(\frac{a}{E}\right)$$

Where

- PC = pumping cost, \$/year, based on 24-h/d pump operation/1,000 m
- H_L = hydraulic gradient or head loss, m/1,000 m
- Q = flow, m^3/s
- a = unit cost of electricity, cost/kW × h
- E = total efficiency of pump system (pump, motor, transmission), %/100

The hydraulic gradient or head loss as given by the Hazen–Williams equation is

$$H_L = 1{,}000\left(\frac{V}{1.318\,Cr^{0.63}}\right)^{1.852} \qquad \text{(Eq 9-2)}$$

Where

- H_L = hydraulic gradient or head loss, ft/1,000 ft
- V = velocity, ft/s
- C = Hazen–Williams C factor (140 for ductile-iron pipe)
- r = hydraulic radius, ft (actual ID in feet divided by 4, for a pipe flowing full)

In metric units,

$$H_L = 1{,}000\left(\frac{V}{0.3546\,Cd^{0.63}}\right)^{1.852}$$

Where

- H_L = hydraulic gradient or head loss, m/1,000 m
- V = velocity, m/s
- C = Hazen–Williams C factor (140 for ductile-iron pipe)
- d = internal pipe diameter, m

The IDs for the minimum available pressure classes of ductile-iron pipe are given in Table 9-1, and the dependence of head loss on ID is illustrated in Tables 9-2 through 9-6 showing the comparative head loss at a given flow.

Velocity is related to flow by the equation

$$V = \frac{Q}{448.8A} \qquad \text{(Eq 9-3)}$$

Where

- V = velocity, ft/s
- Q = flow, gpm
- A = cross-sectional area of pipe, ft^2
  = $\frac{\pi d^2}{4}$, where d = actual ID, ft

In metric units,

$$V = Q/A$$

Where

V = velocity, m/s
Q = flow, m^3/s
A = $\pi d^2/4$, where d = actual ID, m

9.3.1 Calculation Examples

To determine the additional energy cost of pumping a given volume of water through a nominal or smaller ID pipe, the following steps should be employed:

Step 1. Convert flow to gpm

Q (gpm) = Q (mgd) × 694.4

Q (gpm) = Q (ft^3/s) × 448.8

Q (gpm) = 448.8 AV

Where

V = velocity, ft/s
A = flow area, ft^2

Step 2. Calculate velocity V by using Equation 9-3. (Actual IDs for ductile-iron pipe should be used, and the IDs of the lowest available pressure class of cement–mortar-lined ductile-iron pipe can be obtained from Table 9-1.)

Step 3. Calculate the H_L for each pipe ID using Equation 9-2 or from Tables 9-2 through 9-6.

Table 9-1 Inside diameters and flow areas of the minimum available pressure classes of cement–mortar-lined ductile-iron pipe

Nominal Pipe Size	ID, *in.*	Flow Area, *ft²*
6	6.28	0.215
8	8.43	0.387
10	10.45	0.596
12	12.51	0.854
14	14.55	1.16
16	16.61	1.51
18	18.69	1.91
20	20.75	2.35
24	24.95	3.40
30	31.07	5.27
36	37.29	7.58
42	43.43	10.29
48	49.63	13.43
54	56.29	17.28
60	60.28	19.82
64	64.30	22.55

NOTE: To convert inches (in.) to millimeters (mm), multiply by 25.4; to convert square feet (ft^2) to square meters (m^2), multiply by 0.0929.

Table 9-2 Head loss in feet per 1,000 ft of pipe (6 in. to 12 in.), cement–mortar-lined ductile-iron pipe, nominal ID pipe, C = 140

	6 in.		8 in.		10 in.		12 in.	
Q, gpm	Pressure Class 350	Nominal ID	Pressure Class 350	Nominal ID	Pressure Class 350	Nominal ID	Pressure Class 350	Nominal ID
250	3.98	4.96	0.95	1.22	0.33	0.41	0.14	0.17
500	14.36	17.96	3.42	4.41	1.20	1.49	0.50	0.61
750	30.42	37.97	7.25	9.35	2.55	3.15	1.06	1.30
1,000	51.83	64.67	12.35	15.93	4.34	5.37	1.81	2.21
1,500	109.83	137.03	26.18	33.75	9.19	11.38	3.83	4.68
2,000	—	—	44.60	57.50	15.66	19.39	6.52	7.98
2,500	—	—	—	—	23.68	29.32	9.86	12.06
3,000	—	—	—	—	33.19	41.09	13.82	16.90
3,500	—	—	—	—	—	—	18.38	22.49
4,000	—	—	—	—	—	—	23.54	28.80

NOTE: To convert inches (in.) to millimeters (mm), multiply by 25.4; to convert feet (ft) to meters (m), multiply by 0.3048; to convert gpm to cubic meters per second (m^3/s), multiply by 6.308×10^{-5}.

Table 9-3 Head loss in feet per 1,000 ft of pipe (14 in. to 20 in.), cement–mortar-lined ductile-iron pipe, nominal ID pipe, C = 140

	14 in.		16 in.		18 in.		20 in.	
Q, gpm	Pressure Class 250	Nominal ID	Pressure Class 250	Nominal ID	Pressure Class 250	Nominal ID	Pressure Class 250	Nominal ID
1,000	0.87	1.04	0.45	0.54	0.26	0.31	0.15	0.18
2,000	3.12	3.76	1.64	1.97	0.92	1.11	0.55	0.66
3,000	6.62	7.98	3.47	4.16	1.96	2.35	1.17	1.40
4,000	11.28	13.59	5.92	7.09	3.33	4.00	2.00	2.39
5,000	17.05	20.55	8.95	10.72	5.04	6.04	3.03	3.62
6,000	23.90	28.81	12.54	15.03	7.06	8.47	4.24	5.07
7,000	31.80	38.32	16.68	20.00	9.39	11.27	5.64	6.75
8,000	40.72	49.08	21.36	25.61	12.03	14.43	7.23	8.64
9,000	—	—	26.57	31.85	14.96	17.95	8.99	10.74
10,000	—	—	—	—	18.18	21.82	10.92	13.06
11,000	—	—	—	—	—	—	13.03	15.68

NOTE: To convert inches (in.) to millimeters (mm), multiply by 25.4; to convert feet (ft) to meters (m), multiply by 0.3048; to convert gpm to cubic meters per second (m^3/sec), multiply by 6.308×10^{-5}.

Step 4. Calculate the pumping cost (PC) per 1,000 ft from each pipe ID using Equation 9-1 (use known or estimated values for a and E).

Step 5. Multiply the PC times the length of the pipeline divided by 1,000.

Step 6. Multiply the result of Step 5 by the fraction of each day the pump will operate.

Step 7. Take the difference in pumping cost (Step 6) of each pipe ID to obtain the annual savings (AS).

Step 8. Calculate the present worth (PW) of the annual savings adjusted for annual inflation rate of power cost using the following equation:

Table 9-4 Head loss in feet per 1,000 ft of pipe (24 in. to 36 in.), cement–mortar-lined ductile-iron pipe, nominal ID pipe, C = 140

	24 in.		30 in.		36 in.	
Q, gpm	Pressure Class 200	Nominal ID	Pressure Class 200	Nominal ID	Pressure Class 200	Nominal ID
2,000	.23	.27	0.08	0.09	0.03	0.04
4,000	0.82	0.98	0.28	0.33	0.12	0.14
6,000	1.73	2.09	0.59	0.70	0.24	0.29
8,000	2.94	3.53	1.01	1.20	0.42	0.49
10,000	4.45	5.37	1.53	1.81	0.63	0.75
15,000	9.43	11.38	3.24	3.84	1.33	1.58
20,000	16.07	19.39	5.52	6.54	2.27	2.69
25,000	—	—	8.35	9.89	3.43	4.07
30,000	—	—	11.70	13.86	4.81	5.70
35,000	—	—	—	—	6.40	7.59
40,000	—	—	—	—	8.19	9.72

Note: To convert inches (in.) to millimeters (mm), multiply by 25.4; to convert feet (ft) to meters (m), multiply by 0.3048; to convert gpm to cubic meters per second (m^3/s), multiply by 6.308×10^{-5}.

Table 9-5 Head loss in feet per 1,000 ft of pipe (42 in. to 54 in.), cement–mortar-lined ductile-iron pipe, nominal ID pipe, C = 140

	42 in.		48 in.		54 in.	
Q, gpm	Pressure Class 150	Nominal ID	Pressure Class 150	Nominal ID	Pressure Class 150	Nominal ID
10,000	0.30	0.35	0.16	0.18	0.08	0.10
15,000	0.63	0.75	0.33	0.39	0.18	0.22
20,000	1.08	1.27	0.56	0.66	0.31	0.37
25,000	1.63	1.92	0.85	1.00	0.46	0.56
30,000	2.29	2.69	1.19	1.40	0.65	0.79
40,000	3.90	4.59	2.04	2.40	1.10	1.35
50,000	5.90	6.93	3.08	3.62	1.67	2.04
60,000	—	—	4.31	5.07	2.34	2.86
70,000	—	—	5.74	6.75	3.11	3.80
80,000	—	—	7.35	8.64	3.98	4.87

Note: To convert inches (in.) to millimeters (mm), multiply by 25.4; to convert feet (ft) to meters (m), multiply by 0.3048; to convert gpm to cubic meters per second (m^3/s), multiply by 6.308×10^{-5}.

$$PW = AS\left(\frac{(1+i)^n - 1}{i(1+i)^n}\right) \text{ with } i = \frac{j - i'}{1 + i'}$$

Where

AS = annual savings
n = design life of the pipeline in years
i = effective annual investment rate (%/100)
j = annual investment interest rate (%/100)
i' = annual inflation rate of power cost (%/100)

Table 9-6 Head loss in feet per 1,000 ft of pipe (60-in. to 64-in.), cement–mortar-lined ductile-iron pipe, nominal ID pipe, C = 140

	60 in.		64 in.	
Q, gpm	Pressure Class 150	Nominal ID	Pressure Class 150	Nominal ID
10,000	0.06	0.06	0.04	0.05
15,000	0.13	0.13	0.09	0.10
20,000	0.22	0.22	0.16	0.16
25,000	0.33	0.34	0.24	0.25
30,000	0.46	0.47	0.34	0.35
40,000	0.79	0.81	0.58	0.59
50,000	1.19	1.22	0.87	0.89
60,000	1.67	1.71	1.22	1.25
70,000	2.23	2.28	1.63	1.66
80,000	2.85	2.91	2.08	2.13
90,000	3.54	3.62	2.59	2.65
100,000	4.31	4.41	3.15	3.22

NOTE: To convert inches (in.) to millimeters (mm), multiply by 25.4; to convert feet (ft) to meters (m), multiply by 0.3048; to convert gpm to cubic meters per second (m^3/s), multiply by 6.308×10^{-5}.

Example 9-5. Criteria

- 24-in. pipeline 30,000-ft long
- Design flow = 6,000 gpm
- Unit power cost = \$0.10/Kw × h
- Pumping efficiency = Keep at 70 percent
- Pump to operate 24 h/d
- Design life of 100 years
- Annual investment interest rate of 8 percent
- Annual inflation rate for power cost of 4 percent
- C = 140

			Example Calculations	
Step Number			Ductile-Iron Pipe	Pipe With Nominal ID
1	*Q*, gpm	=	6,000	6,000
2	*V*, ft/s	=	3.94	4.26
3	H_L, ft/1,000 ft	=	1.73	2.09
4	PC, \$/year/1,000 ft	=	\$2,447	\$2,986
5	PC, \$/year/pipeline	=	\$73,410	\$89,580
6	Same as Step 5—pump operates 24 h/d			
7	AS (with DIP)*			\$16,170
8	PW*			\$418,395

* This number is the additional cost necessary to pump the above quantity of water through pipe with a nominal ID.

The results of these calculations give the annual savings in energy costs from using pressure class 200 ductile-iron pipe and the present worth of projected annual savings adjusted for the inflation rate of power cost. The present worth is the amount that should be used in comparing alternative bids on a pipeline with respect to energy cost alone.

9.3.2 Equivalent Pipelines

Another way to realize the savings available with ductile-iron pipe because of its larger ID is to consider equivalent head losses in the pipeline. Consider a pipeline of diameter D and length L designed to deliver Q gpm (m^3/s) of water. As previously explained, if the pipe has an ID equal to or smaller than the nominal diameter D, the head loss over length L as a result of the friction will be greater than the head loss in an equal length of the same size ductile-iron pipe. Again, this is because of the larger ID of the ductile-iron pipe.

If a nominal or smaller than nominal diameter pipeline is designed to produce the equivalent head loss as the ductile-iron pipe over the length L at flow Q, it can only be made up of pipe size D for part of the distance L. The remainder of the pipeline will have to be made up to the next larger size pipe. This approach will yield savings by using ductile-iron pipe in the initial installation by not using a larger diameter pipe for part of the pipeline.

To achieve equivalent flow and head loss in this manner, the larger size pipe will have to be used over length S, where S is given by

$$S = \left(\frac{H_D - H_N}{H_{N'} - H_N} \right) \times L$$

Where

H_D = head loss per 1,000 ft (1,000 m) of pipe in the ductile-iron pipe of size D at flow Q

H_N = head loss per 1,000 ft (1,000 m) of pipe in the nominal or smaller than nominal diameter pipe of size D at flow Q

H_N = head loss per 1,000 ft (1,000 m) of pipe in the nominal or smaller than nominal diameter pipe of the next larger size than D at flow Q

Example 9-6.

Find the nominal ID pipeline that is equivalent to a 24-in. (610-mm) pressure class 200 ductile-iron pipeline, 30,000-ft (9,144-m) long, delivering 6,000 gpm (0.378 m^3/s). The equivalent nominal ID pipeline will be made up of 24-in. (610-mm) pipe and 30-in. (762 mm) pipe.

H_D (24-in., pressure class 200 ductile-iron pipe) = 1.73 ft/1,000 ft

H_N (nominal diameter 24-in. pipe) = 2.09 ft/1,000 ft

H_N' (nominal diameter 30-in. pipe) = 0.70 ft/1,000 ft

$$\text{Then } S = \left(\frac{1.73 - 2.09}{0.70 - 2.09} \right) \times L$$

$$S = 0.26\,(30{,}000) = 7{,}800$$

Thus, the equivalent nominal ID pipe would include 22,200 ft (6,767 m) of 24-in. (610-mm) pipe and 7,800 ft (2,377 m) of 30-in. (762-mm) pipe.

9.3.3 Water Hammer or Surge

Water hammer or surge has caused failures in many pipelines of various materials. Water hammer is the result of a velocity change in the pipeline fluid. A rapid change in velocity of the liquid mass converts dynamic energy to elastic energy and sets up pressure waves that are transmitted rapidly through the pipeline system. Water hammer can be caused by rapid closure of valves or fire hydrants, quick-acting valves, check valves, sudden loss of power at pumping stations, and other situations that affect the velocity of the fluid. Surge suppression devices are available, but experience has shown that because of inactivity, rusting, silting, loss of power, and overloading, these devices do not always properly arrest surge forces.

Thus, water hammer should be considered in all pipe thickness calculations. It is estimated that in water transmission and distribution systems, a change in velocity of 1 ft/s (0.3048 m/s) can increase the pressure in a pipe by approximately 50 psi (345 kPa).

The design procedure for ductile iron detailed in ANSI/AWWA C150/A21.50, Thickness Design of Ductile-Iron Pipe, includes a standard allowance of 100 psi (689 kPa) for water hammer. After that, a safety factor of 2 is applied resulting in a surge design pressure of 200 psi. This figure is generally accepted as a conservative value for most water piping systems. However, if higher surge pressures are anticipated, higher values should be used in the design calculations.

Water hammer or surge pressure can be calculated for a simple pipeline using the following equations:

$$P_s = \frac{aVW}{144g} \qquad \text{(Eq 9-4)}$$

In metric units,

$$P_s = \frac{aVW}{g}$$

$$a = \frac{u}{\sqrt{1 + \frac{Kd}{Et}}} \qquad \text{(Eq 9-5)}$$

$$u = \sqrt{\frac{144K}{\left(\frac{W}{g}\right)}} \text{ (normal use for water is 4,660 ft/s [1,420 m/s]} \qquad \text{(Eq 9-6)}$$

$$T_c = \frac{2L}{a} \qquad \text{(Eq 9-7)}$$

$$P_{s'} = \frac{T_c}{T_x}(P_s) \qquad \text{(Eq 9-8)}$$

Where

P_s = surge pressure above initial pressure, psi (kPa) (when $T_c > T_x$)
V = velocity change, ft/s (m/s)
W = weight density of water, 62.4 lb/ft^3 (9.80 kN/m^3)

g = acceleration due to gravity, 32.2 ft/s² (9.81 m/s²)
a = wave velocity, ft/s (m/s)
u = wave velocity of fluid, ft/s (m/s)
K = bulk modulus of elasticity of water, 294,000 psi (2.03×10^6 kPa)
d = pipe diameter (nominal), in. (mm)
E = modulus of elasticity of pipe material: ductile iron = 24,000,000 psi (165.48×10^6 kPa); gray cast iron = 15,000,000 psi (103.43×10^6 kPa)
t = thickness of pipe wall, in. (mm)
T_c = critical time in seconds required for pressure wave to travel 2 L
L = length of pipeline, ft (m)
$P_{s'}$ = surge pressure above initial pressure, psi (kPa) (when $T_c < T_x$)
T_x = valve closure time, s

The formula for P_s is used for near instantaneous velocity changes and quick-closing valves. The value for P_s' from Equation 9-8 is for slow-closing valves and considered by many to be an approximation. To obtain more accurate P_s' values, it would be necessary to use other formulas that are considerably more complicated. Because of the high factors of safety incorporated in the standard design procedure for ductile-iron pipe, it normally is not necessary to go through this process. However, the use of other materials having lower factors of safety requires a more accurate determination of surge pressure. Such analyses can be found in various reference works.

It is emphasized that the formulas listed here are for simple pipelines. The analysis of water hammer in a network of piping, such as that of a conventional water distribution system, is much more complex. For this reason, the standard design for ductile-iron pipe includes a 100-psi (689-kPa) water hammer allowance. Further conservatism is realized considering the fact that the safety factor of 2 used in the hoop stress formula for ductile-iron pipe also applies to the 100-psi (689-kPa) water hammer allowance.

A sample calculation for a ductile-iron pipeline 10,000-ft (3,048-m) long, 36 in. (914 mm) in diameter, pressure class 150, and flowing at 5 ft/s (1.52 m/s) with valve closure times of 5 s and 20 s is as follows:

$$d = 36 \text{ in.}, t = 0.38 \text{ in.}, E = 24 \times 10^6 \text{ psi}$$

From Eq 9-5,
$$a = \frac{4,660}{\sqrt{1 + \frac{294,000(36)}{24 \times 10^6 (0.38)}}} = 3,170 \text{ ft/sec}$$

From Eq 9-7,
$$T_c = \frac{2(10,000)}{3,170} = 6.31 \text{ sec}$$

For a valve closure time of 5 s:

$$T_c > T_x \ (6.31 > 5)$$

Therefore, use Equation 9-4 to calculate P_s:

$$P_s = \frac{3,170 \times 5 \times 62.4}{144 \times 32.2} = 213.30 \ psi$$

For a valve closure time of 20 s,

$$T_c < T_x \ (6.31 < 20)$$

Therefore, use Equation 9-8 to calculate $P_{s'}$:

$$P_{s'} = \left(\frac{6.31}{20}\right) \times 213.30 = 67.30\,psi$$

From this example, it is apparent that the standard surge allowance of 100 psi is not sufficient (100 psi < 213 psi), and 213 psi should be added to the design working pressure before selecting the pipe class. It is also apparent that the critical time to closure is 6.31 s. A slower valve closure would diminish the surge pressure.

REFERENCE

Wagner, E.F. 1974. "Autogenous Healing of Cracks in Cement–Mortar Linings for Gray-Iron and Ductile Iron Water Pipe." *Journal AWWA*, 66:6:358–360.

This page intentionally blank.

Chapter 10

External Corrosion Protection

10.1 INTRODUCTION

For many years, underground pipelines were installed with little attention given to the possible need for external corrosion protection. The causes of corrosion were not well understood, and reliable methods of corrosion prevention were not readily available. As a result, repair and replacement of pipelines where soil conditions promoted extensive external corrosion were accepted practices for utilities, albeit expensive solutions to corrosion problems.

Today, utilities can install underground pipelines in practically any soil environment with confidence that external corrosion can be effectively controlled. Modern technology permits the reliable determination of corrosive conditions and provides effective and economical recommendations for the mitigation of corrosion. Although corrosion is a complex subject, a basic understanding of the principles of corrosion and available methods of prevention will enable utility managers and engineers to seek out the most cost-effective solutions. Much of the material contained in this chapter is covered in AWWA M27, *External Corrosion Control for Infrastructure Sustainability*, and is repeated here for easy reference. However, the reader is encouraged to supplement the information found here by referring to AWWA M27. The soil evaluation procedures and methods of corrosion mitigation presented in this chapter are currently utilized only for ductile-iron pipe and fittings and do not apply to other piping materials.

The specific service conditions for ductile-iron pipe and fittings are variable (buried, exposed, and submerged). Also variable are the ways these products are manufactured. Ductile-iron pipe is cast centrifugally and annealed, while iron fittings are statically cast and do not require annealing. Consequently, the control methods for each service condition and for each production method may be similar in some circumstances but mutually exclusive in others.

10.2 BASIC CORROSION THEORY

Corrosion has been defined as the deterioration of a substance or its properties as a result of a reaction with its environment. Most metals are found in nature in their more stable, oxidized form. For example, iron ores are found as iron oxides. During the refining process, these oxides are chemically reduced to convert the metal to its elemental (unoxidized) form (NACE 1984). When exposed to the elements, the refined metal will revert to its more stable form when conditions permit. This oxidative reversion process is known as corrosion.

Corrosion is an electrochemical process that involves a chemical reaction and the flow of electrical current. A typical corrosion cell consists of four basic components, all of which must be present for oxidation to occur. They are an anode, a cathode, a common electrolyte, and a return current path between the anode and cathode. For corrosion to occur, the anode and cathode must be electrically connected and immersed in a common ionized electrolyte (Peabody 1967). The flow of current results from an electrical potential that has developed between the anode and the cathode, with the magnitude of current being determined by this electrical potential and the overall resistance of the cell.

In a pipeline, the anode and cathode form on the pipe itself, the return current path is the wall of the pipe, and the soil is the electrolyte. For a variety of reasons, potential differences can occur between separate locations on a metallic structure buried in soil (the electrolyte). When this occurs, conventional current flows from the higher potential location (the cathode) through the metallic structure to the lower potential location (the anode). The current leaves the metal at the anode and returns to the cathode through the electrolyte. The mechanisms for the transfer of current from the metal to the electrolyte are chemical reactions. Oxidation reactions occur at the anode where current leaves the metal and enters the electrolyte. Reduction reactions occur at the cathode where the current leaves the electrolyte and enters the metal. Corrosion occurs at the anode as a result of the oxidation reactions. In general, there are two basic types of corrosion that occur on underground pipelines: galvanic corrosion and electrolytic corrosion. A detailed discussion of each follows.

10.2.1 Galvanic Corrosion

A galvanic corrosion cell is one in which current is generated due to a potential difference between two locations on a structure. It may result from two dissimilar metals being immersed in a uniform electrolyte, by similar metals immersed in an electrolyte of uneven composition, or a combination of both. In all cases, the four basic components of a corrosion cell, as described previously, must be present for corrosion to occur.

10.2.1.1 Dissimilar metals. Figure 10-1 illustrates a galvanic cell where the corrosion current is generated by two dissimilar metals (iron and copper) that are electrically connected to each other and immersed in a common, uniform electrolyte. When the return current path is provided, the potential difference between the two metals will cause current to flow from the anode (iron) through the electrolyte to the cathode (copper) and back to the anode (iron) through the return current path. Electrons migrate from the anode (iron) through the return current path to the cathode (copper), which results in a polarization of the cell. The loss of electrons from the anode leaves positively charged iron atoms and creates a negative charge at the cathode. This potential difference results in the dissociation of water in the electrolyte, with negatively charged hydroxyl ions (OH^-) migrating to the

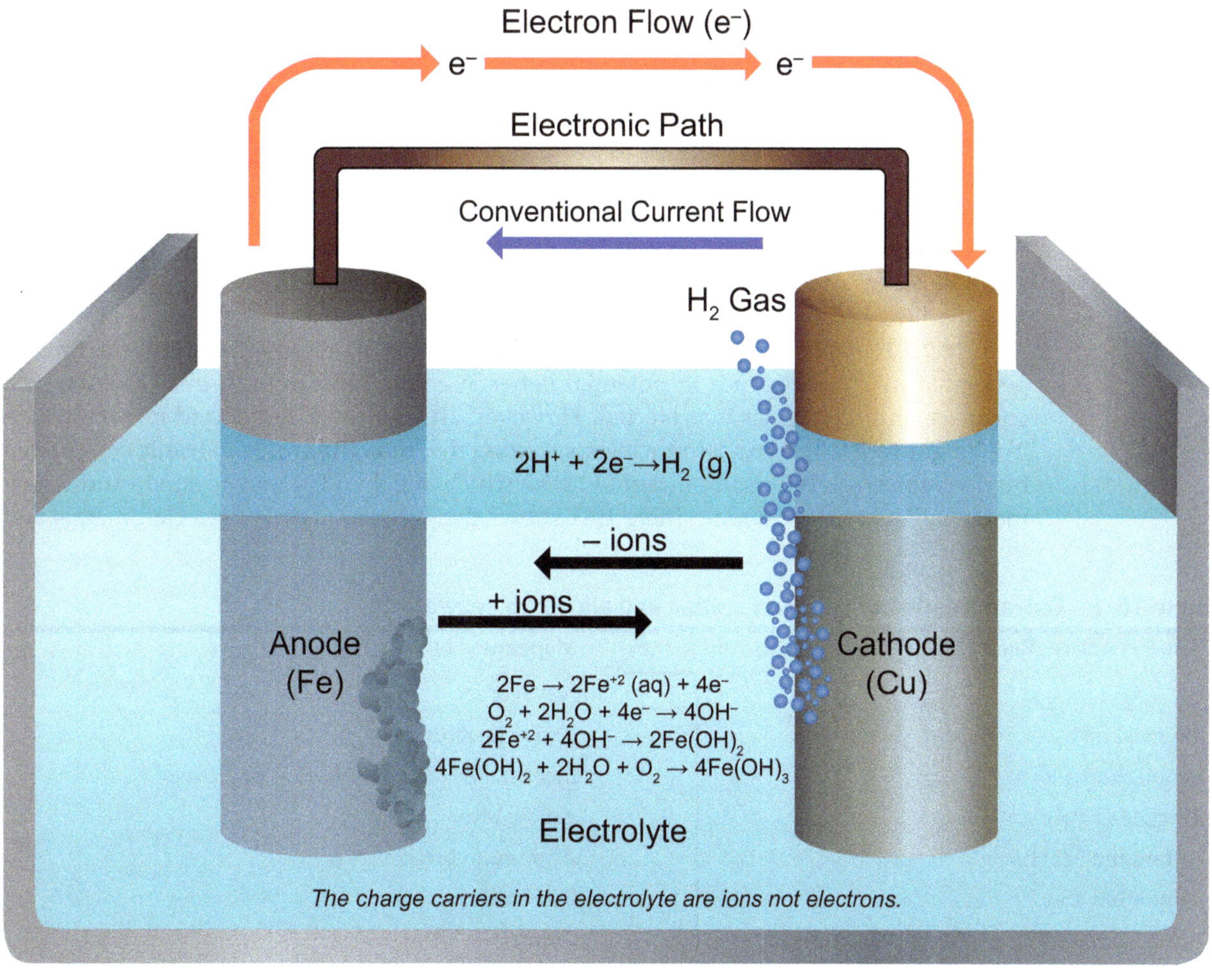

Source: Association for Materials Protection and Performance (AMPP), © 2023 AMPP

Figure 10-1 Chemical reactions in a typical galvanic corrosion cell

anode and positively charged hydrogen (H) ions migrating to the cathode. The hydroxyl ions react at the anode to form $Fe(OH)_2$, which may react further to form $Fe(OH)_3$. This oxidative reaction presents as a deterioration of the metal at the anode surface.

At the cathode, there is an accumulation of negatively charged electrons (migrated from the anode), which attract positively charged hydrogen ions (H^+) from the moisture in the electrolyte. The electrons combine with the hydrogen ions (H^+) to form hydrogen gas (H_2); however, no metal is lost, and thus the cathode is protected. The fact that electron migration is in a direction opposite to conventional current flow can be somewhat confusing. However, one need only remember that conventional current flows from the cathode to the anode through the return current path and from the anode to the cathode through the electrolyte, that metal is consumed where the current leaves the anode to enter the electrolyte, and that the metal (cathode) receiving current from the electrolyte is protected.

It should also be noted that in any galvanic cell, the reactions that occur at the anode and cathode can also influence the rate of corrosion. For example, in the cell illustrated in Figure 10-1, the products of corrosion, $Fe(OH)_2$ and $Fe(OH)_3$, will accumulate on or near the anode while hydrogen gas is generated on or near the cathode. In many soils, particularly relatively dry ones, the buildup of corrosion byproducts can slow or even stop the corrosion process by sealing off the corroding area and preventing the migration of iron ions into the electrolyte. Similarly, the hydrogen gas that forms on the surface of the cathode may accumulate and effectively insulate the cathode from the electrolyte, thereby preventing current flow and rendering the cell inactive. Formation of the hydrogen film on the cathode surface is called *polarization*. As long as this film remains intact, the rate of corrosion will be substantially suppressed. Removal of the hydrogen film is called *depolarization*, which will allow corrosion to proceed at an unsuppressed rate.

Although there are several mechanisms that can affect the removal of the hydrogen film, such as the presence of dissolved oxygen in the electrolyte and the activity of certain types of bacteria, the rate of removal is directly related to the rate of corrosion activity. Thus, the chemical composition of the electrolyte and the materials acting as the cathodic and anodic electrodes in a corrosion cell can greatly affect the severity of corrosion.

10.2.1.2 Galvanic series. In the cell illustrated in Figure 10-1, current flow is generated by the difference in potential between copper and iron. In this case, the iron corrodes, and the copper is protected. However, another metal connected in circuit with iron might well suffer corrosion while protecting the iron. Thus, for a galvanic cell created by dissimilar metals, it is important to know which metal will act as the anode and which metal will act as the cathode. Table 10-1 shows the galvanic series of selected metals and

Table 10-1 Galvanic series of selected metals and alloys (in seawater)*

Anodic, Active (Read Down)†	Manganese bronze
	Naval brass
Magnesium	Nickel (active)
Magnesium alloys	Inconel—76% Ni, 16% Cr, 7% Fe (active)
Zinc	Yellow brass
Aluminum 52SH	Aluminum bronze
Aluminum 4S	Red brass
Aluminum 3S	Copper
Aluminum 2S	Silicon bronze
Aluminum 53S-T	Ambrac—5% Zn, 20% Ni, 75% Cu
Alciad	70% Cu, 30% Ni
Cadmium	Comp. G bronze—88% Cu, 2% Zn, 10% Sn
Aluminum 17S-T	Comp. M bronze—88% Cu, 4% Zn, 6.5% Sn, 1.5% Pb
Aluminum 24S-T	Nickel (passive)
Mild steel	Inconel—75% Ni, 16% Cr, 9% Fe (passive)
Wrought iron	Monel—70% Ni, 30% Cu
Gray and ductile cast iron	18-8 stainless steel, type 304 (passive)
Ni-resist	18-8, 3% Mo stainless steel, type 316 (passive)
13% Cr stainless steel, type 410 (active)	Titanium
50–50 lead–tin solder	Silver
18-8 stainless steel, type 304 (active)	Graphite
18-8, 3% Mo stainless steel, type 316 (active)	Gold
Lead	Platinum
Tin	
Muntz metal ←	**Cathodic, Noble** (Read Up)†

* There is a specific galvanic series for each environment. The relative positions of the various metals and alloys may vary slightly from environment to environment.

† In a galvanic cell of two dissimilar metals, the more active metal will perform as the anode and be corroded, while the more noble metal will perform as the cathode and be protected.

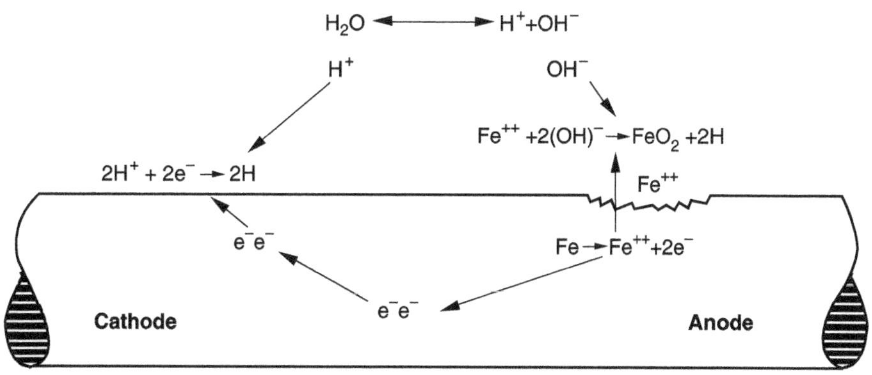

Figure 10-2 Corrosion cell

alloys in seawater. When any two metals shown in this series are electrically connected and placed into a common electrolyte, the metal on the more active side will be the anode and will corrode. The farther apart the two metals are in the galvanic series, the greater the potential difference available to generate current flow and the more rapidly corrosion will occur. Conversely, metals that are close together in the series will generate less corrosion current because of a smaller difference in potential.

It should be noted that in addition to the natural potential that exists between different metals, the rate of corrosion is also controlled by the amount of resistance in the electrolyte and the electrical connection between the anode and cathode. The galvanic series shown in Table 10-1 is accurate when the electrolyte is seawater. In electrolytes of different composition (such as freshwater and soil), there may be slight variations in the order of the metals listed as well as variations in potential difference between any two metals.

10.2.1.3 Nonuniform electrolytes. Galvanic corrosion can also occur when similar metals are exposed to a solid electrolyte of uneven composition. For example, a corrosion cell can develop on a length of pipe exposed to dissimilar soils as illustrated in Figure 10-2. In this case, one area of the pipe acts as the anode and another area as the cathode, with the electrical connection between the two being provided by the pipe itself. The difference in potential necessary for such a cell to develop is a result of the difference in the electrolyte composition. Because of the nature of pipeline construction and because soils are often nonuniform, it is not uncommon for such cells to develop on underground pipelines. Dissimilar soils, differential aeration, differential moisture content, and soil contaminants like cinders and organics are some of the more common causes of this type of galvanic corrosion cell. In general, however, galvanic corrosion as a result of dissimilar electrolytes is less severe than other types of corrosion. Whether or not the formation of such cells on underground pipelines will result in any significant damage is largely dependent on the conductivity of the soil and the type of piping material involved.

10.2.2 Electrolytic Corrosion

An electrolytic corrosion cell is similar to a galvanic corrosion cell in that both require an anode, a cathode, an electrolyte, and a return current path. In a galvanic cell, the electrical energy required to drive the corrosion reaction is generated within the cell as a result of dissimilar metals or dissimilar electrolytes. In an electrolytic cell, the electrical energy is supplied by a direct current source originating outside the cell as illustrated in Figure 10-3. In this case, the corrosion reaction is driven by a direct current source located within the current return path. The end result is the same as that of a galvanic corrosion cell; however, in that corrosion occurs at the anode where current leaves to enter the electrolyte.

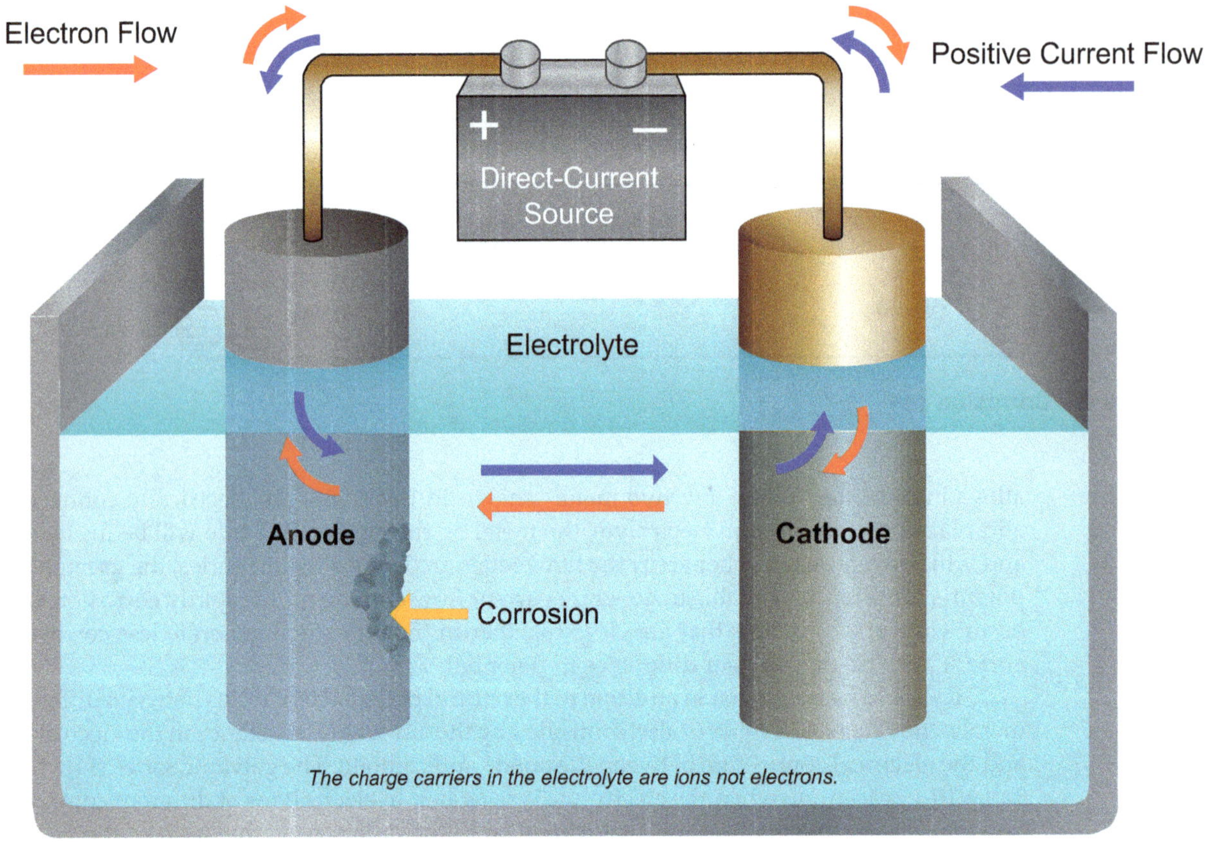

Source: Association for Materials Protection and Performance (AMPP), © 2023 AMPP

Figure 10-3 A typical electrolytic corrosion cell

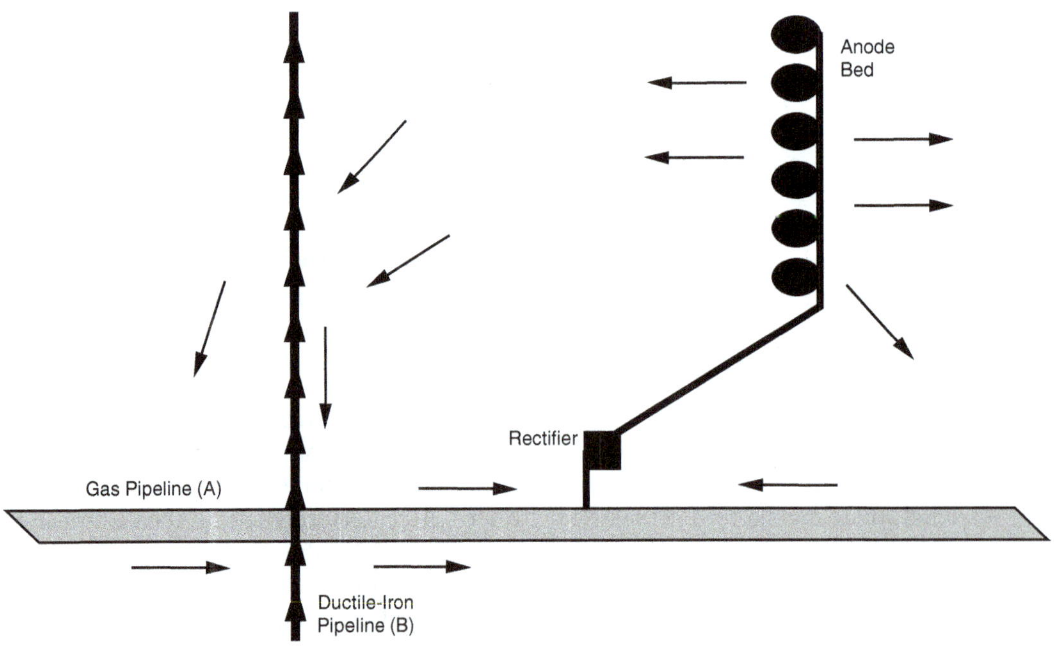

Figure 10-4 Typical stray current environment

Electrolytic corrosion of underground pipelines is often referred to as *stray current corrosion*. This occurs when a direct-current source is located near a pipeline and some of the current from this source strays from its intended path and is picked up by the foreign pipeline as illustrated in Figure 10-4. In such cases, a section of the pipeline becomes part of the current path but is not protected. Corrosion occurs where the current leaves the foreign pipeline to complete its intended circuit.

Ironically, impressed-current cathodic protection systems are a primary cause of electrolytic corrosion on underground pipelines (Bonds 1991). Although designed to protect underground metallic structures from corrosion, these systems can also cause corrosion damage to other underground metallic structures located nearby that are not a part of the system. For example, in Figure 10-4, pipeline A is cathodically protected. Direct current impressed through the ground bed is intended to travel through the earth to pipeline A and return to the rectifier via the metallic connection provided. However, part of the current may be picked up by pipeline B if it offers a less resistant path than the earth. If so, pipeline B will corrode where the current leaves to continue its path to pipeline A.

Electric transit systems and electric welding equipment grounded to underground utilities are other sources of direct current that can cause electrolytic corrosion of pipelines. When electrolytic corrosion does occur, it is usually more severe than other types of corrosion. Fortunately, however, it is probably the least frequent cause of corrosion problems on underground pipelines.

10.2.3 Microbiologically Influenced Corrosion

Microbiologically influenced corrosion is one manifestation of the effect on corrosion by soil bacteria. Certain bacteria that can exist under anaerobic conditions (absence of oxygen) at the pipeline surface will reduce any sulfates present. This releases sulfur ions that combine with the hydrogen ions at the cathode. The consumption of hydrogen at the pipe surface acts to depolarize cathodic areas and permits more rapid consumption of the metal by galvanic corrosion cells. The bacteria, then, do not directly attack the pipe but prevent the effective polarization of the corrosion cell, allowing the rate of corrosion to continue unabated.

10.2.4 Rate of Corrosion

From the preceding discussion, it might seem that corrosion cells on metallic pipelines are ubiquitous in a soil environment. However, as a practical matter, whether corrosion will be a significant problem in a given underground pipeline is determined by the rate of corrosion. The rate of corrosion at an anodic area is directly related to the magnitude of the current leaving the metal at the anode. The magnitude of current is in turn related to such factors as the relative size of the anode and cathode areas, the electrical resistance of the various current paths, the magnitude of the potential difference between the anode and the cathode, and the physical and thermodynamic characteristics at the metal–electrolyte interface at the anode and cathode—all of which will influence the rate of chemical reaction that can occur. The chemical reactions at the anode and cathode are the mechanisms of current transfer from the metal to the electrolyte and vice versa. The specific chemical reactions and resultant compounds created at the anode and cathode often change the environment and limit the rate of corrosion. This phenomenon, known as *passivation*, depends on specific chemical characteristics of the metal and the soil. Thus, the mere existence of corrosion cells does not necessarily mean that corrosion problems will occur.

Galvanic corrosion is the type of corrosion activity most prevalent on underground pipelines. Because galvanic corrosion cells, particularly those created by dissimilar electrolytes, generally have small potential differences that drive the corrosion reaction, the conductivity of the electrolyte has a significant effect on the rate of corrosion. This is why the soil is tested to evaluate its corrosive character. Although other aspects of the soil contribute to the rate of corrosion, the most important soil characteristic is its resistivity. As the name implies, a soil's resistivity is a measure of its ability to conduct an electric current. In relatively high resistivity soils, the rate of corrosion is generally low and would rarely be of concern. For that reason, corrosion engineers generally place a great deal of importance on the resistivity of soils when evaluating the need for corrosion protection of underground pipelines. Additionally, if the cell is a result of dissimilar metals, both the materials involved and the relative size of the anode and cathode areas will affect the corrosion rate. As discussed previously, metals close together in the galvanic series of metals and alloys (Table 10-1) will have a low difference in potential and thus require a low-resistivity electrolyte for corrosion activity to proceed at an appreciable rate. Also, a small cathode in combination with a large anode might only be of concern in a low-resistivity soil, whereas the reverse could conceivably cause a problem in a much wider range of soil conditions.

10.3 CORROSION RESISTANCE OF DUCTILE-IRON PIPE

Before the introduction of ductile-iron pipe, gray cast-iron pipe had been the principal material used for construction of buried pressure pipelines, particularly for water supply and distribution, for more than 150 years.

Ductile-iron pipe succeeded gray cast-iron pipe because of its dramatically improved structural properties (DIPRA 1984). In addition to its greater strength, ductile-iron pipe also possesses the other superior properties of gray cast iron, such as corrosion resistance, ease of casting and of installation, and economy. Both ductile iron and gray iron contain similar amounts of carbon, which contributes to the machinability and corrosion resistance of the metals. In gray cast iron, most of the carbon is present in the form of interlocking flakes of graphite dispersed throughout the metal matrix. In ductile iron, an additional step in the treatment of the molten metal during manufacture causes the graphite to take the form of spheroids or nodules. When corrosion of either material occurs, the graphite present will be left as an integral part of the corrosion byproducts, adhering firmly to the remaining metal substrate. As indicated previously, these graphite-containing corrosion byproducts provide a barrier against further corrosion attack, which can slow or even stop the corrosion process in many soil environments. In addition to their protective value, the graphite byproducts also have considerable mechanical strength, which undoubtedly accounts for the long service life of many unprotected iron pipelines in corrosive environments (see DIRPRA's Cast Iron Pipe Century Club and Cast Iron Pipe Sesquicentennial Club website, dipra.org/about-dipra/century-club).

A number of studies have been conducted to determine the corrosion resistance of ductile-iron pipe. Mostly as a result of conservative interpretation of the data, the early studies (e.g., CIPRA 1968; Sears 1968; Romanoff 1968; Smith 1968) concluded that the corrosion resistance of ductile-iron pipe was at least as good as that of gray-iron pipe. Theoretical considerations also support this conclusion because there is little difference chemically between the two materials. However, it was recognized early on that ductile iron has a fundamental advantage over gray iron with respect to corrosion resistance because of the spheroidal shape of the graphite (LaQue 1964). The surface area of graphite capable of acting as a cathode to support anodic corrosion of the iron matrix is much less when the graphite is in the form of spheres rather than flakes. Additionally, the interlocking flakes of graphite in gray iron (as opposed to the well-dispersed and separate graphite spheres in ductile iron) tend to promote a greater depth of penetration when localized corrosion pitting progresses along the margins of the graphite flakes. Subsequent research and analysis of data from ductile pipe that had been exposed for longer periods, as well as successful experience with ductile-iron pipe in actual field installations, further confirmed the existence of this fundamental advantage and led to a refinement of the earlier conclusions regarding the corrosion resistance of ductile-iron pipe (Fuller 1981). The spheroidal graphite structure of ductile iron encourages a given amount of corrosion attack to spread out over the surface of the metal; this occurs more than in gray iron where the attack usually occurs as deeper pitting. As a result, ductile-iron pipe is less susceptible to deep localized pitting than gray-iron pipe, which is important in evaluating its relative resistance to failure by perforation.

Another important but often overlooked consideration is the fact that ductile-iron pipe is manufactured in 18- and 20-ft (5.5- and 6.1-m) nominal lengths and employs a rubber-gasketed jointing system. These joints offer resistance that may vary from a fraction of an ohm to several ohms but nevertheless is of sufficient magnitude that ductile-iron pipelines are considered to be electrically discontinuous. As a result, long line corrosion current, which is dependent on exposure of a single electrical unit to varying soil conditions, cannot develop. Additionally, any accumulation of stray direct current is limited to short electrical units and usually is of little significance. For these reasons, joint bonding of ductile-iron pipelines is generally discouraged except in cases where electrical continuity is needed for corrosion monitoring and cathodic protection.

10.4 EVALUATION OF CORROSIVE SOILS

Buried service conditions present challenges due to corrosivity of different soil types. Several soil evaluation procedures have been developed for ductile and gray cast-iron pipe installations and used to varying extents.

10.4.1 The 10-Point Soil-Test Evaluation System

One procedure historically used in soil analysis to evaluate potentially corrosive environments to ductile-iron pipe is the 10-point soil-test evaluation system. The 10-point system was originally developed by the Cast Iron Pipe Research Association (CIPRA), which in 1979 became the Ductile Iron Pipe Research Association (DIPRA). This soil evaluation procedure is specific to gray and ductile cast-iron pipe and should not be applied to other materials. The 10-point system and its interpretation are incorporated in an appendix to ANSI/AWWA C105/A21.5, Polyethylene Encasement for Ductile-Iron Pipe Systems. This system is presented in Table 10-2. Providing decades of useful service, the 10-point system is a one-dimensional system that accumulates points based on the results of evaluations of soil to determine whether the soil was corrosive. If the accumulated point count met or exceeded 10 points, polyethylene encasement would be recommended.

10.4.2 The Design Decision Model

More recently, DIPRA cooperated with Corrpro Companies to develop the first two-dimensional risk model for corrosion control of ductile-iron pipe. The Design Decision Model (DDM®) balances the likelihood of a corrosive condition against the consequences of a related failure and makes recommendations that recognize the practicalities of corrosion control for distribution pipe networks.

The DDM® provides five levels of corrosion control based on the results of the likelihood and consequences evaluations. Building on the 10-point system and the statistical evaluation of more than 60,000 corrosion datapoints (Bonds et. al. 2005), the DDM® was designed to provide a minimum average service life of 100 years through increasing levels of corrosion control recommendations.

Table 10-2 Soil-test evaluation for gray or ductile cast-iron pipe (10-point system)*

	Soil Characteristics	Points
Resistivity Ω -cm[†]	<1,500	10
	≥1,500–1,800	8
	>1,800–2,100	5
	>2,100–2,500	2
	>2,500–3,000	1
	>3,000	0
pH	0–2	5
	2–4	3
	4–6.5	0
	6.5–7.5	0[‡]
	7.5–8.5	0
	>8.5	3
Redox potential	> +100 mV	0
	+50 to +100 mV	3.5
	0 to +50 mV	4
	Negative	5
Sulfides	Positive	3.5
	Trace	2
	Negative	0
Moisture	Poor drainage, continuously wet	2
	Fair drainage, generally moist	1
	Good drainage, generally dry	0

* Ten points (See Sec. 10.4.1)—corrosive to gray or ductile cast-iron pipe, protection is indicated.
† Based on water-saturated soil box.
‡ If sulfides are present and low or negative redox-potential results are obtained, three points shall be given for this range.

The evaluation procedure is based on the results of tests and observations made at appropriate points along the proposed pipeline route. At each discrete location, soil is sampled and tested for the likelihood of corrosion, and observations are made regarding the potential consequences of a corrosion-related problem. In this way, corrosion control recommendations can be applied to specific sections of a pipeline as required.

The likelihood factors and associated point counts are shown in Table 10-3. The consequence factors and point counts are shown in Table 10-4

A score is assigned based on the results at each location, with more points resulting from more aggressive conditions. The points for the likelihood factors are totaled, as are the points for the consequence factors. These two results are used to enter the grid shown in Table 10-5 where the intersection of the likelihood (x-axis) and consequence (y-axis) factors reveals a recommended method of corrosion mitigation. These recommendations are shown in Table 10-6.

If the soil exhibits a more aggressive character, alternative recommendations are provided that recognize the distinction between distribution and transmission mains and the preferences of the owner of the pipeline:

1. Installing the pipe as manufactured with no supplemental corrosion control beyond the inherent corrosion resistance offered by its standard shop coating and the annealing oxide layer.

Table 10-3 Design decision model likelihood factors

Likelihood Factor		Points	Maximum Possible Points
Resistivity	< 500 ohm-cm	30	**30**
	≥ 500–1,000 ohm-cm	25	
	> 1,000–1,500 ohm-cm	22	
	> 1,500–2,000 ohm-cm	19	
	> 2,000–3,000 ohm-cm	10	
	> 3,000–5,000 ohm-cm	5	
	> 5,000 ohm-cm	0	
Chlorides	> 100 ppm = positive	8	**8**
	50–100 ppm = trace	3	
	< 50 ppm = negative	0	
Moisture content	> 15% = wet	5	**5**
	5–15% = moist	2.5	
	< 5% = dry	0	
Ground water influence	Pipe below the water table at any time	5	**5**
pH	pH 0–4	4	**4**
	pH > 4–6	1	
	pH 6–8, with sulfides and low or negative redox	4	
	pH > 6	0	
Sulfide ions	positive (>– 1 ppm)	4	**4**
	trace (> 0 and < 1 ppm)	1.5	
	negative (0 ppm)	0	
Redox potential	= negative	2	**2**
	= positive 0–100 mV	1	
	= positive > 100 mV	0	
Bimetallic considerations	Connected to noble metals (e.g., copper)—Yes	2	**2**
	Connected to noble metals (e.g., copper)—No	0	

RELATIVE IMPORTANCE

1 = Highest
5 = Lowest

Known corrosive environments

TOTAL POSSIBLE POINTS 60

Cinders, mine waste, peat bog, landfill, fly ash, coal

21

* Soils with known corrosive environments shall be assigned 21 points or the total of points for likelihood factors, whichever is greater.

Table 10-4 Design decision model consequence factors

Consequence Factor		Points	Maximum Possible Points
Pipe service	3–24″	0	22
	30–36″	8	
	42–48″	12	
	54–64″	22	
Location: construction-repair considerations	Routine (fair to good access, minimal traffic/other utility consideration, etc.)	0	
	Moderate (typical business/residential areas, some right of way limitations, etc.)	8	20
	Difficult (subaqueous crossings, downtown metropolitan business areas, multiple utilities congestion, swamps, etc.)	20	
Depth of cover considerations	0–10 ft depth	0	5
	>10–20 ft depth	3	
	>20 ft depth	5	
Alternate water supply	Alternate supply available—No	3	3
	Alternate supply available—Yes	0	
	Total possible points		50

Table 10-5 Recommendations from the DDM®

	Recommendations
1	As manufactured with shop coat
2	V-Bio®Enhanced Polyethlene Encasement
3	V-Bio®Enhanced Polyethylene Encasement or V-Bio®Enhanced Polyethylene Encasement with Joint Bonds
4	V-Bio®Enhanced Polyethylene Encasement with Metallized Zinc Coating or V-Bio®Enhanced Polyethylene Encasement with Life Extension Cathodic Protection
5	V-Bio®Enhanced Polyethylene Encasement with Metallized Zinc Coating or V-Bio®Enhanced Polyethylene Encasement with Cathodic Protection

NOTE: Recommendations in Zones 4 and 5 recognize a practical difference between distribution and transmission mains. Distribution mains are generally smaller sized pipes, with the final classification to be defined by the pipeline owner. Cathodic protection should be considered where external corrosion is a significant risk or where pipe repairs/replacements would be cost prohibitive.

2. Encasing the pipe in an enhanced polyethylene.
3. Encasing the pipe in an enhanced polyethylene or encasing the pipe and providing bonded joints. The joint bond alternative would allow for retrofit of the pipeline with cathodic protection.
4. Encasing the pipe in an enhanced polyethylene and providing life-extension cathodic protection currents or encasing the pipe in an enhanced polyethylene and installing ductile-iron pipe that is provided with a special metallized zinc coating.

5. Encasing the pipe in an enhanced polyethylene and providing cathodic protection currents or encasing the pipe in an enhanced polyethylene and installing ductile-iron pipe that is provided with a special metallized zinc coating.

10.4.3 Soil Testing

10.4.3.1 Soil resistivity. The soil-resistivity test yields the reciprocal of conductivity and evaluates how well the soil would serve as an electrolyte. A low soil resistivity is a more conductive electrolyte and would indicate a more aggressive soil. Resistivity, reported in ohm-centimeters (Ω-cm), represents the average of the electrical resistances across each cubic centimeter of soil in a given volume. The test may be accomplished using the 4-pin method or a soil box in conjunction with an appropriate resistance meter.

The 4-pin method has limitations because it measures the resistivity of the soils at the time of the test, which may be affected by recent weather, for example. The 4-pin method may also average a dry topsoil with wetter subsoils or contaminated soil with clean soil. Because soil moisture affects resistivity and moisture in the field varies unpredictably, the best way to obtain a repeatable soil resistivity result is to remove discrete soil specimens down to the pipe depth for laboratory testing. The soil box or a similar unit makes it possible to saturate the soil with distilled water, causing available salts to enter into solution and simulating the worst conditions likely to occur after pipe installation. In all cases, the lowest resistivity measured should be used for evaluation purposes.

10.4.3.2 pH. Soils with a pH range below 4.0 generally serve well as electrolytes. They are higher in total acids and have a record of being aggressive. Neutral pH (6.0–8.0) is a benign condition unless other test results indicate the soil supports sulfate-reducing bacteria. Soils with a high pH (greater than 7.0) are generally high in dissolved salts and usually exhibit a low resistivity.

10.4.3.3 Redox potential. The oxidation–reduction (redox) potential of a soil is significant because the most common sulfate-reducing bacteria live only under anaerobic conditions. A redox potential greater than +100 mV shows the soil is sufficiently aerated and will not support sulfate reducers. Soils with redox potentials of 0 to +100 mV may or may not indicate anaerobic conditions. A negative redox potential definitely indicates anaerobic conditions under which sulfate reducers thrive.

10.4.3.4 Sulfides

10.4.3.4.1 Qualitative measurement of sulfides. If sulfides are found in the soil, the presence of sulfate-reducing bacteria is likely. Sulfate-reducing bacteria act as depolarizers by stealing the oxygen from sulfates present in the soil, which forms negative sulfur ions. These negative ions react with the hydrogen ions accumulating at a cathode and prevent a corrosion cell from polarizing, thereby allowing the rate of corrosion to continue unabated. The sulfides test is qualitative and accomplished by introducing a solution of 3 percent sodium azide in 0.1 *N* iodine into a test tube containing a small quantity of the soil being tested. If sulfides are present, they catalyze a reaction between sodium azide and iodine with the effervescent release of nitrogen. The chemical reaction is

10.4.3.4.2 Quantitative measurement of sulfides. A quantitative method of measuring sulfides is also available using a selective ion probe in conjunction with an appropriate meter. The soil is prepared in the laboratory using the probe manufacturer's guidelines and provides a quantitative measurement of the concentration of sulfide ions in the soil sample.

$$\underset{\text{Sodium Azide}}{2NaN_3} + \underset{\text{Iodine}}{I_2} \rightarrow \underset{\text{Sodium Iodide}}{2NaI} + \underset{\text{Nitrogen}}{3N_2}$$

10.4.3.5 Chlorides. Chlorides, like sulfate-reducing bacteria, are also depolarizers. Chloride ions react with any hydrogen ions that migrate to the cathode of a galvanic cell. Chlorides may be measured directly or by use of an indicator strip that provides a positive, trace, or negative result. In general, soils with high chloride content will also exhibit low resistivity results.

10.4.3.6 Moisture. The prevailing moisture conditions in the soil are determined as an observation of the prevailing moisture content of the soil. Moisture is, of course, an absolute necessity for active corrosion to occur. Because the soil moisture may vary throughout the year, general drainage characteristics are recorded rather than specific moisture content.

10.4.3.7 Groundwater influence. In addition to the assessment of relative moisture condition, it is recognized that a fluctuating water table may influence the effectiveness of polyethylene encasement if the encasement is not properly installed. For that reason, added emphasis is placed on the likelihood factors if the groundwater table is expected to intrude into the pipe zone at any time after installation. This factor also triggers the special installation measure for wet trench conditions described in section 10.5.1.

As with all methods for evaluating potentially corrosive soil conditions, the previous procedures should be employed only by qualified personnel experienced in soil analysis and evaluation of potentially corrosive conditions for gray- and ductile-iron pipe. Moisture content, soil temperature, location of soil sample with respect to pipe, time between removal of soil sample and testing, and other factors can significantly affect the soil-test evaluation. For example, the redox test should be done on a sample that has been relatively undisturbed and stored so as to minimize the opportunity for what is a poorly aerated soil to become aerated after sampling.

10.4.3.8 Known corrosive soils. In addition, certain soil environments are generally accepted to be potentially corrosive to gray- and ductile-iron pipe based on experience and do not require evaluation to determine the necessity of providing corrosion protection. Such environments include but are not limited to coal, cinders from burned coal, muck, peat, mine wastes, and landfill areas high in foreign materials. Experience with existing installations and potential for stray direct current corrosion should also be taken into consideration as a part of the evaluations.

10.5 BURIED SERVICE—CORROSION CONTROL

Not all soil or water environments are corrosive to ductile-iron pipe. Ductile-iron pipe, as manufactured, has a protective annealing oxide and is provided with an external shop coat. Although the external shop coating is applied to the outside of ductile-iron pipe in accordance with ANSI/AWWA C151/A21.51, Ductile-Iron Pipe, Centrifugally Cast, to minimize atmospheric oxidation for aesthetic reasons, in many environments, the annealing oxide coupled with the external shop coating is sufficient corrosion protection. However, where soil and water testing, known corrosive conditions, or performance history indicate high rates of corrosion, positive corrosion protection should be applied. Corrosion protection methods for ductile-iron pipe are typically polyethylene encasement, stray current control, and/or cathodic protection.

10.5.1 Polyethylene Encasement

Effective protection of ductile-iron pipe in corrosive soil environments can be achieved economically by encasing the pipe with either 8-mil (0.008-in.) linear low-density polyethylene (Stroud 1989b) or 4-mil (0.004-in.) high-density crosslaminated polyethylene. This system of protection was first used experimentally in the United States in 1951 for protection of gray-iron pipe in corrosive environments.

The first field installation of polyethylene-encased gray-iron pipe was in 1958 in Lafourche Parish, La., where a 4-inch cast-iron water main was encased using polyethylene sheets to protect the pipe against what was known to be an aggressive environment (Barrios 2019). The success of these early installations led to expanded research, which determined that polyethylene encasement provides a high degree of corrosion protection for both gray- and ductile-iron pipe. Since that time, polyethylene encasement has been used to protect gray- and ductile-iron pipe in thousands of installations across the United States and in other countries. Because of its widespread successful use and field-proven corrosion protection capabilities, the first standard for Polyethylene Encasement for Ductile-Iron Pipe Systems (ANSI/AWWA C105/A21.5) was adopted in 1972. An International Standard for Ductile-Iron Pipelines—Polyethylene Sleeving (ISO 8180 2020) has also been adopted since development of the procedure in the United States.

Polyethylene encasement is not a bonded coating, although it offers some of the qualities of a bonded coating, such as dielectric strength. It is mainly an environmental improvement for the pipe, isolating the pipe from the soil and providing a uniform environment between the film and the pipe. At the trench, ductile-iron pipe is wrapped with a tube or sheet of polyethylene, which acts as an unbonded film and prevents direct contact of the pipe with the corrosive soil. The polyethylene film also effectively reduces the environment to a very thin space between the pipe and the wrap, thus limiting the electrolyte available to support corrosion activity.

Although polyethylene encasement is not a watertight system, the weight of the earth backfill and surrounding compacted soil helps to prevent any significant exchange of groundwater between the wrap and the pipe. What must be avoided is an ongoing exchange of the groundwater that seeps under the film, which can be controlled by appropriate installation methods for wet-trench conditions (ANSI/AWWA C105/A21.5). When done properly, an initial seepage of groundwater beneath the wrap may occur, but the corrosive characteristics of the water would soon be depleted by the action of initial corrosion, resulting in superficial oxidation. Once the available dissolved oxygen in the moisture beneath the wrap has been consumed, further corrosion activity is effectively halted, and a uniform environment exists around the pipe. The polyethylene film retards the diffusion of additional dissolved oxygen to the pipe surface and the diffusion of corrosion products away from the pipe surface.

Extensive use of polyethylene encasement has demonstrated its effectiveness for gray- and ductile-iron pipe in the following ways (Stroud 1989a):

- It provides a uniform environment for the pipe that eliminates local galvanic corrosion cells.
- It offers good electrical shielding that resists the assimilation of stray direct current.
- It is less likely to be damaged than factory-applied coatings because it is installed on the pipe at the time of pipeline construction. If damage occurs during installation, it is easily repaired, requiring no surface preparation beyond removing any soil that may have adhered to the pipe surface at the exposed location.
- Pinholes and minor damage to the loose wrapping material do not diminish its protective ability; however, cuts, tears, punctures, or other significant damage to the polyethylene film should be repaired.
- Initial cost of material and installation is very low, and there are no maintenance costs.

Polyethylene encasement has also been found to be effective in mitigating the effects of stray current corrosion. Polyethylene possesses excellent dielectric properties that enable it to effectively shield ductile-iron pipe from low-level direct current. Because ductile-iron pipe is laid in relatively short (18- or 20-ft [5.5- or 6.1-m]) lengths that are electrically isolated by rubber-gasketed joints, stray current is not considered to be a particularly great hazard. The few verified instances of stray current corrosion on ductile-iron pipe that have been observed occurred in unusual circumstances involving extreme current density conditions. Nevertheless, it is always prudent to investigate potential stray current sources so that appropriate mitigation measures can be taken to protect the pipe. In most cases, properly installed polyethylene encasement applied through the area of influence will provide adequate protection of ductile-iron pipe against stray current corrosion. When ductile-iron pipelines are exposed to high-density stray current environments, the pipeline should be rerouted or the anode bed relocated. If neither of these options is feasible, the ductile-iron pipe in this area should be electrically bonded together, electrically isolated from adjacent pipe, polyethylene encased, and appropriate test leads and current drain installed.

The material requirements for polyethylene are covered in ANSI/AWWA C105/A21.5 with four alternate methods of installation (Figure 10-5). Modified method A, illustrated in Figure 10-5 and depicted in Figure 10-6, is the recommended method as a modification to the widely used method A that features polyethylene tubes that overlap at each joint. As with all corrosion-protection methods, proper installation is essential to obtaining the desired level of protection. The single most important installation criterion is that the polyethylene material completely encases the pipe, preventing contact between the pipe and surrounding soil. Care should also be taken to prevent soil or bedding material from becoming trapped between the pipe and the polyethylene. The polyethylene film should be fitted to the contour of the pipe to effect a snug, but not tight, encasement with minimal space between the polyethylene and the pipe. Sufficient slack should be provided in contouring to prevent stretching the polyethylene where it bridges irregular surfaces, such as bell-spigot interfaces, bolted joints, or fittings, and to prevent damage to the polyethylene during backfilling operations. Overlaps and ends should be secured with adhesive tape or plastic tie straps to hold the polyethylene in place until backfilling operations are completed. Any cuts, tears, punctures, or damage to the polyethylene should be repaired with adhesive tape or with a short length of polyethylene tube cut open, wrapped around the pipe, and secured in place. For installation below the water table or in areas subject to tidal actions, it is recommended that both ends of the polyethylene tube be sealed as thoroughly as possible by wrapping them circumferentially with adhesive tape or plastic tie straps at each joint overlap. It is also recommended that circumferential wraps of tape be placed at 2-ft (0.61-m) intervals along the barrel of pipe to help minimize the space between the polyethylene and the pipe. When installing service taps on polyethylene-encased iron pipe, the preferred method is to apply two to three wraps of polyethylene adhesive tape around the pipe and tap through the tape and polyethylene film as described in Section 11.7.

Polyethylene encasement, as described above, is a passive system of protecting ductile-iron pipe in corrosive soils. Once installed properly, polyethylene encasement does not require continued monitoring, maintenance, or ongoing operating costs. Its success has been demonstrated by its popular use around the world, including through case studies of field inspections such as published by the Association for Materials Protection and Performance (AMPP) and the American Society of Civil Engineers (ASCE) (Cox 2010; Barrios 2019; Horn 2021).

Research has shown that polyethylene encasement alone is a viable corrosion protection system for ductile- and gray-iron pipe in most environments. However, other options should be considered for environments where all the following characteristics coexist: (1) soil resistivity ≤ 500 Ω-cm; (2) anaerobic conditions in which sulfate-reducing bacteria thrive (neutral pH [6.5 to 7.5]) with low or negative redox-potential (negative to +100 mV) and the presence of sulfides (positive or trace); and (3) a water table intermittently or continually above the invert of the pipe.

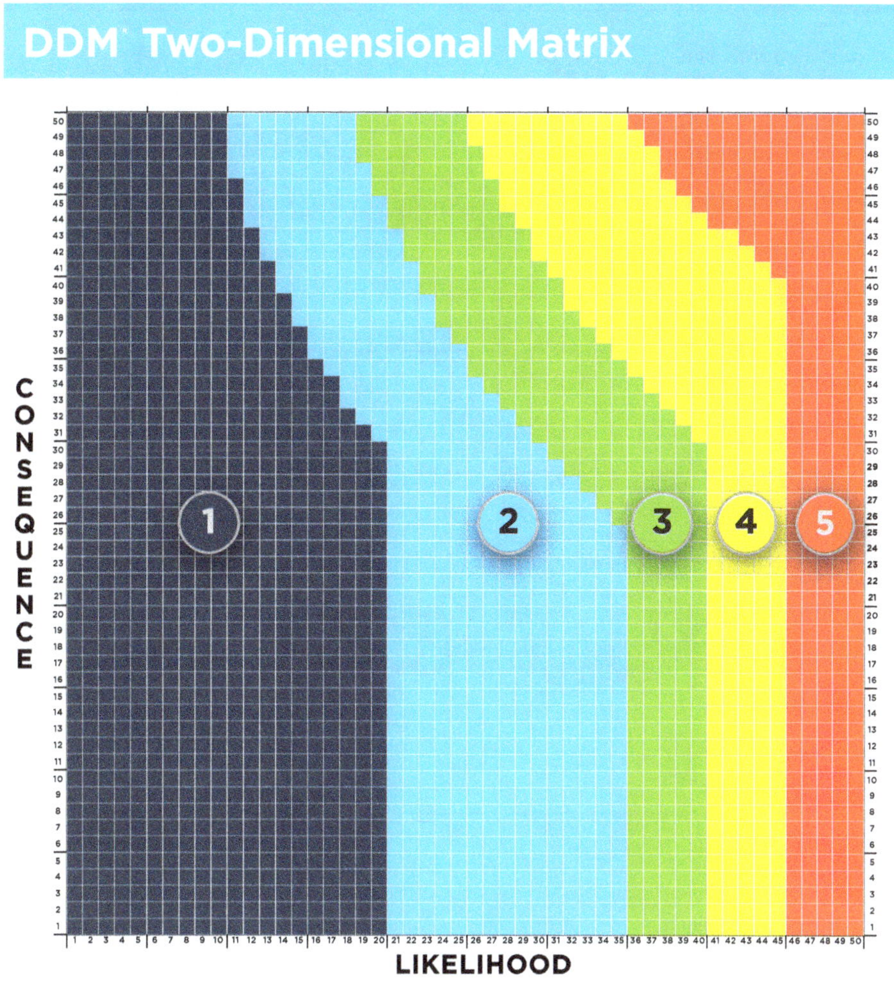

Source: Ductile Iron Pipe Research Association

Figure 10-5 Two Dimensional DDM® Matrix

10.5.2 Enhanced Polyethylene Encasement

In 2013, the ductile-iron pipe industry introduced a linear low-density polyethylene film that had been enhanced by the infusion of a volatile corrosion inhibitor (VCI) and an antimicrobial additive into the film. Taking advantage of a polyethylene film coextrusion manufacturing technology that allows the extrusion of the film simultaneously from separate bins of polyethylene pellets, this enhanced polyethylene film provides an active component to traditional polyethylene encasement by infusing the VCI and antimicrobial additive into the innermost 1.5 mils of the total minimum film thickness of 8 mils.

10.5.3 Metallized Zinc

Metallized zinc has been used as a coating for ductile-iron pipe for more than 65 years (Horton 2014) in ISO standards (ISO 8179-Part 1 2017). The systems include arc-sprayed and flame-sprayed applications of metallized zinc using a high-purity zinc wire to apply the zinc at the rate of 200 g/m^2 over the exterior of ductile-iron pipe with a finishing layer (typically the standard AWWA C151 shop coating) applied over the zinc.

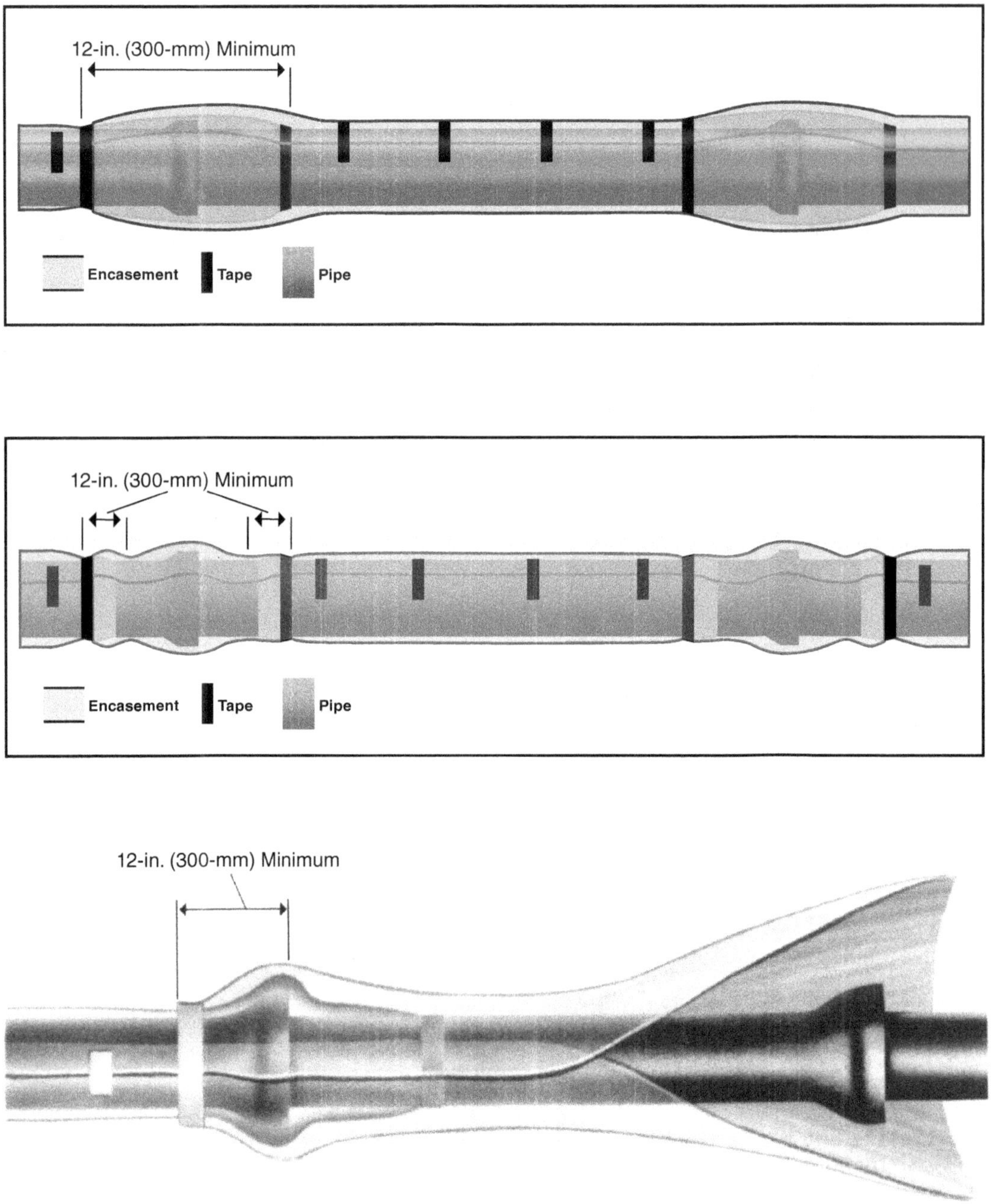

Source: Ductile Iron Pipe Research Association

Figure 10-6 Alternate polyethylene encasement installation methods

Metallized zinc is applied as an alternative to cathodic protection in the most severe soil environments and is used in conjunction with encasement using enhanced polyethylene.

Advantages of the zinc coating include the provision of a uniform layer of anodic zinc to protect the pipe surface from corrosion in a manner akin to cathodic protection. Also, at the recommended concentration, zinc-coated ductile iron is compatible with standard joint tolerances, allowing field cuts and the use of either direct taps or tapping saddles while requiring no special handling or installation procedures.

While the zinc coating is considered to be a betterment in corrosion control, it is not the only method of corrosion control for ductile-iron pipe to be placed in all corrosive soils.

10.5.4 Cathodic Protection

Cathodic protection is a widely used technology for external corrosion control of buried and submerged metallic pipelines. Its use should be dictated by soil and water conditions, by the impact of a corrosion failure in the area of operation, and most importantly, by life-cycle cost–benefit considerations.

10.5.4.1 Corrosion and cathodic protection. The predominant underground corrosion cell is created by differences in the concentration of oxygen along the surface of a pipeline. Where oxygen concentrations are low, anodic areas are formed where the ferrous component of the ductile-iron pipe goes into solution (oxidation reaction). The corresponding cathodic areas are where the oxygen concentration is relatively higher. At the cathode, the metal is protected, and a reduction reaction occurs. Depending on the conditions, the rate of corrosion may be acceptable to realize the desired service life of the pipe, or supplemental corrosion protection may be needed in aggressive environments.

At anodic locations, positive direct current flows from the pipe into the soil or water, causing dissolution of the outside pipe wall. Cathodic protection reduces or eliminates the flow of this corrosion current by providing an opposing current that flows from the cathodic protection system, through the soil or water, onto the surface of the pipe. The pipe is the cathode in the cathodic protection system, thus preventing dissolution of the pipe material at the anodic sites.

Cathodic protection is applied to pipelines by either sacrificial anode or impressed current systems. Sacrificial systems usually use either zinc or magnesium anodes that are placed near the pipeline in the soil or water and connected by a lead wire or metallic bracket. The protective current in sacrificial systems is created by the difference in potential between the more active anode material and the ferrous pipe material. In an impressed current system, a DC power supply (typically a transformer-rectifier) is used to energize ground electrodes that are connected to the positive terminal with the pipe connected to the negative terminal. In both cases, positive DC current flows from the anodes through the soil or water onto the pipe surface. When sufficient current is applied and properly distributed, cathodic protection systems can prevent external pipeline corrosion.

10.5.4.2 Application to ductile-iron pipe. A set of cathodic protection criteria is published in ASTM G218, External Corrosion Protection of Ductile Iron Pipe Utilizing Polyethylene Encasement Supplemented by Cathodic Protection (ASTM 2019). The most common criteria used for pipelines are a polarized potential at least as negative as –850 mV to a copper/copper sulfate reference electrode or a 100-mV polarization shift from the free corrosion potential as determined by the formation or decay of polarization. These criteria were developed for use on all steel and cast-iron pipelines, including those transporting natural gas and hazardous liquids. Under normally encountered soil and water conditions, they are intended to prevent all external corrosion activity. For ductile-iron pipelines, full cathodic protection, according to the NACE criteria, is needed where corrosion conditions are severe (leading to a high likelihood of failure caused by external corrosion) and/or the consequences of an external corrosion failure are not acceptable.

Cathodic protection can be applied to as manufactured shop-coated ductile-iron pipe or polyethylene-encased iron pipe. The use of corrosion-protection currents improves the effectiveness of polyethylene encasement by protecting the pipe where the film may be damaged or not properly applied, resulting in uncovered surfaces. The methods are not exclusive, and there is synergy when they are used in combination (Schiff 1993; Kroon et. al. 2004).

10.5.4.3 Life-extension cathodic systems. When cathodic-protection currents are required for underground or submerged ductile-iron pipe systems, each increment of applied current serves to slow the corrosion process. By slowing the process of corrosion, service life can be greatly extended, even if the corrosion rate is not zero. This has been demonstrated by a 3-year research project involving both electrochemical laboratory testing and field polarization tests to calculate corrosion rates at various magnitudes of applied cathodic protection current (Kroon et. al. 2004). The results indicate that substantial service life extension can be achieved by polarizing the pipe surface to levels well below the NACE criteria. It was also shown that the benefits of life-extension cathodic systems can be realized at cathodic current densities that are a fraction of those typically used by corrosion engineers when designing full cathodic protection for ductile-iron pipelines.

Corrosion-control strategies for ductile-iron pipe may involve either full cathodic protection to NACE criteria or the use of life-extension cathodic systems. A strategy of life-extension cathodic systems is a well-founded engineering approach for typical ductile-iron water main projects where limited corrosion is acceptable relative to pipeline reliability and controlled operating costs. Full cathodic protection requires a notably higher current demand, resulting in more cathodic-protection hardware and greater costs.

10.5.5 Other Corrosion Control Measures

The preceding discussions of polyethylene encasement and cathodic protection were presented in detail. This is because polyethylene encasement is the only recommended method of corrosion protection for ductile- and gray cast-iron pipe that is the subject of an AWWA standard. Also, NACE SP0169, Control of External Corrosion on Underground or Submerged Metallic Piping Systems (2024), includes the application of cathodic protection to as-manufactured iron pipe. Alternative methods of corrosion protection are available and may be appropriate in certain circumstances. Additionally, several other protection methods have been tried over the years with varying degrees of success. A discussion of various protection methods follows.

10.5.5.1 Bonded dielectric coatings. Numerous coating standards for material, application, inspection, and surface preparation on steel surfaces are available through organizations such as AMPP, ASTM, and AWWA. (In 2021, NACE International and The Society for Protective Coatings [SSPC] merged to form The Association for Materials Protection and Performance [AMPP]). These standards and specifications were developed for smooth carbon steel surfaces and are not applicable to ductile-iron pipe. Bonded dielectric coatings for buried ductile-iron pipe may be available on a limited basis but should be considered with extreme caution. Issues such as surface preparation, the depth of the peen surface of the pipe, and joint tolerances are critical and can impact the application and performance of any bonded coating. For these reasons, the DDM® described in Section 10.4.2 was developed to provide alternatives to bonded dielectric coatings. For more information on the use of bonded dielectric coatings, the manufacturers of ductile-iron pipe should be consulted.

In the following discussion, bonded coatings are divided into two categories: (1) high-performance coatings for buried applications and (2) maintenance coatings (i.e., paints and primers) for aboveground service. A separate discussion of maintenance-type coatings is given in Section 10.6.

10.5.5.1.1 Ductile-iron pipe outside diameter (OD) surface characterization. The vast majority of ductile-iron pipe is produced by centrifugal casting by either of two techniques. The first is the deLavaud casting process, in which molten ductile iron is introduced onto a spinning metal mold. In this process, the mold is coated with a thin layer of fine ferrosilicon (FeSi) powder that acts as mold insulation and a parting agent. The other casting method is the wet-spray process, in which a refractory wash or slurry is sprayed onto the inside diameter of the steel mold and dries to form an insulating layer and parting agent. Both casting techniques impart a texture to the pipe as-cast OD surface. In the case of the deLavaud process, the molds are peened as shown in Figure 10-7 to provide friction to accelerate the molten metal to the speed of the spinning mold. Centrifugal forces hold the metal against the mold until it solidifies. During solidification, the pipe OD surface closely replicates the peened surface of the mold, with some of the FeSi powder dissolved or embedded into the pipe OD surface. A similar texturing and replication of the OD surface occurs for pipe cast by the wet-spray process. Depending on the depths of the peens or texturing of the wet-spray, the OD surface of a cast pipe will generally have a roughness profile of from 10 to 30 mil, peak to valley. The typical surface texture of a ductile-iron pipe made by the deLavaud process is shown in Figure 10-8.

Once cast, the pipe is annealed to break down as-cast carbides and soften the microstructure, such that the required mechanical properties (tensile, yield, elongation, and Charpy V-notch) are met. This process forms a silicon-rich annealing oxide that is tightly bonded to the pipe OD surface. The photomicrograph in Figure 10-9 shows the annealing oxide with underlying ductile iron. This photomicrograph is typical of ductile-iron pipe produced by either of the casting methods. The heavy surface texture of 10 to 30 mil and the presence of a 4+ mil–thick oxide layer represent a typical ductile-iron pipe surface and make it characteristically different from that of steel. A schematic diagram of the cross-section of a ductile-iron pipe wall is shown in Figure 10-10. This heavy surface texture makes it difficult, if not impossible, to reliably apply pinhole-free coatings to ductile-iron pipe.

10.5.5.1.2 Ductile- and gray-iron fitting OD surface characterization. Ductile- and gray-iron fittings are cast in green sand or resin-bonded sand molds. They are not produced by the same methods as ductile-iron pipe. As such, their surfaces do not exhibit the heavy textured surface profile and the layer of annealing oxides common to ductile-iron pipe; however, their as-cast condition provides a surface profile greater than most steels.

10.5.5.1.3 Surface preparation. Pipe: Bonded coatings for buried service normally require a high degree of surface preparation. The ductile-iron pipe casting process can result in cosmetic blemishes such as shot, fins, slivers, and splashes that are contained within the 0.08-in. thickness of the service allowance. Overblasting, high nozzle velocities, and excessive blast times can lead to a damaged or more irregular surface than was initially present. Figure 10-12a shows an unblasted surface, and Figure 10-12b shows an overblasted surface of the same pipe that has resulted in large blisters on the OD surface. Overblasting typically occurs when an attempt is made to remove the tightly adherent annealing oxide layer. The blasting media most likely to cause damage include sand, steel shot, and steel grit. Therefore, aggressive blasting is not recommended on the external surface of ductile-iron pipe for buried service. (See Section 10.6.1.1.2 for surface preparation for aboveground coatings.)

Fittings: Ductile- and gray-iron fittings are statically cast in sand molds and therefore do not have the layer of annealing oxides common to ductile-iron pipe. As a result, their surfaces are tolerant of more aggressive surface preparation, if required, without damage to the fitting surface. It should be noted, however, that their as-cast condition provides a surface profile greater than most steels. Consequently, the removal of surface impurities is usually the main concern. This is accomplished by the same methods as those for ductile-iron pipe for exposed applications (see Section 10.6.1.1.2). Coating manufacturers' recommendation for surface preparation (which are normally for carbon steel) will vary depending on the type of coating and on the ultimate service environment. Therefore, the coating and fitting manufacturers should be consulted for specific recommendations concerning surface preparation of ductile- and gray-iron fittings.

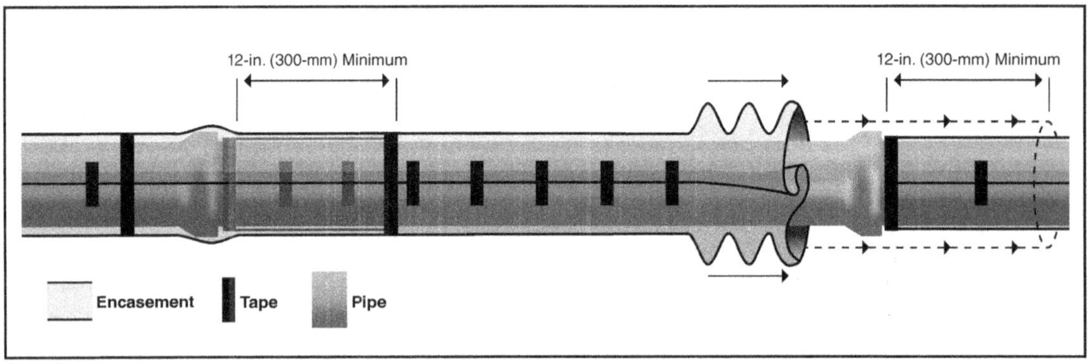

Source: Ductile Iron Pipe Research Association

Figure 10-7 Installation of polyethylene encasement using Modified Method A

Source: American Cast Iron Pipe Company

Figure 10-8 Peened surface of a deLavaud pipe mold

10.5.5.1.4 Surface roughness, coating thickness, and holiday detection. Pipe: The surface texture of ductile-iron pipe is significantly rougher than the smooth surface normally found on steel products (Figure 10-11). As such, the majority of technical datasheets developed for the application of coatings to steel are not applicable for the same coatings on ductile-iron pipe. The rough "orange peel" surface, in addition to the annealing oxide, makes an excessive number of holidays more likely for ductile iron than for a steel surface. Coatings that would normally provide a smooth, uniformly thick, holiday-free surface on steel will not properly cover the textured surface of ductile-iron pipe, as there will certainly be

Source: American Cast Iron Pipe Company

Figure 10-9 Close-up photograph of the cast surface of a typical ductile-iron pipe made by the deLavaud process

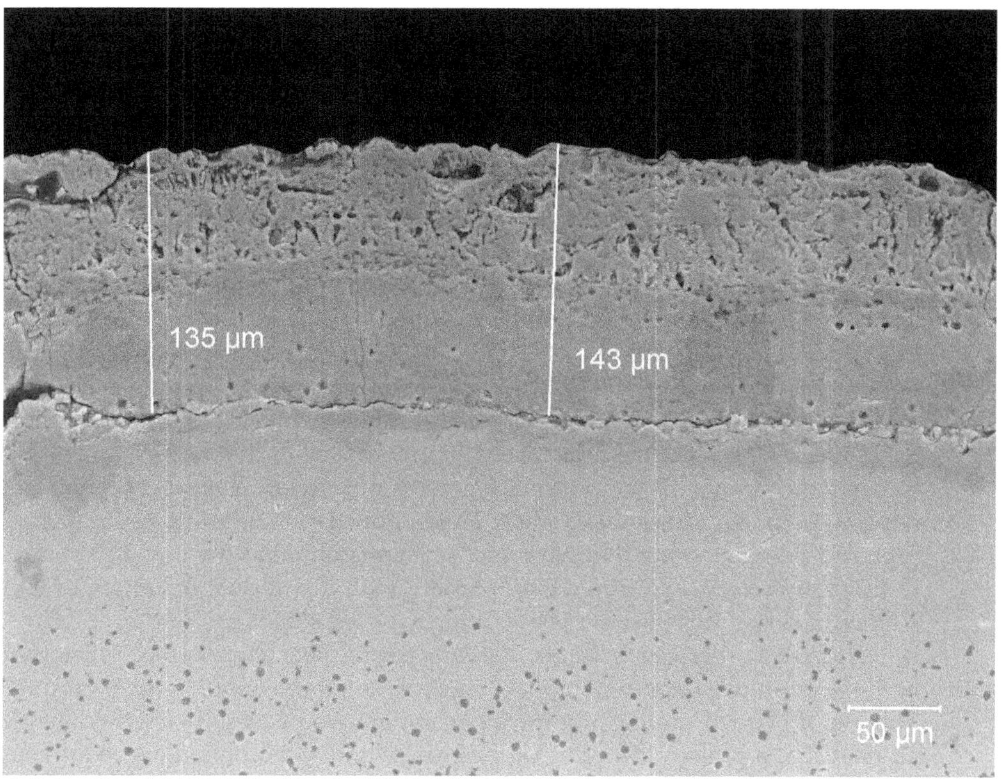

Source: American Cast Iron Pipe Company

Figure 10-10 Scanning electron microscope photomicrograph of a ductile-iron pipe surface profile. The layers consist of a 135- to 143-μm (5-to 6-mil) thickness of annealing oxide with underlying ductile iron

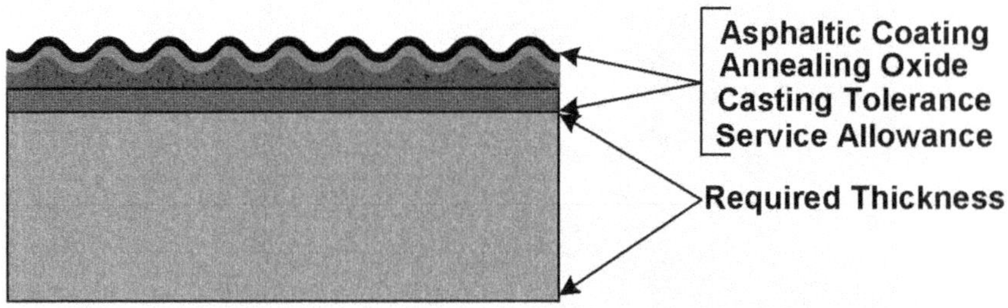

Source: American Cast Iron Pipe Company

Figure 10-11 Schematic diagram of a typical cross section of a ductile-iron pipe

thick and thin areas in the coating. Peen depth patterns of 10 to 30 mil greatly exceed the maximum anchor profile required for some bonded coatings. High-voltage spark testing methods in accordance with ASTM G62 (2022) method B and NACE RP0274 (2004) have been found to damage coatings applied to highly textured ductile-iron pipe if the coating does not follow the surface texture or is of insufficient thickness to cover the roughness of the pipe surface. The coating and pipe manufacturers should be consulted regarding surface preparation, coating thickness, and holiday detection, if required.

Fittings: Ductile- and gray-iron fittings' as-cast condition provides a surface profile greater than most steels. The type and configuration of certain fittings limit the thickness of coating that can be applied. Also, because of casting roughness, grind marks, sharp corners, bolt holes, threaded holes, machined flange faces, and so on, holiday testing is normally ineffective. The coating and fitting manufacturers should be consulted regarding coating thickness and holiday detection, if required.

10.5.5.1.5 Joints. Ductile-iron pipe is manufactured in 18- and 20-foot lengths with several types of gasketed and restrained joints available, many of which also allow for joint movement and flexibility after assembly. Bell and spigot tolerances for proper joint assembly allow an additional coating thickness of only 8 to 10 mil. Coatings that exceed this thickness would have to be ground off or held back during coating application. Pipeline projects with tight rights-of-ways and/or a large number of fittings may require field cuts to accommodate the laying schedule. Coatings on pipe for these locations would have to be adjusted to meet the dimensional requirements of the joint. Restrained pipe joints that rely on weldments and locking segments outside of the gasket seating surfaces will not accommodate a thick OD coating. Field-applied restraint devices for mechanical joint pipe and fittings require that hardened segments bite into the pipe OD surface in order to properly function. Coatings that interfere with the normal operation of these type of restraint devices must be removed, leaving exposed surfaces.

10.5.5.1.6 Shipping, handling, and installation. Ductile-iron pipe is a robust product, and it is not uncommon for shippers and contractors to handle pipe in a rough manner. The use of bonded coatings on ductile-iron pipe would require special attention during all phases of a project. Handling at the coaters and special handling requirements at the jobsite provide multiple opportunities for coating damage.

10.5.5.1.7 Coatings for buried service. Pipe: In accordance with ANSI/AWWA C151/A21.51, an external shop-coating approximately 1-mil thick is applied as the standard exterior coating on ductile-iron pipe unless otherwise specified by the customer at the time of manufacture. This coating, along with the silicon-rich annealing oxide formed during the manufacture of ductile-iron pipe, has been shown to offer some corrosion protection and also reduces current requirements when utilized in combination with cathodic protection systems (Kroon et. al. 2004).

(a)

Source: U.S. Pipe & Foundry

Figure 10-12a Annealed ductile-iron pipe before over-blasting

(b)

Source: U.S. Pipe & Foundry

Figure 10-12b Same ductile-iron pipe resulting in blisters and disbondment from over-blasting

Once applied, the asphalt penetrates the textured surface of the ductile-iron pipe and is extremely difficult, if not impossible, to completely remove. For this reason, ductile-iron pipe that is to receive a special exterior coating other than asphalt (e.g., primer for exposed piping to be topcoated) is usually ordered from the manufacturer bare—that is, without an exterior coating. Polyethylene encasement is unaffected by the asphalt coating and can be applied directly to the asphalt-coated pipe as received from the factory.

Fittings: In accordance with ANSI/AWWA C110/A21.10, Ductile-Iron and Gray-Iron Fittings, and ANSI/AWWA C153/A21.53, Ductile-Iron Compact Fittings for Water Service, an external coating approximately 1-mil thick is also typically applied to the exterior surface of ductile- and gray-iron fittings. Additionally, fusion-bonded epoxy (FBE) coatings may be applied to certain sizes of fittings in accordance with ANSI/AWWA C116/A21.16, Protective Fusion-Bonded Epoxy Coatings for the Interior and Exterior Surfaces of Ductile-Iron and Gray-Iron Fittings. FBE is offered only as a complete inside and outside coating and lining system for fittings. (FBE coatings are not used for ductile-iron pipe because of surface roughness, holidays, and application methods.)

10.5.5.2 Isolation joints. Isolation joints are used to break the metallic electrical connection between two pipelines, pipeline components, or structures, thereby preventing the flow of electrical current between them. This method of corrosion control has limited applications, particularly with respect to ductile-iron pipelines that use a rubber-gasketed jointing system and are considered to be electrically discontinuous. In certain circumstances, however, isolation joints can be used effectively to prevent accelerated corrosion attack caused by dissimilar metals and stray currents. For example, isolation joints are often used to isolate ductile-iron pipelines from pipelines or other structures that are cathodically protected or are part of an electrical grounding system. Isolation joints may also be used at the junction of two dissimilar metals to prevent the formation of a bimetallic corrosion cell. Because of the many variables involved, it is impractical to specify herein every situation in which the use of isolation joints is warranted.

10.5.5.3 Trench improvements. Trench improvement of corrosive soils to reduce their corrosive tendencies is, for the most part, impractical. Extensive studies have been made of bedding and backfilling around underground pipe with sand, limestone, limestone screenings (fines), dolomite, sand plus 10 percent cement, and select soil. Although some reduction in corrosion incidence is realized with most of these treatments, it has been observed that in many instances the substitute fill material eventually takes on the characteristics of the surrounding soil. The long-term result can be inadequate protection. Also, such trench improvements may result in increased groundwater migration through the trench, which replenishes oxygen and other corrosive agents.

10.5.5.4 Sacrificial metal. Increasing pipe wall thickness to allow sacrificial metal loss can extend the service life of a pipeline but is not a long-term solution to corrosion because it doesn't control corrosion so much as it allows corrosion to continue for a longer time and there is no assurance that corrosion will occur uniformly. Corrosion attack may occur in the form of localized pitting, which can result in premature failure of the pipe regardless of wall thickness. In addition to being unreliable, the practice of increasing pipe wall thickness as a safeguard against corrosion is also not cost effective. The availability of more reliable and economical methods of corrosion prevention has generally rendered this practice obsolete.

10.6 EXPOSED SERVICE CONDITIONS

Atmospheric (exposed) corrosion, like corrosion underground or inside of water-containing pipes, is caused by electrochemical reactions. We define atmospheric corrosion as corrosion that occurs anywhere other than in direct contact with a continuous electrolyte such as soil or water. Examples of atmospheric corrosion include

- pipes inside of treatment facilities that are exposed to the air;
- pipes on bridge crossings, either on top of or suspended beneath the bridge;
- aboveground pipe on supports; and
- pipes in manholes or pits where there is no water or soil surrounding the pipe.

Atmospheric corrosion is similar to the corrosion experience by underground pipelines except in magnitude. The major difference is the nature and/or presence of the electrolyte. While the electrolyte in underground and submerged corrosion is typically constant, the electrolyte in atmospheric corrosion is limited and intermittent.

10.6.1 Exposed Corrosion Control

Depending on the environment, ductile-iron pipe has been used successfully in most aboveground applications with no added or special corrosion-control method. Other than loss of gloss associated with aging and exposure to sunlight, the standard external shop coat furnished on ductile-iron pipe has been found to offer sufficient protection in most aboveground applications. This success is attributed in large part to the protective silicon-rich annealing oxide outer surface layer, which is created during the high-temperature annealing process that is part of the manufacture of ductile-iron pipe. As previously noted, this protective oxide has also been shown to offer some degree of corrosion protection in corrosive underground environments compared to nonannealed or coated ductile iron (Kroon et. al. 2004; Bonds et. al. 2005). However, special conditions may exist that involve the use of bonded coatings, such as primers and paints, as described further in the following section.

10.6.1.1 Bonded coatings (primers and paints). Special treatment of ductile-iron pipe for aboveground applications is normally limited to applying a surface tolerant coating. This coating may be applied for aesthetics, for identification, or for added corrosion protection due to more extreme exposures. Due to the absence of a continuous electrolyte, cathodic protection cannot be applied for aboveground piping. Likewise, polyethylene encasement is not useful for aboveground corrosion protection.

10.6.1.1.1 Maintenance coatings versus high-performance coatings. Problems with applying bonded coatings to ductile-iron pipe for underground corrosion protection were discussed in Section 10.5.5.1. As reported, many of these problems are associated with surface preparation and application of high-performance coatings such as high-build epoxies, polyurethanes, and so on. Typically, maintenance coatings or primers are less restrictive with respect to surface preparation and application and are primarily utilized for aboveground service. As reported in the section on "Atmospheric Corrosion" in AWWA M27, External Corrosion—Introduction to Chemistry and Control: "Barrier coatings are easy to apply, less sensitive to surface preparation when compared to high-performance coatings and are initially less costly than high-performance coatings". These maintenance-type primers include alkyds, acrylics, epoxies, moisture-cured urethanes, and latexes. Many of these are primers that will later be provided with a finish coating after installation. As such, holidays (voids or pinholes in the coating) are typically not a concern with the primer coating.

10.6.1.1.2 Surface preparation. Surface preparation is perhaps the most important part of coating. In general, high-performance coatings require better surface preparation than ordinary maintenance coatings. Also, the more severe the environment, the better the surface preparation should be. The purpose of surface preparation is to remove contaminants (dirt, oil, grease, etc.) and to provide a surface profile for the primer to mechanically bond to the pipe. The surface profile on as-cast ductile-iron pipe is generally greater than can be achieved by any degree of blasting.

As cautioned in several publications, AMPP standards were developed for the preparation of steel surfaces and are not directly applicable to ductile-iron pipe (DIPRA 1993; Horton 1995; NAPF 2017). To address this problem, the National Association of Pipe Fabricators developed a surface preparation standard specifically for exposed ductile-iron pipe, NAPF 500-03, Surface Preparation Standard for Ductile Iron Pipe and Fittings Receiving Special External Coatings and/or Special Internal Linings (2017). This standard contains the following warning regarding the specification of AMPP surface preparation standards for ductile-iron pipe:

> "Attempts to apply steel surface preparation specifications to ductile iron (pipe) are inappropriate and may actually result in damage to the pipe surface with subsequent reduced coating effectiveness and life expectancy."

The NAPF standard was developed to provide guidance for application of primers and internal special linings at fabrication shops. Two of the key topics regarding surface preparation and coatings for ductile-iron pipe that are discussed in this standard are (1) damage from blasting and (2) asphaltic standard coating. In the first instance, ductile-iron pipe surfaces can be damaged by overblasting. In the second instance, very few coatings are available that are compatible with the standard external coating provided on ductile-iron pipe. Once the standard coating is applied, it is extremely difficult to completely remove. Ductile-iron pipe and fittings to be specially coated should be ordered uncoated from the factory.

The reader is encouraged to refer to this specification when choosing a bonded coating for aboveground applications of ductile-iron pipe.

Fittings: The same NAPF 500-03 standard includes a surface preparation specifically for exposed gray- and ductile-iron pipe and fittings. This standard should be consulted for surface preparation of exposed ductile- and gray-iron fittings. (Also see Section 10.5.5.1.3.)

10.6.1.1.3 Special considerations. In addition to surface preparation, there are several other areas specific to ductile-iron pipe and fittings that should be considered when specifying a bonded coating for aboveground service. These areas include the following.

Surface Roughness: Most technical datasheets for coatings were developed primarily for smooth steel surfaces. Ductile-iron pipe and fittings have a rough texture on the exterior surface that is normally significantly rougher than those found on steel surfaces. Therefore, coating thicknesses needed to cover the rougher surface on ductile-iron pipe, and fittings are typically thicker than those specified on coating technical datasheets for steel surfaces. If the coating datasheet does not specifically list a recommended thickness for ductile-iron pipe and fittings, then the paint manufacturer and/or pipe manufacturer should be contacted for recommendations specific to ductile-iron pipe and fittings.

Joints: Ductile-iron pipe is available with several types of joints, including push-on, restrained push-on, mechanical, restrained mechanical, flanged, ball and socket, and seismic joints. Most of these joints are proprietary designs, and maintaining critical dimensions in the sealing areas of each joint is critical to providing a proper leak-free assembly. When sealing areas are to be coated, it is recommended that the specific pipe manufacturer be contacted for critical dimensions in those sealing areas.

Holiday Testing: Because of the rough surfaces that are inherent in the manufacture of ductile-iron pipe, holiday test recommendations and procedures developed and written for smooth steel surfaces are not directly applicable to ductile-iron pipe. As with thickness recommendations, if the coating datasheet does not specifically list a recommended holiday test procedure for ductile-iron pipe, the paint manufacturer and/or pipe manufacturer should be contacted for recommendations specific to ductile-iron pipe and fittings.

10.6.1.1.4 Bonded coating selection criteria. The following comments regarding coating selection are taken from the "Atmospheric Corrosion" chapter in AWWA Manual M27, External Corrosion—Introduction to Chemistry and Control.

"Coating selection depends on

- Structure criticality: the effects of corrosion-related failure and aesthetics. It is important to tailor the coating to the environment.
- Exposure environment: severe environments, e.g., marine, heavy industrial, require more durable coatings. In general, the high-performance coating systems perform longer in more corrosive environments.
- Life cycle cost: consider the true cost of the coating system, not just the initial cost. A less expensive, low durability coating will require several recoats within the life expectancy of the structure, while a more durable, more costly coating might require fewer recoats. The more costly coating initially could be the less expensive one in the long run. Life-cycle benefits may also be considered in selecting one of the stainless-steel materials, which do not need coatings for their protection or repeated cleaning, preparation, and recoating.
- Ease of maintenance. This refers to ease of spot repair and recoatability of the coating. A high-performance coating might be justified if maintenance will not be performed. Environmental considerations. If previous coatings contain lead, overcoating may be feasible, or the old coating may have to be removed. If the old primer is intact, a coating suitable for topcoating might be economically justified. However, keep in mind that eventually the old coating will have to be removed. The costs of rigging and scaffolding are often the overriding cost factors. If this is the case, then removal of the old coating might be justified. Where regulatory restrictions on volatile components apply, users can select coatings that conform to the volatile organic compound (VOC) regulations."

The NACE publication *Industrial Maintenance Painting* lists the following criteria that should be used in selection of an aboveground coating (NACE 1976):

1. Abrasion resistance (Note: This is more applicable to linings than coatings.)
2. Adhesion
3. Impact resistance
4. Flexural qualities
5. Resistance to a given media
6. Resistance to sunlight
7. Temperature resistance
8. Drying time
9. Appearance
10. Wetting time
11. Applied cost
12. Antistick properties

10.6.1.1.5 Manufacturer's recommendations. Other factors that should be considered during coating selection for aboveground applications of ductile-iron pipe are as follows:

- Coating manufacturer's recommendation and warranty
- Pipe manufacturer's recommendations and warranty

- Type of ductile-iron surface to be coated (annealed pipe or sand cast fitting)
- Compatibility of ductile-iron pipe surfaces and joint configuration with the surface preparation and application requirements for the coating (maximum anchor pattern of a coating versus peen pattern depths of the pipe)
- Past experience
- Accessibility and requirements for repair/recoating
- Susceptibility to damage during storage, transport, and installation
- Laboratory results in a similar environment
- Resistance to periodic cleaning

Dozens of generic coating materials are available commercially, and each has its own characteristics (NACE 1976). Only by knowing the generic nature of the product and the applicability to ductile-iron pipe can the best coating be selected for the anticipated service. Pipe manufacturers and paint manufacturers knowledgeable with coating ductile-iron pipe should be consulted for recommendations regarding specific applications.

10.6.1.1.6 Factory primer applications. Upon request for exposed installations, the ductile-iron pipe and fittings manufacturers can supply ductile-iron pipe and fittings with a primer coating to which a final coat would be applied. Both the pipe and fittings manufacturer and the coatings manufacturer should be consulted to ensure the primer and final coats are compatible. Special coatings should not be applied over the standard asphaltic coating provided with standard ductile-iron pipe products because of the incompatibility of most such coatings with the paint. Because asphaltic coating is difficult to completely remove, uncoated pipe and fittings should be used for orders requiring special coatings.

10.7 SUBMERGED SERVICE CONDITIONS AND CONTROL

Iron pipe has been serving in submerged installations for more than a century. These installations include both fresh (river and lake crossings and intake lines) and salt (ocean outfall lines) water environments. In the overwhelming majority of these installations, the pipe was provided with standard external coatings with no supplemental corrosion protection. These environments are not particularly corrosive to ductile-iron pipe except in certain situations. In some cases, the pipe is installed in underwater trenches rather than resting exposed on river, lake, or ocean bottoms. In those cases, the underwater soil should be analyzed for its corrosivity potential and additional corrosion protection applied if necessary. Typical service applications have ranged from no additional protection to polyethylene encasement and/or cathodic protection (see Section 10.5).

Another potentially corrosive environment exists in the saltwater splash zone where the transition from submerged piping to exposed piping might occur. In those cases, appropriate corrosion protection should be employed (see Section 10.6).

Surface-tolerant paint systems applied to ductile-iron pipe submerged in potable water tanks or reservoirs should have NSF/ANSI/CAN 61, Drinking Water System Components—Health Effects (2021) approval. Typically, the standard coatings currently provided by ductile-iron pipe manufacturers are NSF/ANSI/CAN 61approved.

REFERENCES

American Water Works Association. 2014. *M27: External Corrosion Control for Infrastructure Sustainability*. Denver: AWWA.

ASTM. 2019. Standard Guide for External Corrosion Protection of Ductile Iron Pipe Utilizing Polyethylene Encasement Supplemented by Cathodic Protection. ASTM G218. West Conshohocken, PA: ASTM.

ASTM. 2022. Standard Test Methods for Holiday Detection in Pipeline Coatings. ASTM G62. West Conshohocken, PA: ASTM.

Barrios, D. and L.G. Horn. 2019. "A 60-year History of the Efficacy of Polyethylene Encasement of an Iron Pipe Installation in an Aggressive Soil Environment." *Pipelines.* Reston, VA: ASCE: 514–520.

Bonds, R.W. 1991. Causes, Investigation, and Mitigation of Stray Current Corrosion of Ductile Iron Pipe. National Association of Corrosion Engineers Conference.

Bonds, R.W., L.M. Barnard, A.M. Horton, and G.L. Oliver. 2005. Corrosion and Corrosion Control of Iron Pipe: 75 Years of Research. *Journal AWWA*, 97:6:88–98.

Cast Iron Pipe Research Association. 1964. Soil Corrosion Test Report: Ductile Iron Pipe. Birmingham. AL: CIPRA.

Cox, A.H. and P.H. Hanson. 2010. "Polyethylene Encasement for External Corrosion Control—A 50 Year History." *Pipelines.* Reston, VA: ASCE: 719–729.

Ductile Iron Pipe Research Association. Cast Iron Pipe Century Club and Cast Iron Pipe Sesquicentennial Club. dipra.org/about-dipra/century-club (accessed September 12, 2024).

Ductile Iron Pipe Research Association. *Handbook of Ductile Iron Pipe.* 1984. Birmingham, AL: DIPRA.

Ductile Iron Pipe Research Association. 1993. "Surface Preparation of Ductile Iron Pipe to Receive Special Coatings." *Ductile Iron Pipe News,* (Fall/Winter):17.

Fuller, A.G. 1981. *Corrosion Resistance of Ductile Iron Pipe.* BCIRA Report 1442. Alvechurch, Birmingham: BCIRA.

Horn, L.G. 2021. "Evaluation of a 60-year old Iron Pipeline in a Corrosive Environment and Flow Test of a 97-year old Cement-mortar Lined Iron Pipeline. Proc. Corrosion 2021, Virtual.

Horton, A.M. 1995. Special Protective Coatings and Linings for Ductile Iron Pipe: Advances in Underground Pipeline Engineering. pp. 745–756 in Proc. Second International Conference Pipeline Division and ASCE-TCLEE. Reston, VA.: ASCE-TCLEE.

Horton, A.M. 2014. "Zinc Metallizing for External Corrosion Control of Ductile Iron Pipe." *Pipelines.* Reston, VA: ASCE: 1307–1318.

International Organization for Standardization. 2017. ISO 8179-1. Ductile iron pipes, fittings, accessories and their joints—External zinc-based coating, Part 1: Metallic zinc with finishing layer. Geneva: ISO.

International Organization for Standardization. 2020. Ductile Iron Pipelines—Polyethylene Sleeving for Site Application. ISO 8180. Geneva: ISO.

Kroon, D.H., D. Lindemuth, S. Sampson, and T. Vincenzo. 2004. 04046 Corrosion Protection of Ductile Iron Pipe. Houston: NACE International.

LaQue, F.L. 1964. "Corrosion Characteristics of Ductile Iron." *Journal AWWA*, 56:11:1433–1442.

National Association of Corrosion Engineers. *Industrial Maintenance Painting*. 1976. Houston: NACE, 46–47.

National Association of Corrosion Engineers. 1984. *Corrosion Basics: An Introduction*. Houston: NACE International.

National Association of Corrosion Engineers. 2004. High-Voltage Electrical Inspection of Pipeline Coatings. NACE RP0274. Houston: NACE International.

National Association of Corrosion Engineers. 2024. Control of External Corrosion on Underground or Submerged Metallic Piping Systems. NACE SP0169. Houston: NACE International.

National Association of Pipe Fabricators. 2017. Surface Preparation Standard for Ductile Iron Pipe and Fittings Receiving Special External Coatings and/or Special Internal Linings. NAPF 500-03. Oklahoma City, OK: NAPF.

National Sanitation Foundation/American National Standards Institute/Canada. 2021. Drinking Water System Components—Health Effects. NSF/ANSI/CAN 61. Ann Arbor, MI: NSF International.

Peabody, A.W. 2018. *Peabody's Control of Pipeline Corrosion*. Houston: National Association of Corrosion Engineers.

Romanoff, M. 1968. "Performance of Ductile Iron Pipe in Soils." *Journal AWWA*, 60:6: 645–655.

Schiff, M.J. 1993. Impressed Current Cathodic Protection of Polyethylene Encased Ductile Iron Pipe. Corrosion Paper No. 583. Houston: NACE International.

Sears, E.C. 1968. "Comparison of the Soil Corrosion Resistance of Ductile Iron Pipe and Gray Cast Iron Pipe." *Materials Performance*, 7:10:33–36.

Smith, W.H. 1968. "A Report on Corrosion Resistance of Cast Iron and Ductile Iron Pipe." *Cast Iron Pipe News*, 35:3:16–29.

Stroud, T.F. 1989a. "Corrosion Control Measures for Ductile Iron Pipe." National Association of Corrosion Engineers Conference.

Stroud, T.F. 1989b. Corrosion Control Methods for Ductile Iron Pipe. *AWWA Waterworld News*, 5:4.

Chapter 11

Installation of Ductile-Iron Pipe

11.1 INTRODUCTION

The installation of ductile-iron water mains and their appurtenances is described in AWWA C600, Installation of Ductile-Iron Water Mains and Their Appurtenances, so the information in this chapter closely parallels that given in C600.

11.2 PIPE INSPECTION, STORAGE, HANDLING, AND DELIVERY

11.2.1 Inspection

The purchaser may arrange with the manufacturer for inspection and acceptance of ductile-iron pipe and appurtenances at the manufacturer's plant. On delivery, the purchaser should inspect the shipment as it is unloaded. Damaged items and shortages should be brought to the attention of the carrier and recorded on the delivery receipt or similar document by the carrier's agent. Tests may be performed as specified in the applicable AWWA standard to ensure compliance with the standard. Pipe or appurtenances that fail to comply with specified tests shall be rejected.

11.2.2 Unloading

During the manufacturing, finishing, testing, handling, and loading processes, the manufacturers of ductile-iron pipe exercise the utmost care to avoid damage to their products. Each length of pipe is hydrostatically tested and inspected before it is loaded for shipment. Every precaution is taken to ensure that pipe arrives at its destination in good condition.

The same care should be taken in unloading pipe and related products from railcars and trucks.

Trucks should be parked on level ground for unloading. Rail sidings are generally level enough for unloading purposes.

Before release of restraint, binder, or strapping around the pipe, the loads should be checked to ensure that all chock blocks are securely in place on both ends of all support timbers. If not, chocks or other suitable wedges should be nailed into position to prevent the pipe from rolling when the other restraints are removed. Under no circumstances should the chocks be removed while there is any possibility of pipe rolling out of control and causing damage or injury. Only after all chocks are in place should the binders or strapping securing the load to the truck or railcar be released. Personnel should never remain on, in front of, or alongside the load of pipe after the restraints are removed.

Unloading should be done by lifting with a fork truck (Figure 11-1) or other suitable lifting devices. Pipe should never be rolled off the truck or railcar. Slings, chains, forks, hooks, or pipe tongs should be padded and used in such a manner as to prevent damage to the coated exterior surface or internal lining of the pipe, fitting, or related product. Nylon slings are particularly well suited for lifting ductile-iron pipe and appurtenances. It is dangerous to lift pipe with a single rope. Precautions should be taken to prevent the pipe from

Source: Ductile Iron Pipe Research Association

Figure 11-1 Unloading pipe with a forklift

rolling or shifting during unloading. Steel bands should be cut with a long-handled bolt cutter or similar tool. The bands should not be cut with an ax, chisel, or other tool likely to cause product damage or personal injury. Personnel not directly involved in the unloading operation should stand clear, and no one should ever stand under a lifted load. Lifting devices should be inspected, repaired, and replaced on a timely basis.

11.2.3 Storage

Materials, if stored, should be safe from damage. The interior of all pipe, fittings, and other appurtenances should be kept free of dirt or foreign matter. Valves, hydrants, and other closed appurtenances should be checked on receipt, drained if necessary, and stored so as to protect them from damage by freezing.

Pipe should be carefully placed onto wood timbers or other suitable supports on level ground, avoiding impact. The end pipe of rows or tiers should always be secured with chock blocks. Pipe should be in parallel rows, with bells and plain ends alternated. When stacking more than one tier, care should be taken that bell ends do not rest on spigot ends or vice versa to avoid load concentration points and possible pipe damage.

At least two rows of timbers should be placed between tiers. Chock blocks should be affixed to each timber to prevent pipe movement. The timbers should be large enough to prevent contact between the pipe in the next tier. For safety reasons and ease of handling, pipe should not be stacked higher than the limits shown in Table 11-1, which will result in stacks approximately 12 to 17 ft (3.7 to 5.2 m) high.

Rubber gaskets should be stored in a cool location out of direct sunlight and out of contact with petroleum products. In cold weather, it may be desirable to warm the gaskets and lubricant prior to their use. Gaskets should be used on a first-in, first-out basis.

Table 11-1 Maximum stacking heights* for ductile-iron pipe

Nominal Pipe Size, in.	Number of Tiers
3	18
4	16
6	13
8	11
10	10
12	9
14	8
16	7
18	6
20	6
24	5
30	4
36	4
42	3
48	3
54	3
60	3
64	3

NOTE: To convert inches (in.) to millimeters (mm), multiply by 25.4.
* For 18- or 20-ft (5.5- or 6.1-m) lengths.

11.2.4 Delivery at Trench Site

To avoid unnecessary handling, each joint of pipe and fitting should be placed as close as possible to the location it will occupy in the finished pipeline. The usual procedure is to place pipe close to the trench on the side opposite the spoil bank. Traffic conditions and the type of excavation and installation equipment will affect this procedure. A desirable safety measure is to keep the spoil bank between the trench and the road traffic. Pipe should be strung along the trench, preferably on timbers and with the bells and spigots out of the dirt with all the bells facing in the same direction.

11.3 TRENCHING, EMBEDMENT, PIPE INSTALLATION, AND BACKFILLING

11.3.1 Trench Alignment and Grade

Water mains must be laid and maintained to the lines and grades established by the contract documents for the project. Fittings, valves, tapped or bossed outlets, and hydrants shall be installed at the required locations unless field conditions warrant otherwise and such changes are approved in accordance with the contract documents. Valve operating stems must be oriented to allow proper operation. Hydrants must be installed plumb.

Prior to excavation, an investigation should be made to determine the depth and location of existing underground structures and other utility lines to identify conflicts. Care should be taken to avoid damage to existing structures during excavation and installation. When unforeseen obstructions are encountered requiring alteration of the contract documents, the owner must approve all changes in the contract documents or arrange for removal, relocation, or reconstruction of the obstructions. When crossing existing pipelines or other structures, approval must be obtained to provide clearance as required by federal, state or provincial, and local regulations as deemed necessary by the purchaser to prevent future damage to other structures and contamination of the water main. Special precautions shall be taken when the water main being installed crosses or is adjacent to a facility that is cathodically protected.

11.3.2 Trench Construction

Excavation of the trench must conform with all federal, state or provincial, and local regulations for the protection of the workers. Trench preparation should precede pipe installation for only as far as stated in the contract documents or as directed by the pipeline purchaser. Discharges from trench dewatering pumps shall be directed away from the trench to avoid altering trench stability and shall be in accordance with federal, state or provincial, and local point discharge requirements.

Excavation material must be placed so as not to obstruct the work, endanger workers, or obstruct sidewalks, driveways, or other structures and so as to be in compliance with federal, state or provincial, and local regulations. Pavement removal is a part of trench excavation, and the amount removed depends on the space requirements for the installation of the pipeline and its appurtenances. Pavement removal should not exceed by more than 6 in. (152 mm) in any direction the dimensions of the opening required for installation of pipe, valves, hydrants, specials, manholes, and other structures. Pavement breakage should be in straight lines using the appropriate tools and methods.

The width of the trench at the top of the pipe should be the same as that afforded by the single-pass capabilities of normally available excavating equipment and must be sufficient to permit the pipe to be laid and joined properly and allow the backfill to be placed as specified (Figure 11-2). Table 11-2 can be used as a guide for trench widths. Additional width may be required to permit placement of timber supports, sheathing, bracing, and appurtenances in accordance with the safety requirements of the agency having jurisdiction.

Source: Ductile Iron Pipe Research Association

Figure 11-2 Installing small-diameter ductile-iron pipe in a narrow trench

Table 11-2 Suggested trench widths at the top of the pipe

Nominal Pipe Size, in.	Trench Width, in.
4	28
6	30
8	32
10	34
12	36
14	38
16	40
18	42
20	44
24	48
30	54
36	60
42	66
48	72
54	78
60	84
64	88

Note: To convert inches (in.) to millimeters (mm), multiply by 25.4.

Source: Ductile Iron Pipe Research Association

Figure 11-3 Because of its ruggedness, ductile-iron pipe is particularly suitable for installation in demanding environments

Holes in the trench bottom or pipe embedment for pipe bells must be provided at each joint and should be no larger than necessary for proper joint assembly and assurance that the pipe barrel will lie flat on the trench bottom. Push-on type joints require only minimum-size bell holes to ensure that the pipe is not resting on the bells. Other than for bell holes, the trench bottom must be true and even provide support for the full length of the pipe barrel. A slight depression may be provided, however, for withdrawing pipe slings or other lifting tackle without damaging the coating or polyethylene encasement.

When rock excavation is necessary (Figure 11-3), the rock must be removed to provide at least a 6-in. (152-mm) clearance below and on each side of all pipe, valves, and fittings for pipe sizes 24 in. (610 mm) or smaller. For larger pipe sizes, the clearance must be at least 9 in. (229 mm). When excavation is completed, a layer of appropriate backfill material should be placed on the bottom of the trench to provide the proper depths after leveling and tamping. These same clearances and bedding procedures must also be used for pieces of concrete or masonry and other debris or subterranean structures, such as masonry walls, piers, or foundations encountered during excavation. These clearances are necessary to eliminate the possibility of causing a fulcrum point that might put excessive stress on the pipe, which could cause damage.

If the trench passes over a sewer or other previous excavation, the trench bottom must be sufficiently compacted to provide support equal to that of the native soil or conform to other regulatory requirements to prevent damage to the existing installation.

Blasting for excavation is permitted only after approval is received. The contract documents will establish the hours of blasting. The blasting procedure, including protection of persons and property, must comply with federal, state or provincial, and local regulations.

Trees, shrubs, fences, and all other property and structures must be protected during construction, unless their removal is shown on the contract documents. Tree roots or branches must be cut in accordance with the contract documents.

Temporary support, adequate protection, and maintenance of all underground and surface structures, drains, sewers, and other obstructions encountered in the progress of the work must be provided. All disturbed properties must be restored as nearly as is practical to their original condition.

When the trench material contains ashes, refuse, organic material, or other unsuitable, corrosive material, or cinders from burned coal, such material must be removed to at least 6 in. (150 mm) below the bottom of the pipe or to the depth required by the contract documents and replaced with clean, stable backfill material. When such corrosive materials are encountered, polyethylene encasement should be used to protect the pipe (Figure 11-4). Additional information on external corrosion protection may be found in Chapter 10 of this manual.

When the trench bottom or the subgrade is found to consist of material that is unstable to such a degree that it cannot be removed, a foundation for the pipe and appurtenances must be constructed using piling, treated timber, concrete, or other materials in accordance with the contract documents.

Appropriate traffic control devices must be provided in accordance with federal, state or provincial, and local regulations to regulate, warn, and guide traffic at the worksite.

11.3.3 Pipe Installation

All materials must be carefully examined for damage and defects immediately before installation. Defective materials must be marked and held for final disposition as called for by the contract documents. Proper implements, tools, and facilities must be provided and used for the safe and proper performance of the work. Pipe, fittings, valves, and hydrants must be carefully lowered into the trench with a backhoe, crane, ropes, or other suitable equipment to prevent damage to the materials and to their protective coatings and linings. Under no circumstances should these materials be dropped or dumped into the trench. If possible, the trench should be dewatered before pipe installation.

All lumps, blisters, and excess coating must be removed from the socket and plain end of each pipe, and the outside of the plain end and the inside of the bell must be clean, dry, and free of dirt, sand, grit, or other foreign materials before the pipe is installed. Foreign material must be prevented from entering the pipe while the pipe is being placed and installed in the trench. No debris, tools, clothing, or other materials should be placed in the pipe at any time.

Source: Ductile Iron Pipe Research Association

Figure 11-4 Polyethylene-encased ductile-iron pipe

As each length of pipe is placed in the trench, the joint should be assembled and the pipe brought to correct line and grade. Bumping the pipe with the backhoe bucket in order to obtain grade is discouraged due to the possibility of such practices causing damage to the pipe and/or lining. The pipe should then be secured in place with approved backfill materials.

It is common practice to install pipe with the bells facing the direction in which work is progressing; however, this is not mandatory. When pipe is being installed on a slope, the pipe bells are frequently faced uphill for ease of installation. Bell direction is not functionally related to the direction of flow in the pipeline.

When pipe installation is not in progress, the open ends of pipe should be closed with a watertight plug or other means as specified. The plug should allow venting and remain in place until the trench is pumped dry. Care must be taken to prevent pipe flotation in case the trench fills with water. Air or water pressure in the pipeline must be released before removing the plug for any reason.

The laying conditions for ductile-iron pipe must comply with ANSI/AWWA C150/A21.50 (2023), as illustrated in Figure 11-5 and required by the contract documents. It should be noted that loosely placed backfill around and/or above the pipe may allow settlement, which could be detrimental to improvements such as roadway surfacing placed over the trench. Figures 11-6 and 11-7 show joint assembly instructions for push-on and mechanical joints, respectively. Mechanical joint bolt torques are provided in Table 11-3.

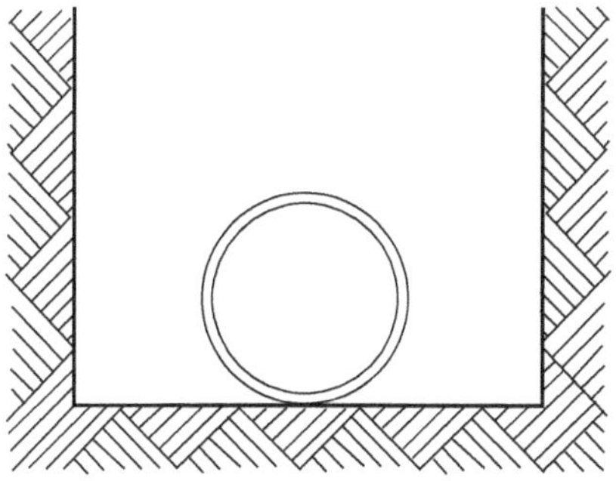

Type 1*

Flat-bottom trench[†] with loose backfill.

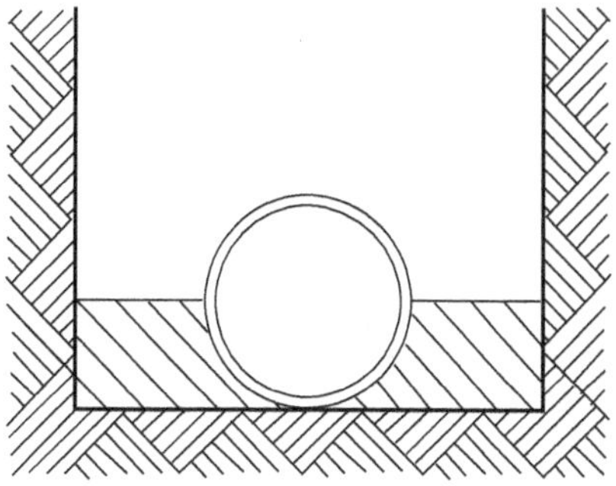

Type 2

Flat-bottom trench[†] with loose backfill lightly consolidated to centerline of pipe.

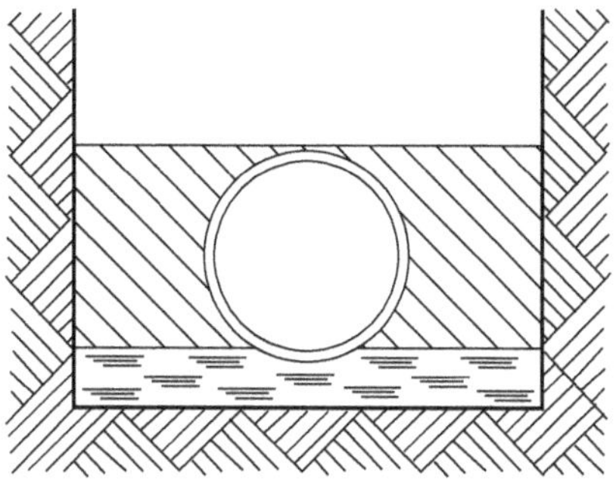

Type 3

Pipe bedded in 4-in. (100-mm) minimum loose soil[‡] with backfill lightly consolidated to top of pipe.

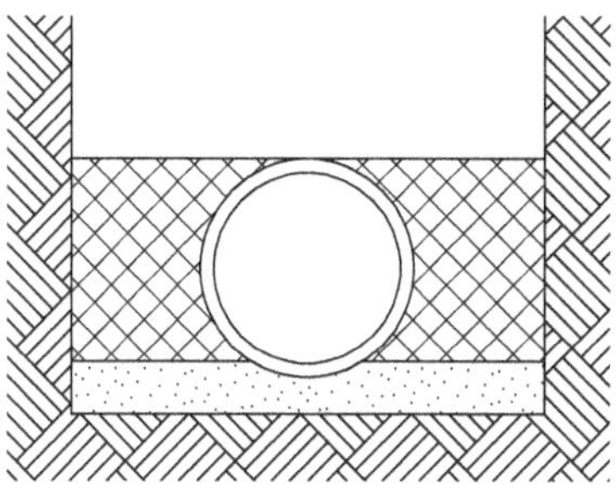

Type 4

Pipe bedded in sand, gravel, or crushed stone to depth of $^1/_8$ pipe diameter, 4-in. (100-mm) minimum, with backfill compacted to top of pipe. (Approximately 80% Standard Proctor, AASHTO[§] T-99)

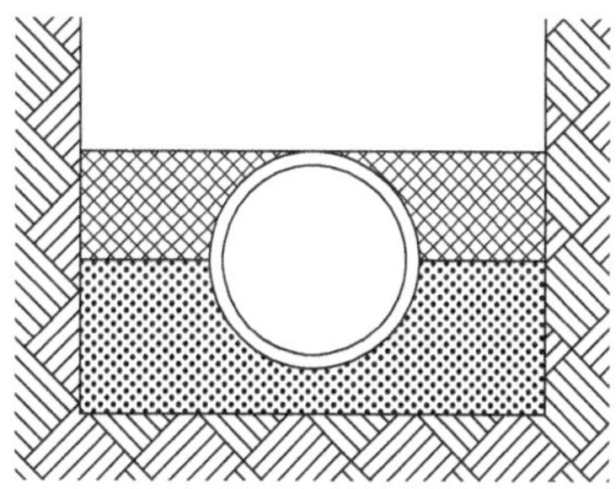

Type 5

Pipe bedded to its centerline in compacted granular material, 4-in. (100-mm) minimum under pipe. Compacted granular** or select material[‡] to top of pipe. (Approximately 90% Standard Proctor, AASHTO[§] T-99)

NOTES: 1) Consideration of the pipe-zone embedment conditions included in this figure may be influenced by factors other than pipe strength. See this chapter for additional information on pipe bedding and backfill.

2) To convert pounds per square inch (psi) to kilopascals (kPa), multiply by 6.895.

* For 14-in. (356-mm) and larger pipe, consideration should be given to the use of laying conditions other than Type 1.

† Flat bottom is defined as undisturbed earth.

‡ Loose soil or select material is defined as native soil excavated from the trench, free of rocks, foreign materials, and frozen earth.

§ AASHTO T-99 "Standard Method of Test for the Moisture-Density Relations of Soils Using 5.5 lb. (2.5 kg) Rammer and a 12 in. (305 mm) Drop." Available from the American Association of State Highway and Transportation Officials, 444 N. Capitol St. N.W., Suite 225, Washington, DC 20001.

** Granular materials are defined per the AASHTO Soil Classification System (ASTM D3282) or the United Soil Classification System (ASTM D2487), with the exception that gravel bedding/backfill adjacent to the pipe is limited to 2-in. maximum particle size per ANSI/AWWA C600.

Figure 11-5 Laying conditions for ductile-iron pipe

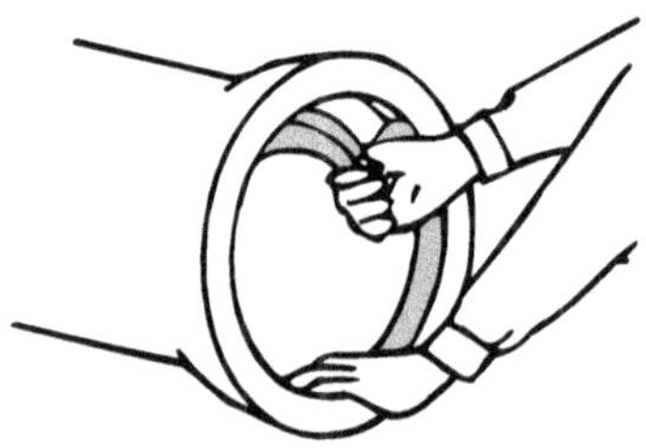

1. Thoroughly clean the groove and the bell socket of the pipe or fitting; also clean the plain end of the mating pipe. Using a gasket of the proper design for the joint to be assembled, make a small loop in the gasket and insert it in the socket. For pipe sizes larger than 20 in., it may be necessary to make two or more loops in the gasket (6 and 12 o'clock). Ensure that the gasket faces the correct direction and that it is properly seated. Note: In cold weather, it is preferable to warm the gasket to facilitate assembly of the joint.

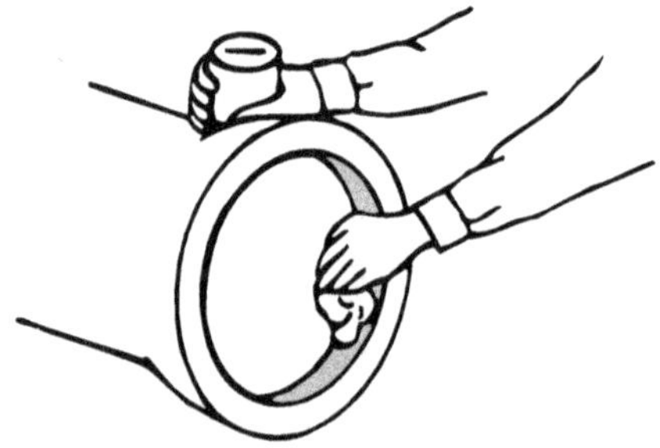

2. Apply lubricant to the exposed surface of the gasket and plain end of the pipe in accordance with the pipe manufacturer's recommendations. Do not apply lubricant to the bell socket or the surface of the gasket in contact with the bell socket. The lubricant is furnished in sterile containers, and every effort should be made to protect against contamination of the containers' contents.

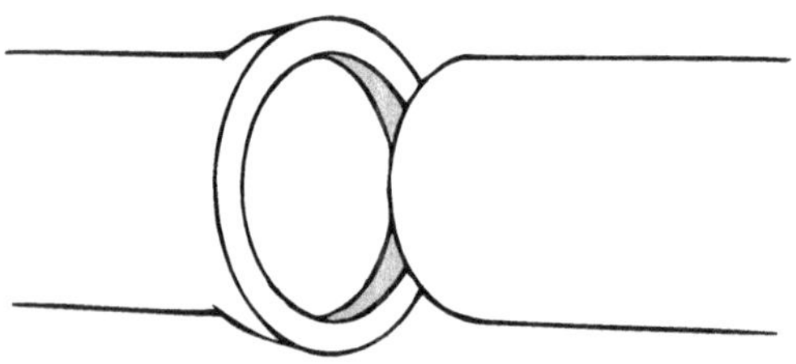

3. Be sure that the plain end is beveled per the manufacturer's recommendations; square or sharp edges may damage or dislodge the gasket and cause a leak. When pipe is cut in the field, bevel the plain end with a heavy file or grinder to remove all sharp edges. Push the plain end into the bell socket of the mating pipe. Keep the joint straight while pushing. Make deflection after the joint is assembled.

4. Small pipe can be pushed into the bell socket with a long bar. Large pipe requires additional power, such as a jack, lever puller, or backhoe. The supplier may provide a jack or lever puller on a rental basis. A timber header should be used between the pipe and the jack or backhoe bucket to avoid damage to the pipe.

Figure 11-6 Push-on joint assembly

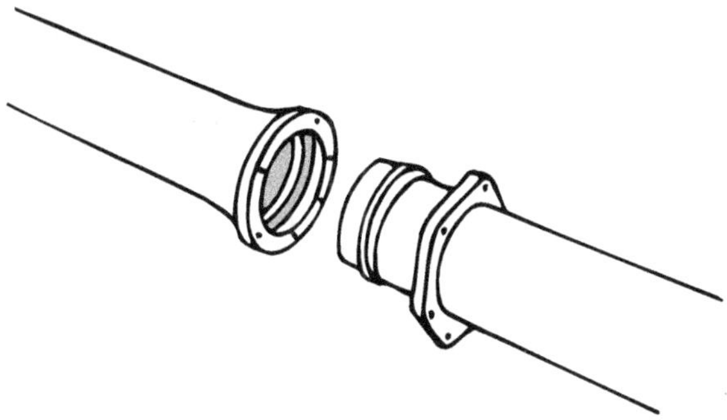

1. Wipe clean the socket and the plain end. Lubrication and additional cleaning should be provided by brushing both the gasket and plain end with soapy water, or an approved pipe lubricant meeting the requirements of ANSI/AWWA C111/A21.11, just prior to slipping the gasket onto the plain end for joint assembly. Place the gland on the plain end with the lip extension toward the plain end, followed by the gasket with the narrow edge of the gasket toward the plain end. Note: In cold weather, it is preferable to warm the gasket to facilitate assembly of the joint.

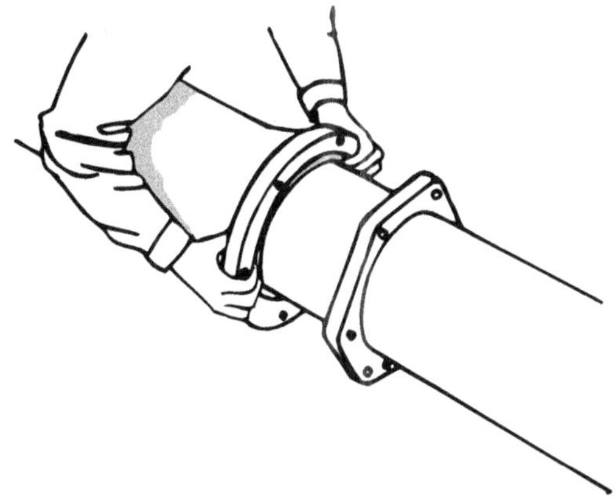

2. Insert the pipe into the socket and press the gasket firmly and evenly into the gasket recess. Keep the joint straight during assembly.

3. Push the gland toward the socket and center it around the pipe with the gland lip against the gasket. Insert bolts and hand tighten nuts. Make deflection after joint assembly but before tightening bolts.

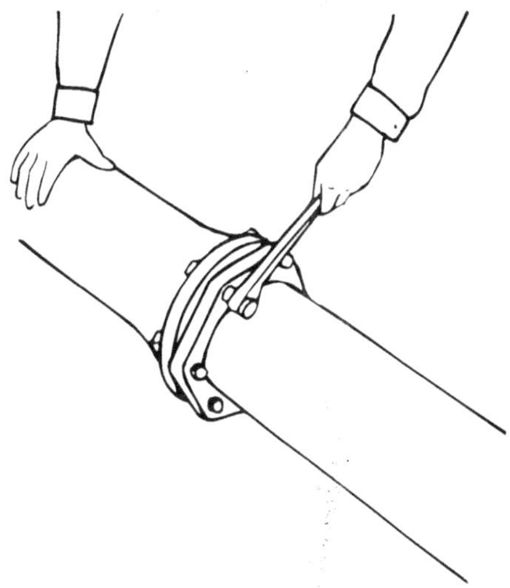

4. Tighten the bolts to the normal range of bolt torque, as indicated in Table 11-3, while at all times maintaining approximately the same distance between the gland and the face of the flange at all points around the socket. This can be accomplished by partially tightening the bottom bolt first, then the top bolt, next the bolts at either side, finally the remaining bolts. Repeat the process until all bolts are within the appropriate range of torque. In large sizes (30 in. through 48 in. [762 mm through 1,219 mm]), five or more repetitions may be required. The use of a torque-indicating wrench will facilitate the procedure.

Figure 11-7 Mechanical joint assembly

Table 11-3 Mechanical joint bolt torques

Nominal Pipe Size, *in.*	Bolt Size, *in.*	Range of Torque, *ft-lb*
3	⅝	45–60
4–24	¾	75–90
30 and 36	1	100–120
42–64	1 ¼	120–150

NOTE: To convert inches (in.) to millimeters (mm), multiply by 25.4; to convert foot-pounds (ft-lb) to Newton-meters (Nm), multiply by 1.356.

When it is necessary to deflect pipe from a straight line (Figure 11-8) in either the horizontal or vertical plane, the amount of joint deflection should not exceed that shown in Tables 11-4 and 11-5 (see also Section 8.12).

The deflections listed are maximum deflections and should not be exceeded. For design purposes, deflection should be limited to 80 percent of the values shown in Tables 11-4 and 11-5. Figure 11-9 illustrates the maximum offset *S* and approximate radius of curve *R*, which are listed in Tables 11-4 and 11-5.

Special transition couplings or gaskets are required for joining ductile-iron pipe to different types of pipe, such as steel pipe, asbestos–cement pipe, and some plastic pipe. Such transition devices are available. When ordering, the actual outside diameter of the pipe should be given.

Cutting of pipe for insertion of valves, fittings, or closure pieces must be done in conformance with all safety recommendations of the manufacturer of the cutting equipment. Cutting must be done in a safe manner without damaging the pipe or the cement–mortar lining. Ductile-iron pipe may be cut using an abrasive wheel saw (Figure 11-10), rotary wheel cutter, guillotine pipe saw, or a milling wheel cutter. Ductile-iron pipe may also be cut with an oxyacetylene torch if recommended by the pipe manufacturer. ANSI/AWWA standards for ductile-iron pipe require factory gauging of the spigot end. Accordingly, pipes selected for cutting should be field gauged. An MJ gland inserted over the barrel might serve as a convenient indicator for this purpose. When glands are not available, pipe can be selected by measuring with a tape in accordance with the manufacturer's recommendation. Cut ends and rough edges must be ground smooth. For push-on joint connections, the cut ends must be beveled.

When the presence of aggressive soil is identified that requires polyethylene encasement for ductile-iron pipe, the encasement must be installed in accordance with ANSI/AWWA C105/A21.5, Polyethylene Encasement for Ductile-Iron Pipe Systems, (see Chapter 10) and as required by the contract documents.

11.3.4 Backfilling

Backfilling must be done in accordance with the laying condition specified. (See description and illustrations in Figure 11-5.) All backfill material must be free from cinders, ashes, refuse, vegetable or organic material, boulders, sharp rocks or stones, frozen soil, or other unsuitable material.

From 1 ft (305 mm) above the top of the pipe to the subgrade of the pavement, materials containing stones up to 8 in. (203 mm) in their greatest dimension may be used, unless otherwise specified. When the type of backfill material is not indicated on the contract documents or not specified, the excavated material may be used, provided that such material consists of loam, clay, sand, gravel, or other suitable materials and does not contain peat or other organic materials.

Source: Ductile Iron Pipe Research Association

Figure 11-8 Sweeping curves without fittings and specials are possible because of liberal joint deflection allowance by ductile-iron pipe push-on joints

Table 11-4 Maximum joint deflection* full-length pipe—push-on type joint pipe

Nominal Pipe Size, *in.*	Deflection Angle—θ *deg.*	Maximum Offset–*S*,† *in.*		Approximate Radius of Curve–*R*† Produced by Succession of Joints, *ft*	
		L† = 18 ft	*L*† = 20 ft	*L*† = 18 ft	*L*† = 20 ft
3	5	19	21	205	230
4	5	19	21	205	230
6	5	19	21	205	230
8	5	19	21	205	230
10	5	19	21	205	230
12	5	19	21	205	230
14	3*	11	12	340	380
16	3*	11	12	340	380
18	3*	11	12	340	380
20	3*	11	12	340	380
24	3*	11	12	340	380
30	3*	11	12	340	380
36	3*	11	12	340	380
42	3*	11	12	340	380
48	3*	12	380	—	—
54	3*	12	380	—	—
60	3*	12	380	—	—
64	3*	12	380	—	—

Note: To convert inches (in.) to millimeters (mm), multiply by 25.4; to convert feet (ft) to meters (m), multiply by 0.3048.
* For 14-in. and larger push-on joints, maximum deflection angle may be larger than shown above. Consult the manufacturer.
† See Figure 11-9.

Table 11-5 Maximum joint deflection full-length pipe—mechanical type joint pipe

Nominal Pipe Size, *in.*	Deflection Angle—θ *deg.– min.*	Maximum Offset–*S*,* *in.*		Approximate Radius of Curve–*R** Produced by Succession of Joints, *ft*	
		*L** = 18 ft	*L** = 20 ft	*L** = 18 ft	*L** = 20 ft
3	8–18	31	35	125	140
4	8–18	31	35	125	140
6	7–07	27	30	145	160
8	5–21	20	22	195	220
10	5–21	20	22	195	220
12	5–21	20	22	195	220
14	3–35	13.5	15	285	320
16	3–35	13.5	15	285	320
18	3–00	11	12	340	380
20	3–00	11	12	340	380
24	2–23	9	10	450	500

Note: To convert inches (in.) to millimeters (mm), multiply by 25.4; to convert feet (ft) to meters (m), multiply by 0.3048.
* See Figure 11-9.

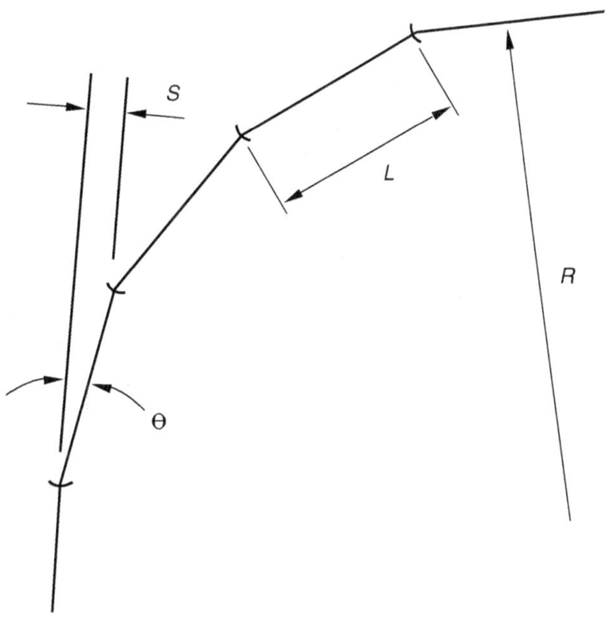

θ = deflection angle
S = joint deflection offset
L = laying length
R = radius of curve

$$R = \frac{L}{2\tan\frac{\theta}{2}}$$

Figure 11-9 Pipeline curve geometry

If excavated material is indicated on the contract documents or specified for backfill and there is a deficiency as a result of rejection of a part of this material, the required amount of sand, gravel, or other approved material must be provided.

For purposes of definition:

- Sand is material graded from fine to coarse, containing less than 10 percent by weight of loam and clay, that all passes a ¾-in. (19-mm) sieve with no more than 5 percent by weight remaining on a No. 4 sieve.
- Gravel is a reasonably uniform combination of stone, containing none larger than 2 in. (51 mm) and not containing excessive amounts of clay and loam.
- Crushed stone is limestone or dolomite ledge-rock material that all passes a ½-in. (13-mm) sieve with no more than 25 percent passing a No. 100 sieve.

When special backfill compaction procedures are required, they must be accomplished in accordance with the contract documents or applicable federal, state or provincial, and local regulations. Particularly in installations of large-diameter pipes where pipe support is more critical, machine-operated roller-compacting wheels are often quite helpful. When these are remotely operated (e.g., by a backhoe arm), they also offer safety advantages over workers in deep trenches.

Newly installed pipelines are normally tested after backfilling. When unusual conditions require that testing allowance and pressure testing be accomplished before completion of backfilling or with pipe joints accessible for examination, sufficient backfill material should be placed over the pipe barrel between the joints to prevent movement, and consideration should be given to restraining thrust forces during the testing. Restrained joint systems that derive their stability from the interaction of pipe and soil should be fully backfilled before testing.

Source: Ductile Iron Pipe Research Association

Figure 11-10 Cutting pipe with an abrasive wheel saw

11.4 PIPELINE ACCESSORIES

11.4.1 Valve and Fitting Installation

Prior to installation, valves should be inspected for direction of opening, number of turns to open, freedom of operation, tightness of pressure-containing bolting and test plugs, cleanliness of pressure-containing bolting and valve ports and seating surfaces, handling damage, and cracks. Defective valves shall be marked and held for final disposition as called for by the contract documents. All bolts and nuts, with the exception of seat-adjusting bolts or screws in butterfly valves, shall be checked for proper tightness. Seat-adjusting bolts in butterfly valves shall only be adjusted on recommendation from the manufacturer.

Valves, fittings, plugs, and caps should be set and connected to the pipe in the same manner as specified for cleaning, laying, and joining pipe, except that valves larger than 12 in. (305 mm) should be supported in such a manner that settling does not affect the pipe and valve joints (Figures 11-11 and 11-12). Valves installed aboveground or in plant piping systems should be supported to prevent bending of the valve end connections as a result of pipe loading. Likewise, valves with heavy offset operator weights may require external support for stabilization to prevent torsional loads/movements in pipe joints that are not designed to support such loads. Valves should be closed before being installed.

Water main valves should be located within or immediately adjacent to street property lines where practical, unless shown otherwise on the plans. Mains should be drained through drainage branches or blowoffs. Drainage branches, blowoffs, air vents, and appurtenances should be provided with control valves and located and installed as shown on the plans. Drainage branches or blowoffs should not be directly connected to any storm or sanitary sewer, submerged in any stream, or installed in any other manner that will permit backsiphonage into the water distribution system. Air release and/or vacuum vents should be provided at high points in the line and in areas of potential negative pressure. The air release and/or vacuum vents should not be connected to any storm or sanitary sewer, and they should be protected from freezing.

Source: Ductile Iron Pipe Research Association

Figure 11-11 Installing valves and fittings

Source: Ductile Iron Pipe Research Association

Figure 11-12 Installation of a 48-in. (1,219-mm) butterfly valve

The underground use of valves with flanged joints is generally not considered desirable because of the rigidity of the joint. If valves with flanged joints are to be used underground, it is good practice to install them in vaults. Swing check valves should normally be installed aboveground or in adequately-sized vaults to allow servicing and room for operation.

A valve box, vault, or other device to allow valve access and operation should be provided for each valve. The valve box must not transmit shock or stress to the valve and must be centered over the operating nut of the valve with the box cover flush with the surface of the finished installation or such other level specified.

When a valve vault is used, care must be exercised to prevent it from settling on the pipe. The operating nut should be readily accessible for operation through the opening in the valve vault. Vaults should be constructed to permit minor valve repairs and afford protection to the valve and pipe from impact where they pass through the vault walls. Valves should never be used to bring misaligned pipe into alignment during installation. Pipe must be supported to prevent stress on the valve. Thrust resulting from closure of valves should be carefully considered in the design of the piping system and vaults.

All dead ends on new mains must be closed with suitably restrained plugs or caps to prevent separation under pressure. If a blowoff valve precedes the plug or cap, it must also be restrained against separation. All dead ends should be equipped with suitable blowoff or venting devices. Additional information on installation of valves can be obtained from AWWA.

11.4.2 Hydrant Installation

Prior to installation, all hydrants should be inspected for direction of opening, nozzle threading, operating-nut and cap-nut dimensions, tightness of pressure-containing bolting, cleanliness of inlet elbows, handling damage, and cracks. Defective hydrants should be marked and held for final disposition as called for by the contract documents. All hydrants must stand plumb and have their nozzles parallel with or at right angles to the curb, with pumper nozzle facing the curb. An exception would be hydrants with two hose nozzles 90° apart, which should be set with each nozzle facing the curb at an angle of 45°. Hydrants should be set to the established grade, with the centerline of the lowest nozzle at least 12 in. (305 mm) above the ground or as otherwise specified. They should be installed away from the curb line a sufficient distance to avoid damage from or to vehicles. Traffic model hydrants (hydrants that are intended to fail at the ground-line on vehicle impact) shall be installed such that the breakaway flange shall be installed not less than 2 in. (51 mm) nor more than 6 in. (152 mm) above established grade.

Each hydrant must be connected to a main with a 6-in. (152-mm) minimum diameter branch controlled by an auxiliary valve, unless otherwise specified by the owner. The valve should be restrained to maintain shutoff when the hydrant is to be removed. When a dry-barrel hydrant is set in impervious soil, drainage should be provided at the base of the hydrant by placing coarse gravel or crushed stone mixed with coarse sand from the bottom of the trench to at least 6 in. (152 mm) above the drain port opening in the hydrant and to a distance of 1 ft (305 mm) around the elbow. Where groundwater rises above the drain port, when the hydrant is located within 8 ft (2.4 m) (or such distance as required by the applicable regulatory agency) of a sanitary sewer main, or when drainage is not permitted by the applicable regulatory agency, the drain port should be plugged and water pumped from the hydrant when freezing occurs. When a dry-barrel hydrant with an open drain port is set in clay or other impervious soil, a drainage pit 2 ft × 2 ft × 2 ft (0.6 m × 0.6 m × 0.6 m) must be excavated below each hydrant and filled with coarse gravel or crushed stone mixed with coarse sand under and around the elbow of the hydrant and to a level of 6 in. (152 mm) above the drain port. To prevent possible contamination of the water supply, hydrant drains should not be connected to a sanitary sewer or storm sewer.

The hydrant should be set on a firm footing like as stone slabs or concrete to prevent settling and strain on the lateral line joints. Provisions should be made for anchoring the hydrant on the end of its lateral by strapping, blocking, or use of restrained joints. Care should be taken not to plug the drain holes of dry-barrel hydrants when pouring thrust blocks. Hydrants should be located as shown on the plans or as specified by the purchaser (Figure 11-13).

In the case of traffic hydrants, specific care must be taken to provide adequate soil resistance to avoid transmitting shock movement to the lower barrel and inlet connection. In loose or poor load-bearing soil, this may be accomplished by pouring a concrete collar approximately 6-in. (152-mm) thick to a distance of 2 ft (0.6 m) at or near the ground line around the hydrant barrel. Additional information on hydrant installation can be found in AWWA M17, *Installation, Field Testing, and Maintenance of Fire Hydrants.*

11.5 THRUST RESTRAINT

Methods for providing restraint from thrust forces, including blocks and restrained joints, are covered in Chapter 8 of this manual. If tie rods or clamps are used for thrust restraint, they shall be of adequate strength and made from corrosion-resistant material (or suitably protected from corrosion).

Source: Ductile Iron Pipe Research Association

Figure 11-13 Fire hydrant installation

11.6 FLUSHING, FIELD TESTING, AND DISINFECTION

11.6.1 Flushing

Foreign material left in the pipelines during installation often results in valve or hydrant seat leakage during pressure tests. Every effort should be made to keep lines clean during installation. Thorough flushing is recommended prior to a pressure test.

To reduce the probability of damaging hydrant sealing surfaces as a result of debris trapped in the hydrant valve during the flushing of a newly installed line, it is important to make sure the hydrant is fully open. After flushing and prior to closing the hydrant, slowly close the gate valve in front of the hydrant to reduce the likelihood of any remaining rocks and debris being trapped in the hydrant seat.

11.6.2 Hydrostatic Testing

The testing methods described in the following section are specifically for water pressure testing. These procedures should not be used for air pressure testing because of the serious safety hazards involved. Hydrostatic pressure and testing allowance checks should be performed at the same time, that is, while pressure is maintained within 5 psi (34.5 kPa) of the test pressure for at least 2 h. Testing allowance is determined by measuring the quantity of makeup water required to maintain the required pressure level.

11.6.3 Hydrostatic Pressure Test

All newly laid pipe or any valved section thereof shall be subjected to a hydrostatic pressure test.

Test pressures should meet the following criteria:

- Be no less than 1.25 times the stated working pressure of the pipeline measured at the highest elevation along the test section
- Be no less than 1.5 times the stated working pressure at the lowest elevation of the test section
- Will not exceed the thrust-restraint design pressure or 1.5 times the pressure rating of the pipe or joint, whichever is less (as specified by the manufacturer)
- Be at least 2 h in duration
- Vary by no more than ±5 psi (34.5 kPa) for the duration of the test
- Will not exceed the rated working pressure of the valves when the pressure boundary of the test section includes closed resilient-seated gate valves or butterfly valves

Hydrants in a test section should only be tested with the main valve of the hydrant closed. Valves shall not be operated in either direction at a differential pressure exceeding the rated valve working pressure. A test pressure greater than the rated valve working pressure can result in trapped test pressure between the gates of a double-disc gate valve. For tests exceeding the rated valve working pressure, the test setup should include a provision independent of the valve to reduce the line pressure to the rated valve working pressure on completion of the test. The valve can then be opened enough to equalize the trapped pressure with the line pressure or fully opened.

Following the installation of a new pipeline, all newly laid pipe or any valved section thereof shall be subjected to a hydrostatic pressure test. Each valved section of pipeline shall be slowly filled with water. When venting air from pipelines, it is important to limit the pipeline fill rate to avoid excessive surge pressures when the water reaches the air venting opening(s). The specified test pressure shall be applied using a suitable pump connected to the pipeline (Note: The specified test pressure shall be based on the elevation of the lowest point of the pipeline or section under test and corrected to the elevation of the test gauge.) Before applying the specified test pressure, air shall be expelled completely from the pipeline section under test. If permanent air vents are not located at all high points, corporation cocks shall be installed at these points to expel any air as the line is filled with water. Use of corporation cocks above rated pressure must be at the risk of the user and authorized specifically by the manufacturer. Following removal of any air, the corporation cocks shall be closed and the test pressure applied. (At the conclusion of the pressure test, the corporation cocks shall be removed and the pipe plugged or left in place as required by the contract documents). The pipeline shall be allowed to stabilize at the test pressure before the hydrostatic test is conducted. This may require several cycles of pressurizing and bleeding trapped air prior to beginning the test. The hydrostatic test shall be of at least a 2-h duration. The test pressure shall not vary by more than ± 5 psi (34.5 kPa) for the duration of the test. Test pressure shall be maintained within this tolerance by adding makeup water through the pressure test pump into the pipeline. The amount of makeup water added shall be accurately measured (in gal/h or L/h) by suitable methods and shall not exceed the applicable testing allowance as specified in Section 11.6.4.

Any exposed pipe, fittings, valves, hydrants, and joints should be carefully examined during the pressure test. Any damaged or defective pipe, fittings, valves, hydrants, or joints that are discovered should be repaired or replaced with reliable material and the test repeated until satisfactory results are obtained.

11.6.4 Testing Allowance

Testing allowance is defined as the maximum quantity of makeup water that can be added into a pipeline undergoing hydrostatic pressure testing (or valved section thereof) in order to maintain pressure within 5 psi (34.5 kPa) of the specified test pressure (after the pipeline has been filled with water and the air has been expelled). Note: The testing allowance may not be reasonable if the pressure boundary of the test section includes appurtenances subjected to pressures above their rated working pressures due to possible leakage by those appurtenances. The testing allowance is exceeded if the amount of makeup water is greater than that determined using the following equation:

In inch-pound units,

$$L = \frac{SD\sqrt{P}}{148{,}000} \qquad \text{(Eq 11-1)}$$

Where

L = testing allowance (makeup water), gal/h (gph)
S = length of pipe tested, ft
D = nominal diameter of the pipe, in.
P = average test pressure during the hydrostatic test, psi (gauge)

In metric units,

$$L_m = \frac{SD\sqrt{P}}{798{,}797} \qquad \text{(Eq 11-2)}$$

Where

L_m = testing allowance (makeup water), Lph
S = length of pipe tested, m
D = nominal diameter of the pipe, mm
P = average test pressure during the hydrostatic test, kPa

These formulas are based on a testing allowance of 10.49 gpd/mi/in. (0.971 L/day/km/mm) of nominal diameter of pipe at a pressure of 150 psi (1,034 kPa). Values of testing allowance at various pressures are shown in Table 11-6.

When testing against closed metal-seated valves, an additional testing allowance per closed valve of 0.0078 gph/in. (1.2 mL/hr/mm) of nominal valve size is allowed. When hydrants are in the test section, the test should be made against the main valve in the hydrant.

Acceptance of an installation is determined on the basis of testing allowance only. If any test of a new pipeline requires a quantity of makeup water greater than that allowed, repairs or replacement shall be accomplished in accordance with the contract documents. All visible leaks must be repaired regardless of the allowance used for testing.

Table 11-6 Hydrostatic testing allowance per 1,000 ft (305 m) of pipeline* (gph†)

Average Test, Pressure	Nominal Pipe Diameter, *in.*																
psi	3	4	6	8	10	12	14	16	18	20	24	30	36	42	48	54	60	64
450	0.43	0.57	0.86	1.15	1.43	1.72	2.01	2.29	2.58	2.87	3.44	4.30	5.16	6.02	6.88	7.74	8.60	9.17
400	0.41	0.54	0.81	1.08	1.35	1.62	1.89	2.16	2.43	2.70	3.24	4.05	4.86	5.68	6.49	7.30	8.11	8.65
350	0.38	0.51	0.76	1.01	1.26	1.52	1.77	2.02	2.28	2.53	3.03	3.79	4.55	5.31	6.07	6.83	7.58	8.09
300	0.35	0.47	0.70	0.94	1.17	1.40	1.64	1.87	2.11	2.34	2.81	3.51	4.21	4.92	5.62	6.32	7.02	7.49
275	0.34	0.45	0.67	0.90	1.12	1.34	1.57	1.79	2.02	2.24	2.69	3.36	4.03	4.71	5.38	6.05	6.72	7.17
250	0.32	0.43	0.64	0.85	1.07	1.28	1.50	1.71	1.92	2.14	2.56	3.21	3.85	4.49	5.13	5.77	6.41	6.84
225	0.30	0.41	0.61	0.81	1.01	1.22	1.42	1.62	1.82	2.03	2.43	3.04	3.65	4.26	4.86	5.47	6.08	6.49
200	0.29	0.38	0.57	0.76	0.96	1.15	1.34	1.53	1.72	1.91	2.29	2.87	3.44	4.01	4.59	5.16	5.73	6.12
175	0.27	0.36	0.54	0.72	0.89	1.07	1.25	1.43	1.61	1.79	2.15	2.68	3.22	3.75	4.29	4.83	5.36	5.72
150	0.25	0.33	0.50	0.66	0.83	0.99	1.16	1.32	1.49	1.66	1.99	2.48	2.98	3.48	3.97	4.47	4.97	5.30
125	0.23	0.30	0.45	0.60	0.76	0.91	1.06	1.21	1.36	1.51	1.81	2.27	2.72	3.17	3.63	4.08	4.53	4.83
100	0.20	0.27	0.41	0.54	0.68	0.81	0.95	1.08	1.22	1.35	1.62	2.03	2.43	2.84	3.24	3.65	4.05	4.32

NOTE: To obtain testing allowance in (L /h), multiply the values in the table by 3.785.

* If the pipeline under test contains sections of various diameters, the testing allowance will be the sum of the testing allowances for each size.

† Calculated on the basis of Equation 11-1.

11.6.5 Disinfection

Newly installed potable water pipelines require disinfection in accordance with ANSI/AWWA C651, Disinfecting Water Mains, and ANSI/AWWA C600, Installation of Ductile-Iron Water Mains and Their Appurtenances. Following chlorination, the main should be flushed as soon as possible (within 24 h) because prolonged exposure to high concentrations of chlorine may damage the asphaltic seal coat. It is important to note that provisions should be made to avoid contamination of existing mains by cross-connection during flushing, testing, and disinfection of newly installed mains.

11.6.6 Pressure Cleaning

The practice of pressure washing ductile-iron pipelines may present some risk of damage to the sealcoat, cement lining, or ceramic epoxy lining. Any such risk is dependent on water pressure, speed, jet design and angle to the lining, distance of the jet from the lining, type of lining, and other factors. Consult the pipe manufacturer for guidelines to minimize the risk of pipeline damage. Damage from pressure washing could reduce pipeline effectiveness and void the warranty.

11.7 Service Taps

Corporation stops may be installed either before or after pipe installation. Generally, they are located at 10- or 2-o'clock positions on the circumference of the pipe and may be screwed directly into the tapped and threaded main without any additional appurtenances. In cold climates with deep frost penetration, freezing of service lines can be a problem. In such locations, installing corporation stops horizontally at 3- or 9-o'clock positions on the pipe circumference will conserve available cover over the service line or reduce the necessary

Table 11-7 Maximum recommended direct tap size

	Pressure Class				
	150	200	250	300	350
Size (in.)	Maximum Recommended Direct Tap Size (in.)				
3	—	—	—	—	¾
4	—	—	—	—	¾
6	—	—	—	—	1
8	—	—	—	—	1
10	—	—	—	—	1
12	—	—	—	—	1¼
14	—	—	1¼	1½	1½
16	—	—	1½	2	2
18	—	—	2	2	2
20	—	—	2	2	2
24	—	2	2	2	2

depth of cover on the main pipeline. Standard corporation stops may be used on all classes of ductile-iron pipe. Torque requirements can be effectively reduced by using two layers of 3-mil TFE tape on the male threads of the corporation stop.

The maximum recommended direct tap sizes to ensure a watertight tap for 3-in. through 24-in. ductile-iron pipe are shown in Table 11-7. All classes of 24-in and larger in diameter ductile-iron pipe can be direct tapped for 2-in. corporation stops. Necessary pipe thicknesses for two, three, and four full-thread engagement of different tap sizes are given in Tables 11-8 and 11-9. Because of the high strength of ductile iron, the number of threads engaged is less critical than with weaker material.

Service taps on gray- and ductile-iron mains encased in polyethylene may be accomplished by making an X-shaped cut in the polyethylene and temporarily folding back the film or preferably, by tapping directly through the polyethylene. The preferred method of direct-tapping through the encasement and reducing the risk of damage to the polyethylene by the tapping equipment is to apply two or three wraps of polyethylene adhesive tape around the pipe and tap through the tape and the polyethylene film (Figure 11-14). After a tap has been completed, cuts or tears in the polyethylene and any other areas of damage to the film shall be repaired with tape as described in ANSI/AWWA C105/A21.5.

Direct connection of copper services to ductile- and gray-iron pipelines has historically been a common practice in the waterworks industry. To minimize the possibility of corrosion, service lines of dissimilar metals and the attendant corporation stop should be wrapped with polyethylene or a suitable dielectric tape for a minimum clear distance of 3 ft (0.9 m) from the main.

Table 11-8 Pipe thicknesses required for different tap sizes per ANSI/ASME B1.20.1 for standard taper pipe threads with two, three, and four full threads

Nominal Pipe Size, *in.*	Number of Threads	Tap Size, in.									
		½	¾	1	1¼	1½	2	2½	3	3½	4
		Pipe Thickness, *in.*									
3	2	0.18	0.21	0.28	—	—	—	—	—	—	—
3	3	0.26	0.29	0.37	—	—	—	—	—	—	—
3	4	0.33	0.36	0.46	—	—	—	—	—	—	—
4	2	0.17	0.19	0.26	0.31	—	—	—	—	—	—
4	3	0.25	0.27	0.35	0.40	—	—	—	—	—	—
4	4	0.32	0.34	0.44	0.49	—	—	—	—	—	—
6	2	0.17	0.18	0.23	0.27	0.30	—	—	—	—	—
6	3	0.25	0.26	0.32	0.36	0.39	—	—	—	—	—
6	4	0.32	0.33	0.41	0.45	0.48	—	—	—	—	—
8	2	0.16	0.17	0.22	0.24	0.27	0.33	—	—	—	—
8	3	0.24	0.25	0.31	0.33	0.36	0.42	—	—	—	—
8	4	0.31	0.32	0.40	0.42	0.45	0.51	—	—	—	—
10	2	0.15	0.17	0.21	0.23	0.25	0.30	0.44	—	—	—
10	3	0.23	0.25	0.30	0.32	0.34	0.39	0.56	—	—	—
10	4	0.30	0.32	0.39	0.41	0.43	0.48	0.69	—	—	—
12	2	0.15	0.16	0.20	0.22	0.24	0.28	0.40	0.48	—	—
12	3	0.23	0.24	0.29	0.31	0.33	0.37	0.52	0.60	—	—
12	4	0.30	0.31	0.38	0.40	0.42	0.46	0.65	0.73	—	—
14	2	0.15	0.16	0.20	0.22	0.23	0.26	0.38	0.45	0.51	0.58
14	3	0.23	0.24	0.29	0.31	0.32	0.35	0.50	0.58	0.64	0.70
14	4	0.30	0.31	0.38	0.40	0.41	0.44	0.63	0.70	0.76	0.83
16	2	0.15	0.16	0.20	0.21	0.22	0.25	0.37	0.43	0.48	0.54
16	3	0.23	0.24	0.29	0.30	0.31	0.34	0.50	0.56	0.60	0.66
16	4	0.30	0.31	0.38	0.39	0.40	0.43	0.62	0.68	0.73	0.79
18	2	0.15	0.15	0.19	0.21	0.22	0.24	0.35	0.41	0.46	0.51
18	3	0.23	0.23	0.28	0.30	0.31	0.33	0.48	0.54	0.58	0.64
18	4	0.30	0.30	0.37	0.39	0.40	0.42	0.60	0.66	0.71	0.76
20	2	0.15	0.15	0.19	0.20	0.21	0.23	0.34	0.39	0.44	0.49
20	3	0.23	0.23	0.28	0.29	0.30	0.32	0.46	0.52	0.56	0.62
20	4	0.30	0.30	0.37	0.38	0.39	0.41	0.59	0.64	0.69	0.74
24	2	0.14	0.15	0.19	0.20	0.21	0.22	0.32	0.37	0.40	0.45
24	3	0.22	0.23	0.28	0.29	0.30	0.31	0.44	0.50	0.52	0.58
24	4	0.29	0.30	0.37	0.38	0.39	0.40	0.57	0.62	0.65	0.70

(continued)

Table 11-8 Pipe thicknesses required for different tap sizes per ANSI/ASME B1.20.1 for standard taper pipe threads with two, three, and four full threads (*Continued*)

Nominal Pipe Size, *in.*	Number of Threads	Tap Size, in.									
		½	¾	1	1¼	1½	2	2½	3	3½	4
		Pipe Thickness, *in.*									
30	2	0.14	0.15	0.19	0.19	0.20	0.21	0.31	0.34	0.37	0.41
30	3	0.22	0.23	0.28	0.28	0.29	0.30	0.44	0.46	0.50	0.54
30	4	0.29	0.30	0.37	0.37	0.38	0.39	0.56	0.59	0.62	0.66
36	2	0.14	0.14	0.18	0.19	0.20	0.21	0.30	0.33	0.35	0.38
36	3	0.22	0.22	0.27	0.28	0.29	0.30	0.42	0.46	0.48	0.50
36	4	0.29	0.29	0.36	0.37	0.38	0.39	0.55	0.58	0.60	0.63
42	2	0.14	0.14	0.18	0.19	0.19	0.20	0.29	0.32	0.34	0.36
42	3	0.22	0.22	0.27	0.28	0.28	0.29	0.49	0.44	0.46	0.48
42	4	0.29	0.29	0.36	0.37	0.37	0.38	0.54	0.57	0.59	0.61
48	2	0.14	0.14	0.18	0.18	0.19	0.20	0.29	0.31	0.32	0.35
48	3	0.22	0.22	0.27	0.27	0.28	0.29	0.42	0.44	0.44	0.48
48	4	0.29	0.29	0.36	0.36	0.37	0.38	0.54	0.56	0.57	0.60
54	2	0.15	0.15	0.18	0.19	0.19	0.20	0.29	0.30	0.32	0.34
54	3	0.22	0.22	0.27	0.27	0.28	0.29	0.41	0.42	0.44	0.46
54	4	0.29	0.29	0.36	0.36	0.36	0.37	0.54	0.55	0.57	0.59
60	2	0.15	0.15	0.18	0.19	0.19	0.20	0.28	0.30	0.31	0.33
60	3	0.22	0.22	0.27	0.27	0.28	0.28	0.41	0.42	0.44	0.46
60	4	0.29	0.29	0.35	0.36	0.36	0.37	0.53	0.55	0.56	0.58
64	2	0.15	0.15	0.18	0.18	0.19	0.20	0.28	0.30	0.31	0.33
64	3	0.22	0.22	0.27	0.27	0.27	0.28	0.41	0.42	0.44	0.45
64	4	0.29	0.29	0.35	0.36	0.36	0.37	0.53	0.55	0.56	0.58

NOTE: To convert inches (in.) to millimeters (mm), multiply by 25.4.

Table 11-9 Pipe thicknesses required for different tap sizes per AWWA C800 for standard corporation stop threads with two, three, and four full threads*

Nominal Pipe Size, *in.*	Number of Threads	Tap Size, in.						
		½	⅝	¾	1	1¼	1½	2
		Pipe Thickness, in.						
3	2	0.21	0.24	0.25	0.33	—	—	—
3	3	0.29	0.32	0.33	0.41	—	—	—
3	4	0.36	0.39	0.40	0.49	—	—	—
4	2	0.19	0.22	0.23	0.30	0.36	—	—
4	3	0.27	0.30	0.31	0.38	0.45	—	—
4	4	0.34	0.37	0.38	0.46	0.54	—	—

(continued)

Table 11-9 Pipe thicknesses required for different tap sizes per AWWA C800 for standard corporation stop threads with two, three, and four full threads* (*Continued*)

Nominal Pipe Size, *in.*	Number of Threads	Tap Size, in.						
		½	⅝	¾	1	1¼	1½	2
		Pipe Thickness, in.						
6	2	0.18	0.20	0.20	0.26	0.30	0.35	—
6	3	0.26	0.28	0.28	0.34	0.39	0.44	—
6	4	0.33	0.35	0.35	0.42	0.48	0.53	—
8	2	0.17	0.18	0.19	0.24	0.27	0.31	0.39
8	3	0.25	0.26	0.27	0.32	0.36	0.40	0.48
8	4	0.32	0.33	0.34	0.40	0.45	0.49	0.57
10	2	0.17	0.17	0.18	0.23	0.25	0.28	0.35
10	3	0.25	0.25	0.26	0.31	0.34	0.37	0.44
10	4	0.32	0.32	0.33	0.39	0.43	0.46	0.53
12	2	0.16	0.17	0.17	0.22	0.24	0.26	0.32
12	3	0.24	0.25	0.25	0.30	0.33	0.35	0.41
12	4	0.31	0.32	0.32	0.38	0.42	0.44	0.50
14	2	0.16	0.17	0.17	0.21	0.23	0.25	0.30
14	3	0.24	0.25	0.25	0.29	0.32	0.34	0.39
14	4	0.31	0.32	0.32	0.37	0.41	0.43	0.48
16	2	0.16	0.16	0.17	0.21	0.22	0.24	0.28
16	3	0.24	0.24	0.25	0.29	0.31	0.33	0.37
16	4	0.31	0.31	0.32	0.37	0.40	0.42	0.46
18	2	0.15	0.16	0.16	0.20	0.21	0.23	0.27
18	3	0.23	0.24	0.24	0.28	0.30	0.32	0.36
18	4	0.30	0.31	0.31	0.36	0.39	0.41	0.45
20	2	0.15	0.16	0.16	0.20	0.21	0.23	0.26
20	3	0.23	0.24	0.24	0.28	0.30	0.32	0.35
20	4	0.30	0.31	0.31	0.36	0.39	0.41	0.44
24	2	0.15	0.15	0.16	0.19	0.21	0.22	0.24
24	3	0.23	0.23	0.24	0.27	0.30	0.31	0.33
24	4	0.30	0.30	0.31	0.35	0.39	0.40	0.42
30	2	0.15	0.15	0.16	0.19	0.20	0.21	0.23
30	3	0.23	0.23	0.24	0.27	0.29	0.30	0.32
30	4	0.30	0.30	0.31	0.35	0.38	0.39	0.41

(*continued*)

Table 11-9 Pipe thicknesses required for different tap sizes per AWWA C800 for standard corporation stop threads with two, three, and four full threads* (*Continued*)

Nominal Pipe Size, *in.*	Number of Threads	Tap Size, in.						
		½	⅝	¾	1	1¼	1½	2
		Pipe Thickness, in.						
36	2	0.14	0.15	0.15	0.19	0.20	0.20	0.22
36	3	0.22	0.23	0.23	0.27	0.29	0.29	0.31
36	4	0.29	0.30	0.30	0.35	0.38	0.38	0.40
42	2	0.14	0.14	0.15	0.18	0.19	0.20	0.21
42	3	0.22	0.22	0.23	0.26	0.28	0.29	0.30
42	4	0.29	0.29	0.30	0.34	0.37	0.38	0.39
48	2	0.14	0.14	0.15	0.18	0.18	0.19	0.20
48	3	0.22	0.22	0.23	0.26	0.27	0.28	0.29
48	4	0.29	0.29	0.30	0.34	0.36	0.37	0.38
54	2	0.14	0.14	0.14	0.17	0.18	0.19	0.20
54	3	0.22	0.22	0.22	0.25	0.27	0.28	0.29
54	4	0.29	0.29	0.29	0.34	0.36	0.36	0.38
60	2	0.14	0.14	0.14	0.17	0.18	0.19	0.20
60	3	0.22	0.22	0.22	0.25	0.27	0.28	0.29
60	4	0.29	0.29	0.29	0.34	0.36	0.36	0.38
64	2	0.14	0.14	0.15	0.17	0.18	0.19	0.20
64	3	0.22	0.22	0.22	0.25	0.27	0.28	0.29
64	4	0.29	0.29	0.29	0.34	0.36	0.36	0.38

NOTE: To convert inches (in.) to millimeters (mm), multiply by 25.4.
* The corporation stop thread is commonly known to the trade as the Mueller thread.

In addition, grounding of household electrical services to the copper water service line may also result in stray current corrosion of the copper service or the ductile- or gray-iron main. AWWA policy opposes the grounding of electrical systems to pipe systems conveying drinking water to a customer's premises. AWWA further states that interior piping systems may be connected to an electrical service neutral and to a separate grounding electrode, provided these systems are electrically insulated from the water utility's pipe system.

11.8 HIGHWAY AND RAILROAD CROSSINGS

When casing pipe is specified for highway or railroad crossings, the project shall be completed in accordance with the applicable federal, state or provincial, and local regulations. In the case of railroad crossings, the project should comply with regulations established by the railroad company. Crossings are normally made by boring, jacking, or tunneling.

The casing pipe should be 6 in. to 8 in. (152 mm to 203 mm) larger than the outside diameter of the ductile-iron pipe bells. Depending on the manufacturer's recommendations,

Source: Ductile Iron Pipe Research Association

Figure 11-14 Direct service connection tapped through polyethylene adhesive tape and polyethylene film

ductile-iron pipe (carrier pipe) may be pulled or pushed through the casing. For integral restraint systems, the joints should be fully extended to take up the joint slack prior to making end connections. Insulating chocks, skids, or spacers should be placed on or under the carrier pipe to ensure approximate centering within the casing pipe and to prevent damage during installation. If timbers are used, they shall be of pressure-treated lumber and secured with stainless-steel strapping. Care must be exercised to avoid metal-to-metal contact between the carrier and casing pipe.

At very long crossings, it is often necessary to partially fill the space between the ductile-iron carrier pipe and the casing pipe (e.g., with sand) to prevent movement. In order to avoid the transfer of earth and live loads to the carrier pipe, the space between the carrier pipe and the casing pipe should not be filled completely. Pressure grouting of the entire annular space between the pipeline and casing pipe is not recommended unless grouting pressure is controlled by pressures below those which could cause buckling failure of the pipeline.

11.9 TRENCHLESS APPLICATIONS

There are several methods of installing ductile-iron pipe in trenchless applications, including directional drilling (see Chapter 12) and microtunneling. Ductile-iron pipe, manufactured in accordance with ANSI/AWWA C151/A21.51, Ductile-Iron Pipe, Centrifugally Cast, can be installed using various pipe pushing/ pulling methods and directional drilling. The methods involve forming a hole slightly larger than the outside diameter of the pipe joint, after which the ductile-iron pipe is pushed or pulled through the bore path. When pipe is pulled into position, restrained joints are utilized. Also, specially designed and manufactured pipe for very high pushing loads, required in microtunneling and other construction procedures, is currently available.

11.10 SUBAQUEOUS INSTALLATIONS

When it is necessary to cross a body of water requiring only a small deflection in the joints and joint restraint is not a consideration, standard push-on or mechanical joint pipe can be used. If the angle of deflection in the joint necessary to follow the contour of the riverbed is great or if changing bottom conditions are anticipated, ball-and-socket pipe or fittings, which will deflect up to 15°, should be used. A combination of restrained and river-crossing joints may be used depending on bottom conditions and service requirements.

There are several methods of installing subaqueous ductile-iron pipe. Ductile-iron pipe can be assembled in sections of three or four lengths, either on shore or on the deck of a barge, attached to a strongback. Then the assembly can be lowered to the bed of the stream where the sections are connected by divers. Ball-and-socket pipes use joints with positive locking devices. This type of pipe can also be assembled on a chute affixed to a barge and lowered into position as the assembly progresses (Figure 11-15).

Another method of installation is to assemble the pipeline on shore and either float it into position by means of barrels or floats attached to the pipe (Figure 11-16), which are punctured or released in a controlled fashion when the pipe reaches the desired position,

a b c

Source: Ductile Iron Pipe Research Association

Figure 11-15 a, b, c Saddle service connection through polyethylene adhesive tape and polyethylene film

Source: Ductile Iron Pipe Research Association

Figure 11-16 Ball-and-socket ductile-iron pipe used in a subaqueous installation

or drag it into position along the bottom. Regardless of the method used, joints should not be allowed to become overly deflected or subjected to excessive beam load during the installation process.

A somewhat similar method is to assemble the pipe on shore, attach floats, and pull the pipe down skids into the water as each length is connected. The line extends further into the water as each successive length is laid, and the finished line is submerged in the manner described previously.

Subaqueous lines laid in navigable streams must be placed in trenches and covered to protect them from damage or displacement by ship or boat traffic. Where applicable, procedures should conform to appropriate governmental regulations.

After a river crossing or other subaqueous crossings are installed, it is normally suggested that the line be backfilled before testing and allowed to settle for a reasonable period before end connections are made.

11.11 OTHER INSTALLATIONS

Installations that require special attention, techniques, and materials have not been considered in this section. These installations require special considerations based on many influencing factors. This is best accomplished by competent engineering design in consultation with representatives of the materials manufacturers. Such installations include the following:

- Piping through rigid walls
- Certain piping on supports aboveground or underground (see Chapter 4 of this manual and Figure 11-17)
- Piping requiring insulation
- Piping across bridges (Figures 11-18 and 11-19)
- Treatment plant or pumping station piping (Figure 11-20)
- Pipe in unstable soil (Figure 11-21)
- Industrial piping
- Piping through geologically hazardous areas
- Piping in high-density stray current environments
- Trenchless installations (sliplining of larger mains, horizontal directional drillings, microtunneling, pipe bursting, etc.)

Source: Ductile Iron Pipe Research Association

Figure 11-17 Subaqueous pipe being floated into position

Source: Ductile Iron Pipe Research Association

Figure 11-18 Ductile-iron pipe installed on a bridge

Source: Ductile Iron Pipe Research Association

Figure 11-19 Ductile-iron pipe spanning waterway

Source: Ductile Iron Pipe Research Association

Figure 11-20 Pump station piping

Source: Ductile Iron Pipe Research Association

Figure 11-21 Ductile-iron pipe being installed in a deep trench

REFERENCES

American National Standards Institute/American Society of Mechanical Engineers. 2013. Pipe Threads, General Purpose, Inch. ANSI/ASME B1.20.1. New York: ASME.

American National Standards Institute/American Water Works Association. 2023. Disinfecting Water Mains. ANSI/AWWA C651. Denver: AWWA.

American National Standards Institute/American Water Works Association. 2021. Underground Service Line Valves and Fittings. ANSI/AWWA C800. Denver: AWWA.

American Water Works Association. 2016 (5th ed.) M17: *Installation, Field Testing, and Maintenance of Fire Hydrants.* Denver: AWWA.

This page intentionally blank.

Chapter 12

Trenchless Installation—Horizontal Directional Drilling

12.1 INTRODUCTION

Since its first use in the early 1960s, horizontal directional drilling (HDD) has become a fast-growing trenchless technology that provides certain benefits over traditional open-cut installations for pipelines transporting water, wastewater, and other liquids. HDD minimizes disruption to vehicular and maritime activities, can require less time and working space, and reduces surface disruptions and restoration requirements without interference with existing infrastructure (DIPRA 2022). The actual methodology involves three major phases. Using a slant drill rig, specifically designed for HDD, the first phase consists of drilling a small-diameter pilot bore (Figure 12-1) entering the ground surface on an angle that normally ranges between 8° and 20°. The oilfield style drill rods are surface guided along a predetermined designed directional path (Directional Crossing Contractors Association 1995).

The second phase consists of enlarging the borepath to a diameter suitable for installation of the pipe and may actually require multiple passes of sequentially larger-diameter cutters to enlarge the borepath gradually (Figure 12-2). The cutting and enlarging actions result in a homogeneous mixture of native soil and drilling fluid, creating a lubricated path for the pipe to follow while providing some support for the surrounding soil. Recommendations for the reamed interior diameters for the borepath are given in Table 12-1 (Bennett et al. 2004). Generally, multiple reaming passes to gradually enlarge the borepath produce the best results in most soil conditions.

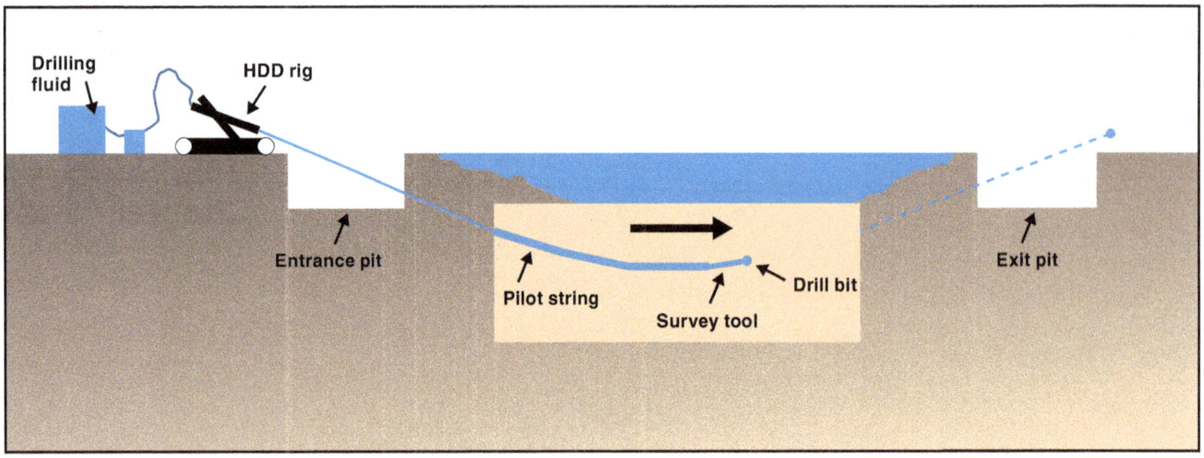

Source: Ductile Iron Pipe Research Association

Figure 12-1 Pilot bore

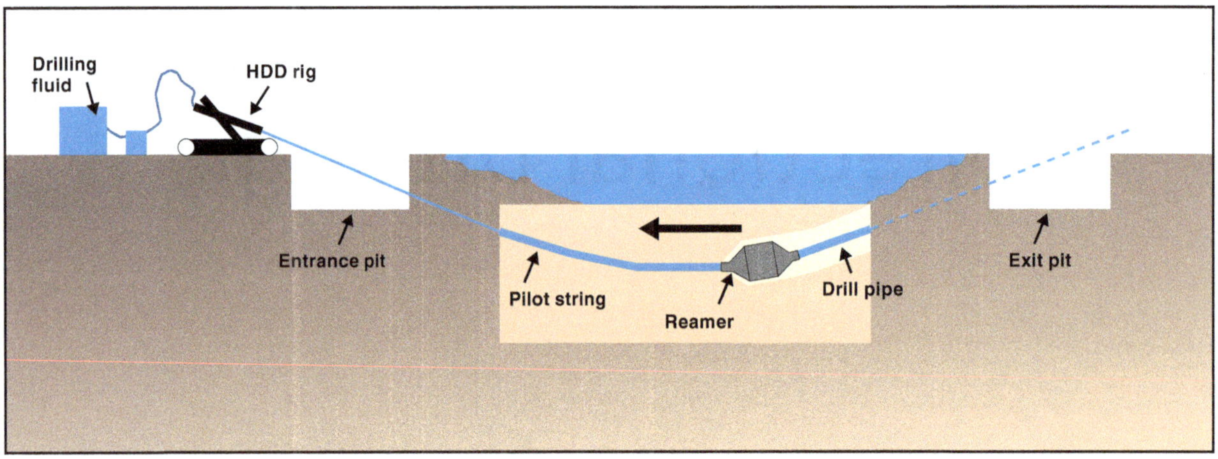

Source: Ductile Iron Pipe Research Association

Figure 12-2 Prereaming

Table 12-1 Recommended borepath reamed inside diameters

Nominal Diameter	Reamed Diameter
<8 in.	Bell outside diameter +4 in.
8–24 in.	Bell outside diameter × 1.5
>24 in.	Bell outside diameter +12 in.

The third phase is the pipe pullback (Figure 12-3), in which the pipe is assembled using either the cartridge (one pipe section at a time) or assembled-line (all joints are pre-assembled and pulled in as a single unit) technique. The cartridge method is preferred in locations where right of way or easements are limited. It allows for reduction in pulling loads and ease of handling, especially for larger diameter pipes, and allows for assembly in restricted project sites that have limited layout area, a requirement of the assembled-line technique.

This chapter will provide recommended guidelines based on industry best practices for the safe and proper installation of ductile-iron pipelines for raw water, potable water, and wastewater using HDD.

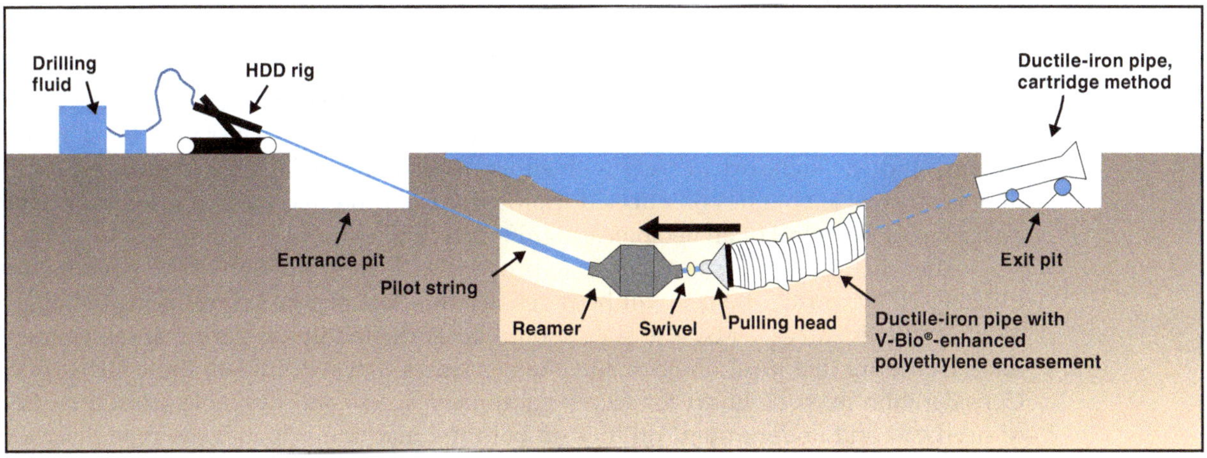

Source: Ductile Iron Pipe Research Association

Figure 12-3 Pipe pullback

Table 12-2 HDD applications and rationales

Application	Rationale
Applications with high working pressures	Ductile-iron pipe has pressure ratings up to 350 psi, material is elastic, and strength does not degrade over time.
Congested, commercialized urban areas with restricted site allocation	Ductile-iron pipe with standard joint lengths of 18 ft and 20 ft can be installed using the "cartridge" assembly method, reducing socioeconomic costs (e.g., over the space required for welded or fused pipe systems).
Deep-bury conditions and high external pressures	Ductile mechanical properties allow for very high, long-term pipe stiffness values.
Water quality/permeation concerns due to contaminated soils	Ductile-iron pipe wall is impermeable to aromatic hydrocarbons. Optional joint gaskets are available.
Extended service life	Life-extending cathodic protection can be applied to critical ductile-iron pipelines. Viscoelastic materials, with a design based on a 50-year extrapolated hydraulic design basis, have no life-extending options.
Ability to locate from the surface	Made of metallic material, ductile-iron pipe is easily located by conventional pipe locators.

HDD using flexible restrained-joint ductile-iron pipe is believed to have occurred for the first time in 1995. Hemlock Directional Drilling in Torrington, Conn., installed approximately 280 ft of 8-in. flexible restrained-joint ductile-iron pipe under a busy intersection in Avon, Conn. Since that first installation of flexible restrained-joint ductile-iron pipe, owners and engineers are specifying ductile-iron pipe for applications where the inherent strengths of ductile-iron pipe are best suited (see Table 12-2).

12.2 PREDESIGN INVESTIGATION

HDD inherently has a higher degree of risk associated with the methodology than conventional open-cut construction. Because of this added risk, it is prudent for the owner and engineer to determine as much information as possible about surface and subsurface features that could potentially impact the method of installation, the risk, and ultimately the cost for each method.

12.2.1 Surface Features

Surface features are most often the primary rationale for owners and engineers to consider HDD or other trenchless methodologies. The most common surface features that suggest consideration of HDD construction include waterway crossings: e.g., lakes, bays, rivers, streams, and wetlands; urban areas with either dense commercial or residential construction; heavy traffic volumes on affected thoroughfares; protected natural growth or habitats; areas of historical or archeological significance; and others.

After the decision has been made to proceed with an evaluation of the potential for utilizing HDD construction, it will be necessary to select one or several potential alignments and exit/entry points. The sites should be evaluated by preparing an assessment for each, noting the most obvious site characteristics that will impact the construction. Consideration must be given for heavy equipment access and available work area free of overhead and underground utilities on both the machine side and the pipe insertion side of the proposed borepath. Other surface details worthy of notation include elevation and grade changes at both ends of the project; any visible evidence of geology, e.g., rock ledges, visible rocks, or boulders; and, wherever possible, information from the owners' geographic information system (Horizontal Directional Drilling Design Guideline Task Committee 2014).

12.3 HDD ASSEMBLY METHODS FOR FLEXIBLE RESTRAINED-JOINT DUCTILE-IRON PIPE

12.3.1 Assembled-Line Method

Flexible ductile-iron pipe can be assembled using two distinctively different methods, assembled-line or cartridge. The first method, the assembled-line (Figure 12-4), is the only method available for pulling in continuously fused or welded pipe materials, such as high-density polyethylene, fusible polyvinyl chloride, or steel pipe. The process involves assembling all the pipe joints prior to the start of pipe pullback. This assembly method requires a staging area that is substantially aligned with the centerline of the borepath extending from the entrance to the borepath (on the pipe insertion side), a distance equivalent to the length of the crossing. The pipe is generally placed on pipe rollers, albeit other methods—using either polyethylene-lined or unlined pipe ramps with no rollers—have proven successful in a number of installations, and is pulled back as a single unit. The assembled-line method allows the slant drill rig to pull continuously, with the exception of the downtime required to remove drill rods every 12 to 40 ft depending on the size of the drill rig. This method also allows the pipeline to be pressure tested before the pipe is pulled into the reamed hole.

12.3.2 Cartridge Assembly Method

The second method, which has proven invaluable in congested urban areas and areas where alignment prevents the use of the assembled-line method, is the cartridge method (Figure 12-5). This method of assembly takes advantage of the quick assembly features of contemporary flexible restrained-joint ductile-iron pipe. By assembling one joint at a time, the staging area required for assembly is reduced to as small as a single length of pipe. After the joint is assembled, the drill rig pulls back the equivalent of that joint length, and then the next joint is assembled. This process is repeated until the entire line is completely assembled and pulled back. It sometimes takes little more time to assemble ductile-iron

Source: Ductile Iron Pipe Research Association

Figure 12-4 Assembled-line method

lines in this fashion than it does to remove and place the drill rods as the drill string is withdrawn.

Because of the differences between the length of drill rods and standard joint lengths of ductile-iron pipe, it may thus sometimes be advantageous to provide a staging area that can accommodate the assembly of multiple joints that will come closer to matching lengths of drill rods; for example, 2 each, 30-ft drill rods are equivalent in length to 3 each, 20-ft joints of ductile-iron pipe.

12.4 SURVEY

When the final route has been selected, the owner and engineer need to provide a survey of the planned horizontal and vertical alignment. Cognizant of the desirability of reducing project risk and ultimately also cost, this detail survey needs to provide plan and profile sheets that offer comprehensive details of the crossing including but not limited to the following:

- Site topography
- Existing utilities

Source: Ductile Iron Pipe Research Association

Figure 12-5 Cartridge method

- Right of way/easement
- Identified wetland
- Waterway—high/low
- Site size and access
- Streets—traffic counts
- Nearby structures
- Boring locations

12.5 SUBSURFACE ANALYSIS

A subsurface analysis should always start with the identification of existing utilities. There are a number of methods utilized for identifying buried pipelines, including pipe locators, ground-penetrating radar, vacuum excavation, and seismic survey. It is good practice that any existing pipeline that encroaches within 25 ft of the proposed centerline alignment of the new pipeline be examined to confirm its elevation. The most common method for verifying the location of buried pipelines is by using vacuum excavation. This system uses compressed air to loosen the in-situ soil, which is then removed by suctioning into an industrial vacuum truck.

12.5.1 Geotechnical

Geotechnical discovery is extremely important when considering HDD. To understand the importance of the geology of the borepath, it is essential to understand the dynamics of the HDD process. The drilling fluid used in HDD has several critical functions that include the following.

1. In a rock borepath, the drilling fluid is the power source for running the mud motor, which is cutting the rock as the bits rotate and the drill string is advanced.
2. In either soil or rock, the drilling fluid acts as a coolant for the tool and for clearing up any residual buildup on the drill bits and reamer.
3. In either soil or rock, the drilling fluid acts as a transport medium for the spoil, which is blended with the cuttings at the reamer.
4. In a soil borepath, the fluid is used to maintain borepath stability by forming a slurry cake on the interior wall of the borepath. This minimizes the chance of the borepath collapsing to bind on the pipe.

HDD contractors need accurate information about the geology along the centerline of the borepath, as it affects the selection in cutting tools/reamers and more importantly, the design of the drilling fluid. Although drilling fluid is mostly water, without properly designed drilling fluid that includes bentonite and/or polymer additives, it may be very difficult to maintain a quality borepath in some soil environments (Table 12-3).

In general, clay soils are considered to provide the best geology for HDD. When clay is anticipated, the contractor can add a polymer to the drilling fluid. The polymer binds to the exterior of the clay, thus preventing the clay from binding together to form larger clay

Table 12-3 Use of HDD in various soil types

Soil Type	HDD Drillability	Drilling Fluid
Soft to very soft clays, silt, and organic deposits	yes	Water/bentonite/polymer
Medium to stiff clays and silts	yes	Water/bentonite/polymer
Hard clays and highly weathered shales	yes	Water/bentonite/polymer
Very loose to loose sands	yes	Water/bentonite
Medium to dense sands (below water table)	yes	Water/bentonite
Gravels and cobbles less than 2 to 4 in.	marginal	Water/bentonite
Soils with significant cobbles and boulders or obstructions larger than 4 to 6 in	marginal	Water/bentonite
Weathered rocks, marls, chalks, and firmly cemented soils	yes	Water/bentonite
Significantly weathered to unweathered rocks	yes	Water/bentonite

Source: Portions of Table 12-3 reproduced from Najafi and Gokhale 2021.

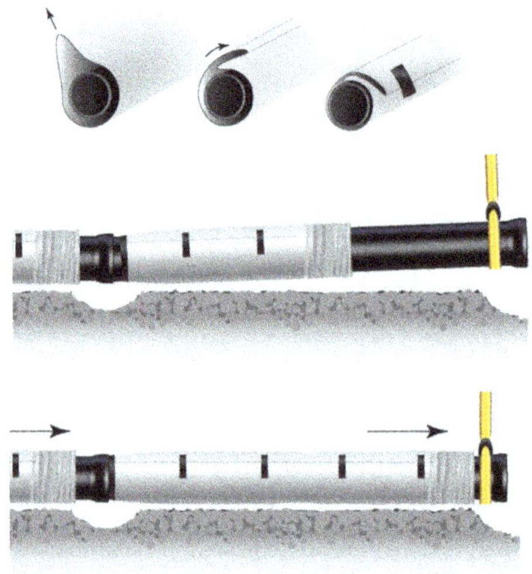

Source: Ductile Iron Pipe Research Association

Figure 12-6 Polyethylene encasement

clogs. In sharp contrast to clay, material containing gravel and cobble is much more difficult. During the drilling process, the drilling fluid should be flowing at both ends of the borepath. This flow indicates that there is an unobstructed borepath (Najafi and Gokhale 2006).

12.5.2 Other Subsurface Considerations

Contaminated soils are generally not an issue with ductile-iron pipe as the entire pipe surface is impermeable to aromatic hydrocarbons. The only components of the pipe that are not impermeable to contaminated soils are the pipe gaskets. By substituting either hydrocarbon-resistant nitrile or least permeable fluorine rubber gaskets in place of the regular styrene butadiene rubber gaskets in such locations, the owner and engineer can have confidence that their potable water delivery systems are not compromised.

Chapter 10 discusses the 10-point soil evaluation procedure (Section 10.4) and design risk models for selecting the appropriate level of protection for ductile-iron pipelines. Utilizing these tools, the owner and engineer can arrange for soil tests and discuss the consequences of failure.

Proper application of polyethylene encasement per the requirements of ANSI/AWWA C105/A21.5, Polyethylene Encasement for Ductile-Iron Pipe Systems, method A for installation below the water table and modified per the illustration shown in Figure 12-6 has proven to be successful in HDD installed ductile-iron pipelines (Bonds 2022a).

Using modified method A, there will be full circumferential wraps of tape around the barrel of the pipe at 2-ft intervals. An alternative fastening method would be to continuously wrap tape spirally around the polyethylene on about 1-ft centers (Figure 12-7). This permits air to escape at the overlap as the hydrostatic pressure of the drilling fluid applies a uniform pressure around the pipe.

12.6 BOREPATH DESIGN

The design of the borepath is dependent on the obstacle that the HDD installed ductile-iron pipeline needs to avoid, as well as the right-of-way/easement obtained by the owner for the installation. When a waterway is involved, consideration needs to be given to the dynamic changes that occur with the alignment of the flow or channel and scour of the bottom. It is best practice to generally keep the HDD installation as straight and as short as possible.

Source: Ductile Iron Pipe Research Association

Figure 12-7 HDD of ductile iron pipe encased in polyethylene showing spiral wrap of tape securing the film to the pipe

12.6.1 Radius of Curvature

When using flexible restrained-joint ductile-iron pipe, the borepath radius is, in most sizes, not the limiting factor. The restrictions on the radius are self imposed by the HDD contractors, who prefer to limit the radius based on the size of the drill rods used. This restriction is placed on their drill rods because of the dynamic stresses that can strain age the steel drill rods. These dynamic stresses are a result of the combination of tension from the pulling force, flexural tension and flexural compression due to bending, and shear stresses resulting from the rotation of the drill rods.

Independent of the contractors' concern for tight radii causing dynamic fatigue of their drill rods, some drilling contractors prefer to limit the design to a radius based on the nominal diameter of the new product pipe. HDD contractors suggest using a radius of between 50 and 100 ft for each diameter-inch. For example, for 36-in. ductile-iron pipe, the minimum radius is 1,800 ft based on 50 ft per nominal inch; for the 100 ft per diameter-inch basis, the 36-in. pipe would require a radius of 3,800 ft. However, borepath curvature should not be less than the value allowed by the manufacturers.

12.7 PIPE DESIGN: INTRODUCTION

Pipe design for HDD pipeline installations can be different from conventionally buried pipelines because of high pulling/tension loads, bending, and external fluid pressures exerted on pipelines during the installation process. In some cases, these loads can be higher than direct-buried design service loads. Pipe and joint strengths, flexibility, and wall thickness must be selected such that the pipeline can be installed and operated with reasonable safety and security. Analysis of the loads and stresses for HDD pipe can be most easily accomplished by examining the two potentially different scenarios of installation and pressure testing/operation.

During HDD installation of other pipe materials with fully welded or fused segments (e.g., continuously welded joint steel or continuously fused polyethylene), the pipe segments are subjected to tension, bending, shape-changing effects due to bending, and external pressure as they are pulled through curved, prereamed borepaths. The stresses and failure potential of the pipe are a result of a complex interaction of these loads and conditions. In the case of HDD with flexible restrained-joint ductile-iron pipes, however, the design situation is somewhat more simplified in that the flexible joints in effect act as hinges that essentially remove the complication of significant bending loads on pipes and joints, as well as stress and shape effects. Additionally, flexible restrained-joint ductile-iron pipes have a somewhat greater effective unit weight or bulk density than some other material-type pipes in a heavy fluid borepath that, per the Archimedes principle, helps in reducing or neutralizing upward buoyancy forces of particularly empty or partially filled pipes exerted against the path in the pulling operation. This condition has resulted in the observation of actual pulling loads for some ductile-iron HDD pulls that are less than encountered for lighter alternate material pipes in comparable installations.

In order to determine if a given ductile-iron pipe specification is adequate, estimated HDD pulling and external pressure installation loads and deflections should be determined and compared to the manufacturer's allowable recommendations so that the adequacy of the pipe and joint system can be reasonably ensured before the operations proceed. The purpose of this section is to describe the loads that act on a pipeline during installation by HDD and to present methods that can be used to estimate these loads.

12.7.1 Estimation of Pulling Force

Tension on the pull section results from friction between the pipe and the borepath, any fluidic drag or cohesion from viscous drilling fluid otherwise surrounding the pipe, and a component of the (submerged) weight of the pipe as it is pulled through the borepath. In addition to these forces that act within the borepath, friction from the portion of the pull section remaining on the surface (supported on ground or sometimes on rollers) also contributes to the tensile load on the pipe. Additional loads that the horizontal drilling rig must overcome during pullback result from the length of the drill string in the hole and the cutting/reaming assembly that precedes the pulled pipe section. These loads do not act on the pull section and therefore have no impact on pipe stresses. Nonetheless, if a direct correlation with the overall rig pulling force is desired, loads resulting from the reaming assembly and drill string must be estimated and added to the tensile force acting on the pull section.

Attempts to calculate the tensile load required to install even a ductile-iron pipeline by HDD can get complicated because of varying geometry of approaches and borepath, variations in the cartridge or assembled-line methods chosen to install the pipe, properties of the pipe being installed, subsurface materials, drilling fluid, and so on. Because of characteristics of the unique and chain-like ductile-iron pipe HDD system as explained in Section 12.7, the previous methods, calculations, or computer software developed based on HDD experience with welded pipe strings, by their nature only forcibly bent against a borepath, are not necessarily applicable to HDD with flexible restrained-joint ductile-iron pipe. Regardless of the method used to calculate HDD pulling load, the design engineer should be aware that pulling loads are affected by numerous variables, many of which are dependent on site-specific conditions and individual contractor practices. These include prereaming diameter, stability of the borepath, removal of cuttings, soil and rock properties, drilling fluid properties, and the effectiveness of buoyancy control measures. Such variables cannot easily be accounted for in a theoretical calculation method designed for use over a broad range of applications. For this reason, theoretical calculations are of limited benefit unless combined with engineering judgment derived from experience in HDD construction. Also, of course, if the borepath is inadequately prepared or unstable and results, for example, in collapse around any piping material, very high or prohibitive pulling loads can result.

Fortunately, there is some history and experience over many years of pulling ductile-iron pipelines in wet, muddy HDD installations that have resulted in some helpful rules of thumb. Basically, if the effective unit weight or unit buoyancy of the pipe in a reasonably well prepared and stable liquid soil-filled borepath is multiplied by a factor in the range of 0.4 to 0.6, and that multiplicand is in turn multiplied by the total length of the pulled pipe, a good approximation of actual pulling load required is obtained.

The effective weight (or unit buoyancy) of the pipe is in turn the unit weight of the pull section minus the unit weight of any drilling fluid and soil cuttings displaced by the pull section. This is typically expressed in lb/ft. The unit weight of the pull section should include not only the product pipe and lining but also any contents (ducts, weighting, internal water used for ballast, and so forth). Calculating the weight of drilling fluid and the like displaced by the pipe requires that the density of the fluid either be known or assumed. For HDD installations, drilling fluid density will range, for example, from approximately 9 to 11 lb/gal, representing a specific gravity (Sp. Gr.) range of approximately 1.1 to 1.3 (some authorities thus suggest an assumption of 1.2 for Sp. Gr. of fluid to be used in buoyancy calculations). Where use of a high-end value for fluid density is warranted for a conservative analysis, 12.0 lb/gal or Sp. Gr.~ 1.4 probably represents a reasonable upper limit.

12.7.2 Limitations on Length of Pull and Pulling Load

With a minimum material yield strength of 42,000 psi and the dimensions/cross-sectional areas (based on outside diameter and thickness) of typical pressure class ductile-iron pipe used for HDD, the tensile yield strength of the pipe barrel itself can be calculated to be far beyond what is normally required for even very long HDD pulls. However, it is necessary also to consider the strength of the specific flexible restrained joints to be utilized in straight alignment and also at the maximum joint deflection angle designed/anticipated per the design of the pulling approach (in the case of "assembled-line pulls") and borepath. Proprietary flexible push-on restrained joints for ductile-iron pipe from many manufacturers—which differ somewhat in dimensions, substantial pressure/pull ratings, and deflection capabilities—are depicted in Appendix A. It is necessary for the designer and installer to compare the specific manufacturer's recommendations for allowable pressure, deflection or curve radius, pulling loads, and so forth of the products proposed to the field requirements for the specific application involved to ensure adequacy.

The actual maximum length of the borepath may be controlled by any of the following key components of the HDD process. These components include the drill rig fluid requirements and usage, drill rig capacities (thrust and torque), ground conditions, and the allowable pulling capacity of flexible restrained-joint ductile-iron pipe. The approximate maximum borepath length based on the allowable pulling loads can be calculated using the equation below and solving for the length as follows:

$$F = L \times W \times \varphi \qquad \text{(Eq 12-1)}$$

Where F = force required or maximum allowable pulling loads
L = length of pull
W = buoyant weight of pipe/unit length
φ = coefficient of friction (0.4 to 0.6)

$$\text{Solve for } L \text{ — length: } L = \frac{F}{W \times \varphi}$$

12.7.3 Bending

The curvature and slopes of the approaches, borepaths, and exits should be so designed and installed such that the flexible restrained ductile-iron joints are not deflected beyond the manufacturer's maximum allowable limits. This being the case, stresses and forces induced by bending are not a significant concern for ductile-iron pipe with flexible restrained joints.

12.7.4 External Pressure

During HDD installation, the pull section can be subjected to external pressure from several sources:

- hydrostatic pressure from the weight of the drilling fluid surrounding the pipe in the annulus of the borepath,
- hydrokinetic pressure required to produce drilling fluid flow from the reamer assembly through the annulus of the borepath back to the mud recirculation pits at the machine end and the pipe insertion end of the HDD installation, and
- hydrokinetic pressure produced by surge or plunger action involved with pulling the pipe into the reamed borepath.

Hydrostatic pressure is dependent on the height of the drilling fluid column acting on the pipe and the density of the drilling fluid that surrounds the pipe. Drilling fluid density values are discussed in Section 12.7.1. The height of the drilling fluid column at any given location along the borepath is typically equal to the elevation difference between that location and the point at which there is no drilling fluid in the reamed hole. Typically, but not always, drilling fluid extends to the entry or exit point, whichever is lower.

External hydrostatic pressure is thus the product of the height of the elevation difference (e.g., converted to inches) multiplied by the density of the fluid in comparable units (e.g., a 1.2 Sp. Gr. drilling fluid would have a density equal to 1.2 × [62.4 lb/ft^3]/[1,728 in.^{3}/ft^3] = 0.0433 lb/in.^{3}), giving the external pressure result in units of lb/in.^{2}

Hydrokinetic pressure required to produce drilling fluid flow might be calculated using annular flow pressure loss formulas. However, these results are dependent on complex drilling fluid properties, flow rates, and hole configuration, and because of uncertainties involving these parameters, they often require a substantial application of engineering judgment to determine a reasonable value. In most cases, annular flow during pullback can be assumed to be very low velocity with low pressure losses.

Hydrokinetic pressure caused by surge or plunger action talked about by some authorities likewise cannot be readily calculated and can only be estimated using engineering judgment and experience.

Critical Buckling Pressure for Ductile Iron Pipe (Bonds 2022b) can be used to evaluate the potential for a pipe structure becoming suddenly unstable and collapsing or buckling because of external pressure and/or vacuum. Ductile iron pipe's strength and high long-term modulus of elasticity provide resistance to collapse compared to thinner steel or plastic pipes with very low long-term elastic moduli. However, it may be advisable to check for the impact of external hydrostatic pressure from deep crossings, especially in larger-sized pipes. Of course, the greatest risk of collapse due to external pressures for any HDD-installed pipeline is when there is no offsetting internal service pressure inside the pipeline.

The "earth load" on pipelines installed by HDD is generally considered to be a "tunnel load," where the resulting soil pressure is less than geostatic stress. While ductile-iron pipes again have greater long-term stiffness than other materials commonly used, some references (e.g., Bizier 2007 or ASTM 2020) are available to determine earth loads on flexible pipes installed by HDD methods if such determination is desired for extreme circumstances.

12.8 TESTING AND SERVICE PRESSURES/LOADS

The testing and operating loads imposed on a ductile-iron pipeline installed by HDD are not significantly different from those imposed on a conventionally installed pipeline. As a result, conventional design procedures for calculating loads and limiting stresses as explained elsewhere in this manual can be applied. The internal hydrostatic pressure from the depth of the HDD installation should be considered when determining the maximum internal design and test pressures, as well as the adequacy of the chosen pipe and joints. Because of the HDD installation process, most joints in an HDD pull will be effectively "extended" to thrust resisting position; however, joint extension may be necessary, particularly in connection areas, to minimize thrust movements (see Chapter 8 for more information).

12.9 THERMAL EXPANSION/CONTRACTION AND PULLING LOAD "RECOIL"

Again because of the aforementioned very high, long-term elastic material modulus and a very low thermal coefficient of expansion, ductile-iron HDD-installed pipelines generally undergo relatively little thermal and pulling recoil movement and stressing compared to some other pipe materials used in HDD. However, the coefficient of thermal expansion of ductile-iron pipe that can be used for any desired thermal analyses based on specific known jobsite conditions is 0.0000062 (0.62 × 10^{-5}) per degree F or 0.0000112 (1.12 × 10^{-5}) per degree C.

REFERENCES

ASTM. 2020. Standard Guide for Use of Maxi-Horizontal Directional Drilling for Placement of Polyethylene Pipe or Conduit Under Obstacles, Including River Crossings. ASTM F1962. West Conshohocken, PA: ASTM International.

Bennett, D., S.T. Ariaratnam, and C.E. Como. 2022. Horizontal Directional Drilling: Good Practices Guidelines. HDD Consortium.

Bizier, P., ed. 2007 (2nd ed.). *Gravity Sanitary Sewer Design and Construction*. ASCE Manual and Reports on Engineering Practice, No. 60. Reston, VA: ASCE.

Bonds, R.W. 2022a. *Horizontal Directional Drilling with Ductile Iron Pipe*. Birmingham, AL: DIPRA.

Bonds, R.W. 2022b. *Critical Buckling Pressure for Ductile Iron Pipe.* Birmingham, AL: DIPRA.

Directional Crossing Contractors Association. Guidelines for a Successful Directional Crossing Bid Package. 2006. Dallas, Texas: DCCA.

Ductile Iron Pipe Research Association (DIPRA). 2022. *Applications: Horizontal Drilling.* chrome-extension://efaidnbmnnnibpcajpcglclefindmkaj/https://dipra.org/wp-content/uploads/2025/02/Horizontal-Directional-Drilling-with-Ductile-Iron-Pipe.pdf. Accessed June 9, 2025.

Horizontal Directional Drilling Design Guideline Task Committee. 2014 (2nd ed.). *Pipeline Design for Installation by Horizontal Directional Drilling*. Reston, VA: ASCE.

Najafi, M. 2006). *Trenchless Technology: Pipeline and Utility Design, Construction, and Renewal.* New York: McGraw-Hill.

Chapter 13

Seismic Design Guidelines for Ductile-Iron Pipe

13.1 INTRODUCTION

One of the most critical resources we have is water. Water is crucial to public health, firefighting, and societal development. People can survive for only a few days without water. A reliable and resilient water supply is necessary to sustain the economy and support responsible growth within a region. A resilient water supply will benefit both current and future generations. Earthquake-resistant ductile-iron pipe (ERDIP) is a component of a resilient water supply.

Presidential Policy Directive (PPD) 21 defines the term *resilience* as "the ability to prepare for and adapt to changing conditions and withstand and recover rapidly from disruptions. Resilience includes the ability to withstand and recover from deliberate attacks, accidents, or naturally occurring threats or incidents" (White House Office of the Press Secretary 2013). Today, approximately 91 percent of Americans live in areas subject to natural disasters or terrorism (Ripley 2009). There are many kinds of disasters, both natural and manmade, that can impact a water supply. Water infrastructure is one of the top four most important types of critical infrastructure following disasters (FEMA 2014). Of all types of natural disasters or other events that can impact water systems, perhaps the most significant are earthquakes. The impact from large earthquakes on water systems and the related communities they serve can be significant. The US Geological Survey (USGS) estimates that over 50 percent of the land mass in the contiguous United States would be affected by at least strong shaking (modified Mercalli intensity [MMI] index VI)

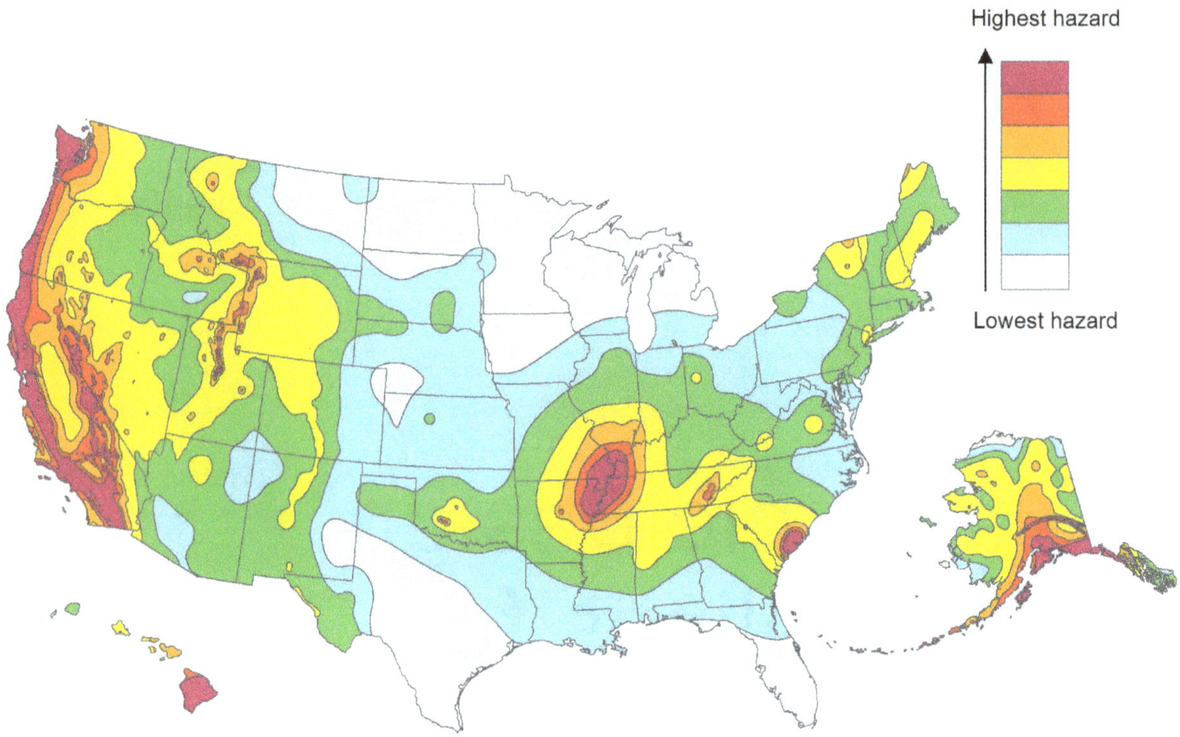

Figure 13-1 2018 Long-term National Seismic Hazard Map. This earthquake hazard map shows peak ground accelerations, having a 2 percent probability of being exceeded in 50 years, for a firm rock site. The map is based on the most recent USGS models for the conterminous United States (2018), Hawaii (1998), and Alaska (2007) (USGS 2019).

from relatively infrequent ground motions from earthquakes (Jaiswal et al. ND). It is estimated that 12 percent of fire stations, 12 percent of hospitals, and 15 percent of all schools are located in regions of the contiguous 48 states where there is a 2 percent chance that severe shaking (MMI VIII intensity) or higher could occur in the next 50 years. As Jaiswal et al. note, "At such levels of shaking, even well-designed earthquake-resistant construction could experience some damage, while unreinforced masonry or nonductile concrete constructions are expected to perform extremely poorly, that is, experiencing partial or complete collapses" (2015). The 2018 Long-term National Seismic Hazard Map, based on work by Peterson et al. (2020), is shown in Figure 13.1. Peterson et al. indicate that one in three Americans may be exposed to very strong shaking (MMI VII intensity). They also state that due to population growth and increased hazard estimates, "more people live or work in areas of high or moderate seismic hazard than ever before, leading to higher risk of undesirable consequences from forecasted future ground shaking."

Earthquakes impact water systems in a variety of ways, including damaging pipelines, pump stations, storage reservoirs, and water treatment plants. For example, in 1994, an earthquake in southern California affected three major transmission systems and damaged pipes in more than 1,000 locations (DHS 2015), and in Baja California, Mexico, a magnitude (M) 7.2 earthquake damaged two water treatment plants in 2010.

A reliable water supply is also vital for firefighting. The second largest fire loss in the history of the United States occurred with 1906 San Francisco Earthquake (NFPA 2016). Firefighting capabilities following the earthquake were rendered useless due to the significant failures in the water system infrastructure (CREW 2013).

This chapter presents a proposed seismic design framework, describes each element of the framework, and concludes with a discussion of implementation considerations and strategies.

13.2 PROPOSED SEISMIC DESIGN FRAMEWORK

The seismic hazard and its impact on infrastructure is a complex problem with many expert opinions on various aspects associated with this problem. However, few frameworks exist on how to effectively approach it. Seismic hazards and related design for infrastructure require involvement by multiple disciplines of science and engineering. The following proposed framework is intended to serve as a bridge between high-level and detailed analysis to help utilities improve the seismic resilience of their water infrastructure. The proposed framework provides an overall approach that helps address the multidisciplinary aspects of the problem and the need for an overall system engineering approach. This framework is not intended to be a comprehensive guide but identifies key issues, provides a methodology to help mitigate seismic hazards and risks, and directs users to other types of identified issues that require more detailed analysis and additional expert involvement.

The proposed seismic design framework consists of the following key steps:

Step 1 – Identify service priorities

Step 2 – Establish level of service (LOS) goals

Step 3 – Establish design earthquake

Step 4 – Evaluate project-specific seismic hazards

Step 5 – Establish design standards and methods

Step 6 – Design for seismic risk mitigation

These six steps were adapted from the American Lifelines Alliance (ALA) seismic design guidelines (ALA 2005) and simplified further as part of the development of seismic standards for a large regional water supply system in the Pacific Northwest that is subject to a major subduction zone earthquake (Britch et al. 2020). This framework is also based on other pertinent work, including water master plan activities (TVWD 2015), a community resilience planning guide (NIST 2020), similar efforts by an ASCE/UESI task committee developing pipeline seismic design guidelines (ASCE 2018), and other emergency framework activities. The six steps for proposed seismic design framework are discussed in the following sections.

13.3 IDENTIFY SERVICE PRIORITIES

The first step in the proposed seismic framework is to identify service priorities and the associated project(s) that include infrastructure that is most critical or vulnerable to earthquake hazards or does not meet the desired level of seismic resilience. Project identification can be straightforward if it is clear that an existing critical deficiency exists. Other projects may require a comprehensive investigation to review, assess, and prioritize infrastructure across an entire system. This section aims to provide concepts that can be used to help identify which infrastructure is more critical to the overall system performance. This requires developing an understanding of the water infrastructure system components, identifying critical customers, identifying a critical backbone system to serve them, and a process that supports the evaluation.

13.3.1 Understanding Infrastructure Components

One must first understand the various components of the system including pipe, appurtenances, and facilities and how they operate together as a system. A geographic information system (GIS) is a useful tool to support the acquisition and management of this information. Basic information to compile includes pipe sizes, materials, joint type(s), length, and age. Condition assessment and asset management information help to provide a more complete understanding of the overall system. There may also be different cohorts of installation, for example, with a certain material, a particular installation process that occurred during a specific time period, or some other consideration that may warrant consideration in your system.

It is also imperative to know how the system operates hydraulically, what the nuances related to it are, operational alternatives, ways to optimize operations, and specific situations that must be avoided. Hydraulic models that have been calibrated to the system and system monitoring data coupled with comprehensive and accurate operator knowledge are the cornerstone of this element of understanding.

13.3.2 Identifying Critical Customers

It is also important to establish which customers are most critical to the community and region following a major seismic event. The staff of water agencies likely have a good initial understanding of some of these customers; however, full development requires diverse stakeholder involvement from emergency response staff, elected officials, and other key groups within the community.

The most obvious critical customers are those involving life safety, such as hospitals and emergency response facilities. Multiple community stakeholders should be involved in identifying a comprehensive list of the appropriate critical customers within a community. This can include disaster response groups to identify planned emergency shelter locations or local firefighting/rescue teams to plan response strategies. Less obvious but also important customers may include other critical infrastructure systems like those reliant on water for cooling. Water infrastructure is one of many interconnected and interdependent infrastructure systems discussed further in this section.

Another way to identify potential critical customers is to consider the overall needs of the community. The National Institute of Standards and Technology (NIST) has prepared a resilience planning guide, *Community Resilience Planning Guide for Buildings and Infrastructure Systems* (NIST 2020), which focuses on understanding the broader context for the community. NIST (2020) states,

> The built environment is an essential part of a community's resilience. Social functions and institutions, including family/kinship, education, health, government, economy, media, and other community-based organizations rely on buildings and infrastructure systems before, during, and after a hazard event occurs. Key buildings and infrastructure systems must be functional to support neighborhood restoration, to care for vulnerable populations, and to enable the community's economy to recover and thrive.

Planning for water systems as an important part of the built environment should consider both the broader and specific needs of the community as part of the identification of critical customers and associated critical water infrastructure.

Another component to creating a comprehensive list of critical customers is to identify and manage the interdependence of the various infrastructure systems to avoid cascading failures that may affect the community. The Department of Homeland Security (DHS) has analyzed the complex interrelationships and interdependence that exist

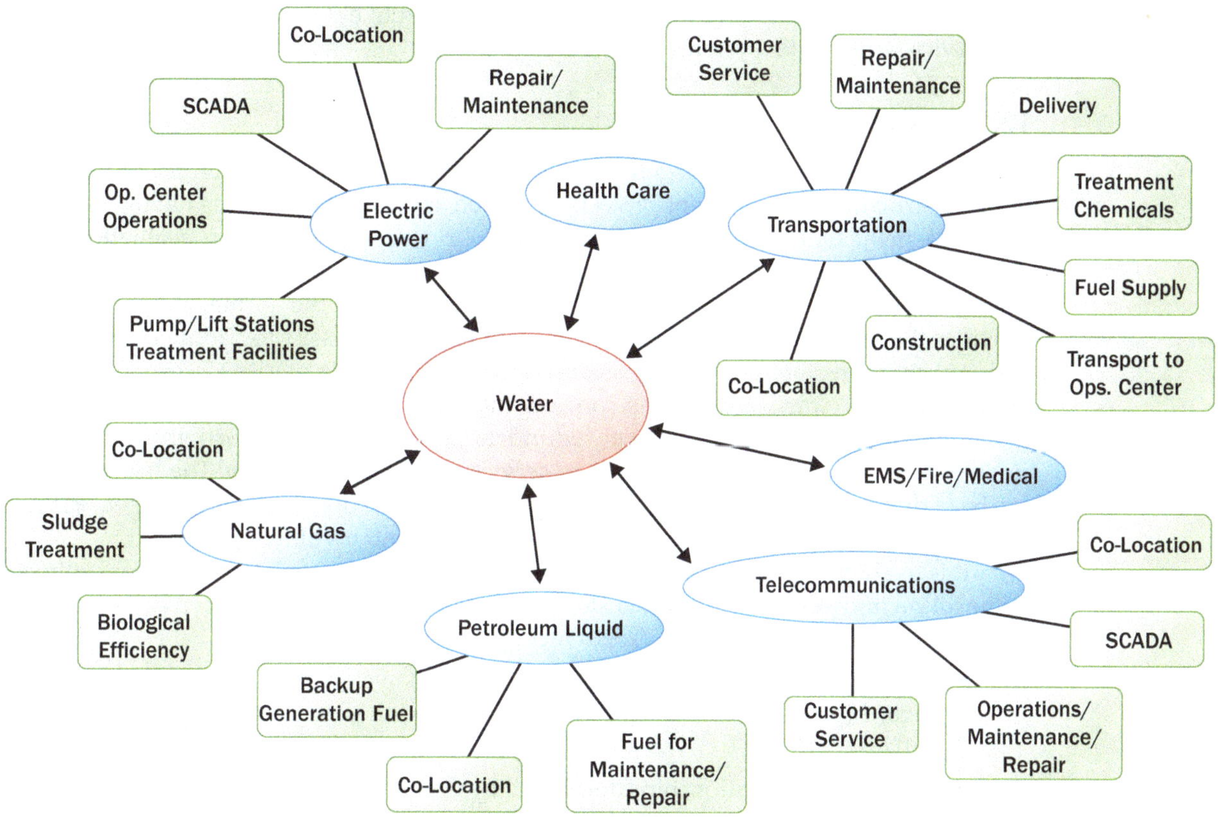

Figure 13-2 Water sector interdependencies (Britch et al. 2020. Adapted from 2010 Water Sector Specific Plan [DHS 2015])

between water infrastructure and other related systems. Figure 13.2 shows this complex set of water sector interdependencies.

A final consideration pertains to those parts of the community that are important to support the recovery process after the earthquake. Following an earthquake, areas that provide key resources are important. These include retail locations to provide resupply of food and medicine and banking locations for access to cash, as well as those areas that provide other key resources to help support community recovery. Finally, to support economic recovery, people will need to go back to work. For people to go back to work, they will need to know that their families are safe, which includes schools for children. These factors should be considered as part of a complete process to identify critical locations to support recovery of the community following a natural disaster.

13.3.3 Identify Infrastructure to Serve Critical Customers

Once critical customers have been identified, the infrastructure necessary to serve them must be identified next. The process needs to take into consideration the complete system of infrastructure required to convey water from the source to the service connection. Hydraulic modeling when coupled with other operator system knowledge is a good tool to help determine which pieces of infrastructure should be considered part of this critical delivery system. Agencies should decide how they will factor system redundancy into this evaluation.

A secondary consideration related to critical infrastructure following a major event pertains to critical inventory items and supplies. Following a major earthquake, access to critical replacement parts will likely be limited. Preplanning to develop a list of those critical items and stockpiling them at key locations will likely help hasten the recovery to full service of the water system.

13.3.4 Evaluation Process

There are several ways that an agency can approach the overall evaluation process to identify infrastructure improvement projects. It can be as straightforward as having a critical system element that is known to be vulnerable up to complete system evaluations that require a much greater level of effort and resources to complete. A practical approach to complete that work is to evaluate seismic resilience as part of the normal master planning activities that occur for water systems. This work lends itself easily to adding elements that address seismic resilience planning as described in this section (TVWD 2015). As part of a master planning effort, one can also consider interdependency of infrastructure, potential cascading effects, and overall system redundancy. The master plan activity should consider system performance from source to service locations holistically. The use of ERDIP can be of diminished value if the water supply upstream of the system is damaged. Master planning can help identify system deficiencies and potential vulnerabilities.

It is likely the evaluation process will be iterative after completing the steps outlined in the following sections. Additional steps as part of a master planning activity for buried pipelines recommended by NIST (2020) include the following:

- In GIS, superimpose the pipeline distribution system onto maps of the scenario hazard (peak ground velocity [PGV], liquefaction potential, and landslide potential).
- Use empirical relationships developed by the ALA to predict the number of breaks and leaks in the pipeline system.
- Estimate the time required to repair the predicted number of breaks and leaks and restore system functionality based on historical crew productivity data.
- Consider the anticipated damage states of dependent systems (transportation, liquid fuel, etc.).

It is typically important to engage elected officials for approval of related long-term financial planning and the development of a related capital improvement plan associated with the work. Because the scope and resources required to accomplish the desired seismic resilience improvements are large and costly, long-term implementation planning may be appropriate. The Oregon Resilience Plan (ORP) suggests a 50-year planning period to accomplish the desired seismic resilience improvements (OSSPAC 2013).

13.4 ESTABLISH LEVEL OF SERVICE GOALS

LOS pertains to the response expected from infrastructure following the design earthquake event selected by an agency (design earthquake discussion covered in the Section 13.5). In the following sections, two primary considerations pertaining to establishing the LOS are discussed. These include the formal classification of the infrastructure into different functional groups or categories and the level of operational capacity required over time.

13.4.1 Classification of Infrastructure

Levels of criticality can be used to organize infrastructure into different categories of importance. The system of categorizing infrastructure can be simple or complex, depending on the needs of the system and the community it serves. Three examples of how infrastructure can be categorized are described further in this chapter, as well as how a utility might consider system redundancy in the categorization process.

The ORP describes two tiers of criticality. The most critical tier is the system's backbone water system, which is described as follows:

> The backbone water system would be capable of supplying key community needs, including fire suppression, health and emergency response, and community drinking water distribution points, while damage to the larger (non-backbone) system is being addressed. (OSSPAC 2013)

Like the ORP, the plan put forth by the Japan Water Works Association (JWWA) considers two levels of importance ranking for water facilities: those facilities with a high level of importance (Rank A) and other facilities (Rank B). In the JWWA approach to determine the level of importance, authorities encourage water utilities to consider the location of the facilities within their system, whether loss of the facilities would create serious collateral impacts, potential threats upstream of the supply source that could impact system resilience, system redundancy, whether these facilities serve important entities, the relative difficulty in replacing the facilities if damaged, as well as other factors (JWWA 1997).

A third example related to categorization of infrastructure is provided by the ALA. The ALA (2005) proposes four classes of pipeline functionality related to different levels of infrastructure criticality. The four classes and a description of the infrastructure they are intended to cover are shown in Table 13-1.

A water utility may consider redundancy during their classification of infrastructure. Where redundancy of supply exists, it may be appropriate to reduce the pipe function classification of redundant sources if at least one of the supplies would meet the higher classification of supply as if it were the only source providing that supply. In consideration of the identified LOS goals, however, no reduction in pipe function classification might be warranted depending on the facility or area being served (e.g., if a single supply does not meet an agency's specific LOS related to the necessary flow capacity). Other water utility objectives or policies may further influence pipeline function classification.

Table 13-1 Pipeline function classes*

Pipeline Function Class	Description
I	Pipelines that represent a low hazard to human life and have a low economic impact in the event of failure. These pipelines are not required to be functional immediately following an earthquake and can endure longer restoration times without impact to the water utility. These pipelines primarily serve agricultural or irrigation usage, certain temporary facilities, or minor (nonwater) storage facilities that do not have a significant role in local or regional economy.
II	Pipelines that provide water for typical use within the utility where only a limited impact would be realized in the event of failure. These pipelines require less restoration time than Class I pipelines to limit the impact to the surrounding community. This category provides water for typical domestic use within the system and includes all pipelines not identified in Class I, III, and IV.
III	Pipelines that represent a higher criticality than the typical pipelines within a utility. These pipelines deliver water to many customers and may also result in significant social or economic impacts in the event of failure and outage. Pipeline restoration times would need to be minimal following a major event.
IV	Pipelines that provide water to essential facilities for postearthquake response, public health, and safety. These pipelines are intended to remain functional during and after a designed earthquake without an immediate required restoration time. These pipelines provide water for postearthquake firefighting and emergency support.

* Originally derived from Table 3-1, ALA (2005) and as modified by ASCE/UESI seismic task committee (ASCE 2018).

Table 13-2 ORP operational level of service guidance—target states of recovery following an earthquake*

Water System Component	0–24 h	1–3 days	3–7 days	1–2 weeks	2 weeks–1 month
Supply source/treatment	20–30%	50–60%	—	80–90%	—
Main transmission system (backbone)	80–90%		—	—	—
Water supply to critical facilities	50–60%	80–90%	—	—	—
Water for fire suppression at key supply points	80–90%	—	—	—	—
Water for fire suppression at hydrants	—	—	20–30%	50–60%	80–90%
Water at community distribution points	—	50–60%	80–90%	—	—
Distribution system	—	20–30%	50–60%	80–90%	—

* The ORP includes two figures with target states of recovery based on a major off-shore rupture of the Cascadia Subduction Zone fault. The two figures represent target states of recovery for the coastal and the valley regions. The values shown in Table 13-2 represent the valley region as presented in Figure 8.19 in the ORP.

13.4.2 Level of Capacity Required Over Time

The amount of flow that is required at a given service connection depends on the level of criticality of the customer served by that service connection plus consideration for when that flow is needed following a major event. Infrastructure serving the most critical customers as indicated in Table 13-1 (e.g., hospitals) will need to be operational following an earthquake. Recognizing the limited resources that agencies have available to implement desired system improvements, some parts of the system such as sections of the distribution system may have to remain inoperative for a period of time until other more critical sections of the infrastructure are repaired following a major event.

Guidance from the ORP includes levels of operational capacity for different water system components in the periods of time immediately following a seismic event as shown in Table 13-2. The ORP considers 90 percent operational to be the normal full operational condition. Agencies and their policymakers can adjust the desired levels in Table 13-2 as necessary to accommodate individual needs. For example, a regional water supply system adopted the transmission goals of the ORP but required its intake, pumping, and treatment components to achieve 50-percent capacity within 48 h (Britch et al. 2020). Full water system recovery, depending on the nature of the seismic event, may not be achievable until other interdependent infrastructure systems like transportation, electric power, and liquid fuel systems have been restored to some minimal level of operation.

The above guidelines should be considered as an example. Each water utility should determine for their own system what LOS is appropriate to serve their customers and community. The above guidelines can be influenced by many factors. For example, the amount of storage and its location within the system can have a bearing on what LOS goals are established for different parts of the system. The size of the system; its components; the geographical layout of source, treatment, transmission, and storage facilities; the location of seismic hazards; and the location of critical customers or points of distribution can influence how LOS goals are established. Finally, the word *transmission* should be applied as appropriate for each system. In some cases, it may not be size specific but more importantly linked to the function provided by the pipeline within a given system.

13.5 ESTABLISH DESIGN EARTHQUAKE

The next step of the seismic framework is to establish the project-specific design earthquake. This is consistent with the guidance provided by NIST. NIST (2020) recommends that for a given system, appropriate earthquake scenarios be established with associated ground motions. This section provides an overview of pertinent information required to establish an appropriate design earthquake. Topics covered in this section include a general background on earthquakes, return intervals to consider, and various sources of information for design earthquakes.

13.5.1 Earthquake Background Information

Seismic waves are the vibrations from earthquakes that travel through the ground. There are several identification systems used to describe the size of earthquakes. In the past, the Richter magnitude scale determined the magnitude of an earthquake based on the logarithm of the amplitude of waves recorded by seismographs. Moment magnitude is a common measure, typically used for larger earthquakes (O'Rourke and Lui 2014). The MMI scale indicates the severity of shaking to the observed effects (Jaiswal et al. 2015). Table 13-3 describes the severity for MMI values. The descriptions in Table 13-3 were derived for and relate to building systems and not water systems, though they provide a good sense for the variability associated with different levels of ground shaking. Similar attribution of the impacts of ground shaking to water systems is not as readily distinguishable.

Table 13-3 MMI and description of corresponding shaking*

MMI level	Description (Source: USGS, https://www.usgs.gov/media/images/modified-mercalli-intensity-scale.)
I	**Not felt.** Not felt except by a very few under especially favorable conditions.
II	**Weak shaking.** Felt only by a few persons at rest, especially on upper floors of buildings.
III	**Weak shaking.** Felt quite noticeably by persons indoors, especially on upper floors of buildings. Many people do not recognize it as an earthquake. Standing motor cars may rock slightly. Vibrations similar to the passing of a truck. Duration estimated.
IV	**Light shaking.** Felt indoors by many, outdoors by few during the day. At night, some awakened. Dishes, windows broken. Unstable objects overturned. Pendulum clocks may stop.
V	**Moderate shaking.** Felt by nearly everyone; many awakened. Some dishes, windows broken. Unstable objects overturned. Pendulum clocks may stop.
VI	**Strong shaking.** Felt by all, many frightened. Some heavy furniture moved; a few instances of fallen plaster. Damage slight.
VII	**Very strong shaking.** Damage negligible in buildings of good design and construction; slight to moderate in well-built ordinary structures; considerable damage in poorly built or badly designed structures; some chimneys broken.
VIII	**Severe shaking.** Damage slight in specially designed structures; considerable damage in ordinary substantial buildings with partial collapse. Damage great in poorly built structures. Fall of chimneys, factory stacks, columns, monuments, and walls. Heavy furniture overturned.
IX	**Violent shaking.** Damage considerable in specially designed structures; well-designed frame structures thrown out of plumb. Damage great in substantial buildings, with partial collapse. Buildings shifted off foundations.
X	**Extreme shaking.** Some well-built wooden structures destroyed; most masonry and frame structures destroyed with foundations. Rails bent.

* The MMI scale is also shown with two additional higher levels of shaking intensity in other documents as part of the 12 original levels that were included when it was first introduced.

Table 13-4 Probability of exceedance and recurrence intervals*

PE	RI
50 percent PE in next 50 years	72 years
10 percent PE in next 50 years	475 years
2 percent PE in next 50 years	2,475 years

* The values identified in Table 13-4 are commonly used in seismic related literature. Examples include ALA (2005) and Jaiswal et. al (2015). Other return intervals may also be used.

Specific engineering measures of earthquake intensity include peak ground acceleration (PGA) and PGV. Sources include the local crustal earthquakes, deep intraplate earthquakes, and earthquakes at plate boundaries that include subduction zone earthquakes, which exhibit the largest magnitudes and long-duration ground shaking.

13.5.2 Probability of Exceedance and Recurrence Intervals

To establish the design earthquake and related LOS, the appropriate design recurrence of the event needs to be selected. This is commonly described in terms of probability of exceedance (PE). Typical PE and associated recurrence intervals (RIs) are shown in Table 13-4.

Critical water infrastructure, such as that related to water source infrastructure, treatment plants, and the parts of the system associated with major transmission, are commonly designed for the 2 percent PE in the next 50 years (2,475-year event). The RI of the design earthquake to be used coupled with level of criticality or importance of the infrastructure or facilities must be determined by each agency and may involve policy-level decisions.

The JWWA (1997) takes a similar approach with their seismic design and construction guidelines for water supply facilities. When subjected to earthquakes with more severe seismic ground motions (Level 2), facilities identified as having a high level of importance should experience only "light damage, but. . . still function." Level 2 events have a larger RI than the lower seismic ground motion but are more frequent than Level 1 events. Level 2 seismic ground motions are "equivalent [to] the seismic motion generated in areas with faults or areas with big scale plates bordering inland areas."

Each agency must determine what level of operability is appropriate for their system. As an example, a regional water supply adopted the goal to have its system operable following a major subduction zone seismic event (Britch et al. 2020). Some components may sustain slight damage as long as it doesn't affect the overall system operability or pressure integrity of the pipelines. Simply put, nothing important can break. This is particularly important in this case, given the expected regional impact of a full fault rupture associated with this event and its impact on other interdependent infrastructure systems that will likely compromise potential repair activities.

13.5.3 Source Information for Earthquakes

A seismic hazard analysis can be performed using either a deterministic seismic hazard analysis (DSHA) or a probabilistic seismic hazard analysis (PSHA). The ALA (2005) describes DSHA and PSHA as follows:

- A DSHA is useful when examining the performance of a complete pipeline network over a spatially large area. A DSHA considers the effects from a particular earthquake scenario on a pipe. This method "would specify similar ground motion levels for similar ground conditions at similar fault distances, but has difficulty accounting for uncertainties in the evaluation and risks associated with the accumulation of all seismic sources potentially affecting the pipe" (Davis 2008).

- A PSHA simultaneously considers the effects from multiple earthquake source hazards on a pipe and the probabilities of a likely range of magnitudes over the length of each seismogenic source. A PSHA is more difficult to carry out and understand than deterministic methods, but it does account for uncertainties in the evaluation and can account for the risks associated with the accumulation of all seismic sources potentially affecting the pipe.

Since 1996, the USGS has produced National Seismic Hazard Models (NSHMs) on a six-year cycle, which coincides with building code updates. The USGS uses PGA data on a uniformly spaced grid across the United States and converts the data into MMI results for the three different RIs identified in Table 13-4. The USGS provides links for tools that may be used for design on the USGS website. These results are derived from a PSHA (Jaiswal et al. 2015).

Project-specific design earthquake results can also be developed. This includes site-specific site-response seismic analyses and a project-specific design earthquake. Site-specific site-response seismic analyses are used for critical facilities where such analysis is warranted. Project-specific design earthquakes can be developed to include supplemental information beyond what is provided in USGS NSHMs. For instance, a large regional water supply program deemed it necessary to include additional ground motion models and source earthquakes other than what was provided by the USGS (Britch et al. 2020). In this case, additional regional specific information was incorporated in the design earthquake developed for that program.

13.6 EVALUATE PROJECT-SPECIFIC SEISMIC HAZARDS

The next step in the seismic framework is to identify the specific seismic hazards. Most seismic hazards are related to ground motions from an earthquake and their impact on infrastructure, but there are other hazards associated with earthquakes. The following sections describe the different types of seismic hazards, discuss associated investigations, and identify the required gap analyses to evaluate the earthquake hazard relative to an infrastructure system.

13.6.1 Types of Seismic Hazards

An earthquake can impose several types of seismic hazards on infrastructure. Some of the things to consider that can influence the behavior of the soils related to seismic hazards include topography, geomorphology, soil stratigraphy, and groundwater levels. Pipeline alignments that cross or parallel a "free face" toward a stream, ocean, cliff, or any face that will allow for landslides should be carefully evaluated as part of the design with ERDIP. The topography for these areas may also have soils that may be more prone to liquefaction. Streambed soils are typically high in fines and have a high water table (or a potentially high water table in a flood condition). These soil characteristics are potentially susceptible to liquefaction and permanent ground deformation.

The types of seismic hazards are outlined in Table 13-5. A description of each hazard follows the table.

Ground shaking. Ground shaking represents transient ground motions that propagate through the ground due to seismic fault movement. This hazard includes loading on the infrastructure that only exists while ground shaking is ongoing. Once the ground shaking stops, the transient loading imposed on the infrastructure subsides.

Table 13-5 Types of seismic hazards (Britch et al. 2020)

Hazard Category	Subcategory/Description	
A. Ground shaking	Transient ground motions	
B. Permanent ground deformation	**1.** Liquefaction	**a.** Settlement
		b. Lateral spreading
	2. Soft or weak soils below pipe	
	3. Seismically induced landslides	
	4. Abrupt offsets	**a.** Differential movement
		b. Soil transitions
		c. Fault ruptures
C. Nearby infrastructure designed by others to lesser standards		
D. Other applicable hazards		

Permanent ground deformation. Permanent ground deformation (PGD) represents ground movements that can impose permanent significant loading on infrastructure. The loading from the different subcategories of PGD hazards remains following the end of the transient ground shaking from seismic waves. PGD is the "irrecoverable movement that persists after the shaking has stopped" (O'Rourke et al. 2015). The different subcategories of PGD may act separately or in combination, depending on the characteristics of the hazard area under investigation. A description of each follows.

Liquefaction. Liquefaction occurs when loose, granular, or certain types of other noncohesive soils below the groundwater table lose their full structural capacity, resulting in behavior comparable to a viscous fluid rather than a solid. Contained soils that are not free to move laterally and subject to liquefaction can experience settlement. Liquefied soils that have a free face such as along a river bank will move toward the free face, which is known as lateral spreading. Liquefaction-induced lateral movement may continue after shaking has ceased, which is known as flow failure. Depending on the sensitivity of soils to cyclic shearing, cyclic strength reduction can occur in plastic silts or clay soils resulting in decreased soil strength.

Soft or weak soils below pipe. Soils with marginal strength located beneath the pipeline or other infrastructure may result in undesirable load-bearing conditions due to seismic ground motions.

Seismically induced landslides. Slopes that are stable in a static condition may fail due to the additional imposed energy from seismic ground motions or from reduced strength of the soils due to liquefaction or cyclic shearing.

Abrupt offset. Abrupt offsets in movement can focus a significant amount of loading in small sections of the infrastructure, potentially resulting in failures. There are three types of abrupt offsets to consider. Differential movement can occur as infrastructure transitions through different types of support conditions, such as a pipe entering a vault. Soil transitions can create conditions of abrupt offsets. The greater the variation in embedment conditions, the more pronounced the effect, for example, with a rock to soft soil transition. The most significant type of abrupt offset condition comes from fault rupture or fault displacement. This is typically the most complicated type of rupture and requires substantial analysis.

Nearby infrastructure designed by others to lesser standards. Water infrastructure is part of the built environment, meaning it is in close proximity to other infrastructure systems that may pose an additional hazard. Nearby infrastructure systems designed to lower seismic resilience standards are a threat to the water infrastructure. For example, consider a water transmission line designed for a seismic event with a high RI, such as a

2,475-year event. If the transmission line is connected to or supported by a bridge or located behind a retaining wall designed to a lesser standard, the water transmission line may fail as the result of the proximate failure of the bridge or retaining wall.

Other applicable hazards. The previously discussed categories are not intended to be a comprehensive list of all potential hazards. Agencies and designers are encouraged to consider the specific environment surrounding their infrastructure and consider whether additional seismic vulnerabilities exist, like tsunamis in coastal regions. Nonseismic hazards also pose a risk to water infrastructure, including flash floods from heavy rainfall or rapid snow melt, excessive flooding from dam breaks, water source contamination from wildfires, and other natural or manmade disasters.

13.6.2 Seismic Hazard Investigations

To assess the potential seismic hazards that exist, an understanding of the classifications of soils along with their characteristics and potential vulnerabilities is necessary. NIST (2020) recommends that agencies and designers obtain liquefaction and landslide hazard maps from the state geology departments. Field explorations such as geotechnical borings, cone penetrometer tests, and specialized cyclic shear tests should also be performed to analyze the soil and associated seismic hazards, such as liquefaction potential.

13.6.3 Infrastructure Resilience Gap Analyses

Gap analyses aid in determining the deficiencies that exist between the desired infrastructure LOS and the current condition and expected performance. (This gap analysis concept is explained further by NIST's Community Resilience Guide [2020]). For the analysis of large systems, GIS and related mapping tools are useful. Establishing an understanding of the current condition and expected level of performance for the infrastructure may require a variety of condition assessment activities and related analyses. Long-term planning integrated with a long-term capital improvement plan to replace aging infrastructure may be an appropriate strategy to improve the expected seismic performance of a water infrastructure system over time.

13.7 ESTABLISH DESIGN STANDARDS AND METHODS

For ductile-iron pipe to accommodate ground deformation resulting from seismic events, considerations need to be made regarding the strength of the pipe, the strength of the joint, the ability of the system to shorten or elongate in response to seismically induced axial ground displacements, and the capacity of the system to promote curvature that accommodates ground movement directed perpendicular to the pipeline orientation. For the latter consideration, segmented pipe like ductile iron predominately promotes curvature through deflection/rotation at each individual pipe joint, and the magnitude of lateral displacement the system can sustain is a function of both the joint deflection capacity and joint spacing. Design considerations related to the pipe system and design performance standards that have been developed within the industry are discussed in the following sections.

13.7.1 Design Considerations

The characterization of pipe systems for the purposes of seismic design typically involves two classifications, segmented pipe and continuous pipe systems. These two classifications are defined as follows (O'Rourke and Lui 2012):

Segmented pipe. A pipeline with lower axial and rotational stiffness at the joints than the rest of the pipeline: ductile-iron and cast-iron pipe are examples of a segmented pipe.

Continuous pipe. A pipeline joined by connections that exhibit axial and rotational stiffness comparable to the rest of the pipe: welded pipe (i.e., with welded joints) and fused pipe (i.e., with fused joints) are examples of a continuous pipe.

Normal pipeline design considerations can include the internal working pressure, external loading, deflection, and hydraulic transients, as well as other conditions. In seismic design, additional external loading must be considered. For seismic design, transient loads are imposed due to ground shaking. Permanent loads also may exist, including those associated with the various contributing components of permanent ground deformation. When seismic loads are considered, transient and permanent loading are not typically considered to be additive. Industry practice considers these sources of ground movement independently but in addition to the normal design loads (O'Rourke and Honegger 2017). Cyclic loading may also be a consideration in areas prone to liquefaction and subject to long duration seismic events.

13.7.2 Pipe Design Performance Standards

Current seismic design performance standards for ductile-iron pipe are limited. The International Organization for Standards (ISO) Standard 16134 (2006) is one of the few standards available. Standard 16134 identifies three performance standard parameters for the seismic design of ductile-iron pipelines: expansion/contraction performance, slip-out resistance, and joint deflection angle. As shown in Table 13-6, each parameter is classified based on component performance requirements. Other standards are in development by ASCE/UESI and others.

13.8 DESIGN FOR SEISMIC RISK MITIGATION

Ductile-iron pipe is considered segmented pipe. The design for risk mitigation with ductile-iron pipe primarily focuses on the performance characteristics of the joint, as this is where the damage is usually found to occur. Singhal and Benavides (1983) state,

Table 13-6 Ductile-iron pipe seismic performance classifications (adapted from ISO 2006*)

Parameter	Class	Component Performance
Expansion/contraction performance	S-1	±1% of *L* or more
	S-2	±0.5% to less than ±1% of *L*
	S-3	Less than ±0.5% of *L*
Slip-out resistance	A	3*d* kN or more
	B	1.5*d* kN to less than 3*d* kN
	C	0.75*d* kN to less than 1.5*d* kN
	D	Less than 0.75*d* kN
Joint deflection angle	M-1	±15° or more
	M-2	±7.5° to less than ±15°
	M-3	Less than ±7.5°

L is the component length in mm.
d is the nominal diameter of the pipe in mm.

* This table was recreated to include missing *d* terms from original table for component performance shown for classes B and C.

The most frequent and severe damage usually occurs at or near a pipeline joint. Most failures result from ground strains that develop axial or shear forces in the pipelines and at the joints.

Four topics related to design of ductile-iron pipe systems for seismic risk mitigation are presented in this section. They include a discussion of limit states, performance categories, industry test results related to ductile-iron pipe joints, and other considerations for risk mitigation.

13.8.1 Limit States for Ductile-Iron Pipe

Limit states establish design thresholds to consider as part of seismic risk mitigation and achieving desired levels of service performance. The primary goal is to maintain pipeline pressure integrity and avoid loss of containment. For ductile-iron pipe systems, there are few guidelines to follow related to limit states. Wham et al. (2018) proposed two limit states based primarily on internal pressure integrity, quantified by leakage rates, associated with large deformation performance of ductile-iron pipe joints: the serviceability limit state and ultimate limit state.

The serviceability limit state for a segmented pipe relates to a leakage allowance that is "typically defined as sustained leakage greater than one drop per second (0.05 mL/s or 4.32 L/day)," according to Wham et al. (2018). This serviceability limit is intended to reflect a condition where no repair would be necessary. From a practical standpoint following a seismic event, other serviceability limits may be considered. As long as a pipeline system maintains its pressure integrity such that it continues to provide adequate delivery of water to critical customers, support firefighting activities, or other identified post–seismic event goals, then its serviceability limit state has been met on an emergency condition basis. This may also be thought of as a pressure integrity performance goal for the system. Following guidance by the ORP for post–seismic event target states of recovery, 80 to 90 percent capacity is identified as the highest postevent LOS goal. In any case, each water utility or water provider must determine its own specific serviceability limit based on the characteristics of its system and the critical needs of the community it serves. In the case of pipelines where post–seismic event access to effect repairs is expected to be limited or difficult, a lower leakage limit may be considered. With the previous consideration of pipeline accessibility coupled with pipeline criticality, a serviceability limit of zero may be warranted in some cases.

The ultimate limit state according to Wham et al. (2018) is defined as "structural failure or fracture of the pipeline resulting in the inability of the system to sustain adequate levels of internal pressure." In this case, the pipeline system would be unable to provide sufficient flow and/or pressure to meet the established minimum LOS goals for the water agency to meet its community's needs.

The limit states for segmented ductile-iron pipe are different from those for continuous pipe. Karamanos et al. (2017) describe four limit states for continuous welded steel pipelines, including tensile strain capacity, local buckling, overall buckling due to compressive loading, and resistance of pipeline joints and fittings.

13.8.2 Performance Categories for Ductile-Iron Pipe

The classifications provided in Table 13-6 are intended to represent general parameters for the overall pipeline performance as a system. Additional design considerations will be needed for a given project. Designers must assess the requirements of a particular project and develop a complete list of specific design criteria. To select appropriate performance criteria, expected ground movements and anticipated loading must be identified for the project site.

Performance limits for typical ductile-iron pipe systems vary widely depending on the specific performance characteristics of a particular joint (or combinations of joints). Joint configurations include standard joint types such as nonrestrained push-on joints, mechanical joints, and flanged joints; special joint types, including ball-and-socket components; typical restrained joints, such as those with an integral restraint gasket, restrained boltless push-on joints, and restrained mechanical joints; and seismic joints, such as pipe with restrained push-on joints, restrained push-on couplings, and seismic coupling ball-and-socket joints. Ductile-iron pipe products may be used in a variety of combinations to achieve the desired pipeline system performance. The designer will need to determine which products to use to achieve the site-specific project requirements.

For the purpose of helping designers select appropriate types of joints or combinations of joints to mitigate seismic risks, four performance categories that represent increasing levels of performance characteristics associated with ductile-iron pipe joints are listed here.

- **Category 1: Nonrestrained joints.** These include standard push-on joints not designed to provide pull-out resistance. An older subcategory involving cast-iron pipe would also include more rigid joints with lead packing that allow for little to no movement whatsoever. Better performance is expected with gasketed joints compared to joints with lead packing. Also, joint geometry affects performance. Longer joints that allow the gasket to be engaged within the joint for a greater distance while maintaining pressure integrity are expected to perform better than shorter depth joints.
- **Category 2: Joints with gripper gaskets/gripping wedges.** This category includes those restrained joint systems that rely on wedges that grip the pipe. These joints are expected to provide higher performance than nonrestrained push-on joints. These systems are designed for static axial loading. It is not clear how they will perform though in relation to cyclic loading from seismic wave action. Also, their thrust restraint may not be bidirectional, designed instead for loading only in a single direction.
- **Category 3: Joints with integral restraint bead and boltless locking segments.** This category provides improved performance due to integrated joint restraining mechanisms. For example, restraint beads welded to the pipe spigot can interact with ductile-iron locking components to provide resistance to joint pullout. They are expected to provide higher performance in tension than the previous categories.
- **Category 4: Joints specially designed for combined seismic performance.** These joints provide the highest level of seismic performance. By design, they are intended to provide a high level of joint axial movement, axial joint strength, and joint deflection/rotation, either individually or when used in combination with other joint systems. Older styles of joints, such as ball-and-socket types, may also fall into this category depending on usage.

The next section provides a brief overview of industry testing associated with the four performance categories. The designer should be aware that the highest category may still not be able to provide the level of seismic performance required for a given location. Multiple joints used in combination, pipeline routing strategies, ground or other structural improvements, or other means may be needed to achieve the desired seismic risk mitigation for a particular installation.

13.8.3 Testing to Establish Ductile-Iron Pipe Performance Characteristics

This section describes some of the documented testing related to ductile-iron pipe performance characteristics relative to the four categories previously described. Designers are encouraged to read published documentation related to pipe joint performance; however, they should always contact the various pipe manufacturers to ensure they are using the most appropriate values in their design calculations. Testing related to the four performance categories is described below.

Nonrestrained joint testing. Laboratory testing of push-on joints related to seismic performance dates to the 1980s. Singhal and Benavides (1983) and Singhal (1984) report on tests addressing the axial and bending strength of push-on joint for 4, 6, 8, and 10 in.-diameter ductile-iron pipe. The tests provided results that described the axial and rotational stiffness of nonrestrained push-on joints. More recently, Wham and O'Rourke (2015) expanded on previous studies to investigate the combined effect of varying levels of imposed joint rotation and axial displacement on joint leakage. Through the combination of full-scale testing and numerical modeling, they showed that while ductile-iron push-on joints are highly susceptible to pull out, they have considerable capacity to accommodate joint deflection before reaching a conservative serviceability limit state for joint leakage. The study provides a comprehensive assessment of 6 in.-diameter ductile-iron pipe joints and can be used to predict serviceability performance for typical seismically induced ground movement.

Gripping gaskets/gripping wedge testing. Meis et al. (2003) summarize testing to establish axial strength associated with restrained joints with gripping gaskets. Tests were performed on 6- and 8-in. ductile-iron pipe for both static and dynamic conditions.

Integral restraint bead/boltless segment testing. Meis et al. (2003) found that for tensile testing, joint with integral restraint bead/boltless restraint had 212% and 147% of the capacity over the joint with the gripping gaskets for 6- and 8-in. pipe, respectively. Both the joints with gripping gaskets and those with the integral restraint bead/boltless restraint experienced reduced capacity between static ultimate load testing and dynamic ultimate load testing. Joints with the integral restraint bead/boltless restraint saw a much lower reduction in capacity between static and dynamic testing. In both static and dynamic testing, integral restraint bead/boltless restrained joints showed higher ultimate capacity than joints with the gripping gaskets.

Specially designed seismic joint testing. Since 2015, testing has been performed at Cornell University, Ithaca, N.Y., to identify performance limits for specially designed seismic joints. Joints developed by multiple manufacturers were tested. Comprehensive physical testing was typically performed on 6-in. pipe, though additional testing was conducted for select 12 and 16 in–diameter joints. A variety of experimental procedures were performed including four-point bending tests, joint compression tests, joint tension tests, and large-scale testing of fault rupture effects. Additionally, some two-dimensional finite element analyses were performed. Results can be found at Cornell University's Geotechnical Lifelines Large-Scale Testing Facility website (https://lifelines.cee.cornell.edu/).

Current practice for joint testing includes procedures that measure loads and displacements associated with axial tension and compression to evaluate performance and system continuity for axial ground movements (oriented parallel to the pipeline), as well as four-point bending tests that quantify the joint deflection (rotation) and pipe deflection capacity of pipelines subjected to lateral ground displacements (Wham et al. 2018). These tests are performed under typical internal water pressure conditions to address water loss and leakage during large, imposed deformation.

While these tests are an important step toward quantifying pipeline response to seismic ground movement and can serve to inform predictive models assessing overall system resilience, it is imperative to recognize that successful performance is highly dependent on the weakest link in the system. An ERDIP connected to a valve or coupling of insufficient capacity will not provide the desired level of performance during a seismic event. Wham et al. (2018) suggest that "each component of a hazard-resilient system (including couplings, valves, tees, etc.) can, and should, be evaluated for seismic performance. . . and qualified for expected levels of ground movement."

13.8.4 Other Considerations for Risk Mitigation

Agencies can take a more programmatic approach to the implementation of performance requirements for the assigned pipe function classes according to the seismic framework proposed in this chapter. For example, an agency could classify the entire pipeline system as pipe function class III (see Table 13-1) and design the system to a moderate level of component performance per the ISO categories shown in Table 13-6. However, even for portions of the system identified as having the highest function (pipe function class IV) and designed to the highest ISO component performance levels, it is not necessarily certain that the design requirements are adequate or that all levels of service goals will be met. Pipelines that require the highest pipe function class may not directly correspond to the highest component performance category. Required performance is subject to both the pipe function class and the specific seismic hazards for that specific part of the system, such as steep slopes with high landslide potential or others listed in Table 13-5. A more in-depth investigation and analysis would be required. The in-depth analysis may show that even greater performance requirements than those associated with the highest ISO component performance levels may be required. Conversely, an agency may overspend on infrastructure where the site-specific ground motions and expected seismic loading are nominal, even with a high pipe function class portion of the system where the highest performance level of the components may not actually be required for the agency to achieve its seismic resiliency goals. In seismically active areas, a minimum performance requirement for restrained joints is recommended.

It should be noted that the expansion/contraction performance and joint deflection angle performance of ductile-iron pipes are interdependent—i.e., the published component performance values are only fully available when the pipe joints are loaded only in that particular mode. When combined performance of the joints (i.e., axial and rotational) is desired either during the seismic loading or during installation to achieve the desired alignment or profile, then the available axial expansion/contraction and rotational parameter values for the combined seismic performance should be estimated by the designers using the joint configuration data, installation practice, and the maximum allowed deflection during installation.

A final consideration is that designers should use appropriate factors of safety in their designs. Information produced by performance test results represents testing of joints in controlled laboratory settings. It does not consider other variabilities that may have important impacts on the ultimate loading conditions and response of the pipeline system. Some of these include the uncertainty and variability of the backfill and native soils surrounding the pipeline; the condition of and connections to existing pipe; the influence of system components, fittings, and other connections that can cause the pipeline system to behave differently than expected or introduce additional vulnerabilities; or other unknown hazards that could affect system performance. As such, designers are strongly encouraged to consider these and other potential variabilities in the design conditions and use appropriate factors of safety in their designs to achieve their LOS goals.

13.9 IMPLEMENTATION CONSIDERATIONS AND STRATEGIES

In conjunction with implementing a seismic resilience improvement project, consideration should be given toward seismic policies, existing infrastructure, other design issues, installation, and critical repair parts. Each of these is described below.

13.9.1 Seismic Policies

Decisions for seismic improvements and standards will require involvement from both agency employees and elected officials to ensure that existing policies are not violated or to establish new policies where appropriate. To set and obtain levels of service goals, support from upper management and elected officials is crucial. Other policy-related issues include restrained joints, seismic valves, and critical backbone and customers.

Restrained joints. For systems in a seismically active area, restraining all joints of the pipeline during installation is a relatively easy and cost-effective means of improving the seismic resilience of the overall system. This can be done with varying levels of restraint systems, from gaskets with embedded restraining features to more substantial system components that achieve greater system performance. Particularly in highly developed urban environments, relying on thrust blocks for restraint provides many challenges, while restrained piping systems have proven to be advantageous.

Seismic valves. Seismic valves within a system are an important component of a seismically resilient water system. Typically, these valves are placed on the outlets of key reservoirs. Devices designed to sense ground shaking can be integrated with controls to actuate the valve in response to varying levels of seismic action. When actuated, these valves will deliberately shut off water supply to specific portions of the system if programmed to do so. The following considerations must also be made:

- If a pressure zone has only one reservoir, it might not be desirable to install a seismic valve on that reservoir to maintain pressure in the system for firefighting and other emergency services.
- If there are multiple reservoirs in a pressure zone, seismic valves can be installed on at least one of the reservoirs but not all. This will help to provide a supply of water for postearthquake needs.
- If seismic valves are installed on a tank, it may be desirable to have emergency connections with standardized fittings so that the water can be withdrawn as needed.

The use of seismic valves should be carefully considered by agency staff, management, and elected officials to make sure their application is consistent with adopted policy.

Critical backbone and customers. As an agency discusses its overall system resilience goals, sensitivity is required when identifying critical customers and developing the associated critical backbone system. Information pertaining to critical customers may be sensitive for a variety of reasons, including overall system security. Revealing critical customers, their locations, and the associated critical backbone water system on publicly distributed documents is not advisable. A general map portraying the proposed critical backbone for the water system can be publicly distributed, but details should be vague to limit potential vulnerability. Established points of public emergency water distribution may be included as applicable and as deemed appropriate. It may be useful to utilize separate appendices in documents containing sensitive maps, such as those with critical customers, critical infrastructure systems, hazard areas, vulnerable facilities, and other confidential information.

13.9.2 Existing Infrastructure

In general, water systems will rely on the current infrastructure. Over time, system upgrades will occur, but understanding the existing infrastructure and associated vulnerabilities and deficiencies is an integral part of developing an overall long-term plan to improve system seismic resilience. Possible system vulnerabilities in need of consideration for the long term include, but are not limited to, rigid joints, deficiencies due to corrosion, or brittle elements of the infrastructure due to the age of the infrastructure and the types of materials used at the time of installation. Smaller investments at key locations within a system may result in improved near-term seismic resilience.

13.9.3 Other Design Issues

Ductile-iron pipe has inherent strength and resilient properties. Much of the focus of the design with ductile-iron pipe for seismic risk mitigation is related to the joints. Appropriate pipe and joint type selection to match expected site conditions is required. The length of pipe, frequency of joints, and use of specially designed seismic joints are design factors to consider related to seismic performance. Ductile-iron pipe products can be used in a variety of combinations to achieve the desired system performance. Consult manufacturers for the performance characteristics of specific ductile-iron pipe products.

Geotechnical investigations are a critical part of identifying and quantifying the potential surface and subsurface hazards along an existing or proposed pipeline alignment. The scope of investigation can vary greatly and should increase with the criticality of the pipeline. However, a designer or municipality may be able to improve their ability to identify hazard areas for further investigation by performing a study of the geomorphology. Maps for topography, geology, and satellite imaging can be evaluated for certain land features that may indicate susceptibility to transient and permanent ground deformations. These deformations materialize in the form of ground amplification, faulting, liquefaction, blocking, slope failure, and settlement. Features to target for further investigation include former river channels, alluvial plains, natural levees, flood plains, deltas, reclaimed land, dunes, fill areas, and steep and potentially unstable slopes. Additionally, the geomorphology should be studied to determine approximate contacts between differing geologic materials. Geologic boundary conditions can cause large amounts of strain on a pipeline due to how the differing soils react under seismic wave propagation. This list of areas of potential hazards is not meant to be comprehensive but intended to provide a general understanding of the types of conditions where hazards to buried pipelines are common. Once hazards are determined, field investigations should be performed to further evaluate the potential hazards. Depending on the criticality of the pipeline, the designer or agency should determine next steps of investigation, which will likely include more advanced field and geotechnical investigation to better quantify surface and subsurface hazards.

Designers can also consider methods to reduce or manage loads experienced by a pipe system. Buried pipelines are subject to frictional forces from the surrounding soils. Different coatings have different frictional resistance coefficients. For a pipeline at rest in a cohesionless soil, $\tan\delta$ is used to estimate frictional resistance where δ is angle of interface friction. A pipe with polyethylene or fusions bonded epoxy coating has $\tan\delta$ values between 0.4 and 0.6 (O'Rourke et al. 2015). Pipelines with mortar coatings and metallic pipelines without coatings have $\tan\delta$ ranges between 0.7 and 1.0 (O'Rourke et al. 2015). While axial pull tests on jointed ductile-iron pipes in partially saturated, medium-dense sand showed $\tan\delta$ values ranging from 0.6 to 0.9 along the pipe barrel and that the frictional resistance provided by the bell contributes significantly to pipeline response under relative soil-pipeline displacement, a specimen wrapped in polyethylene encasement exhibited notable reductions of approximately 15 percent* in frictional resistance at large imposed displacements (Wham et al. 2017). Elhamdi and O'Rourke (1989), however, suggest that the frictional resistance for ductile-iron pipe with a "plastic covering" is equal to $\tan\phi$, where ϕ is the angle of shearing resistance of the soil. They state that "for pipes

* These tests where completed using 6 in.–diameter ductile-iron pipe. In this case, the bell contributed substantially to the resistance of the pipe due to its relative size to the pipe barrel. With larger pipe, it is expected that the contribution of bell resistance to overall resistance of the piping system will diminish as the overall area of the pipe bell to the pipe barrel diminishes. As the overall impact of the bell diminishes compared to the exposed frictional resistance associated with the pipe wall, the reduction in overall frictional resistance of using polyethylene encasement is expected to improve.

wrapped with plastic covering, slippage happens in the soil near the pipe rather than at the interface. The reason is that the soil grains become partially embedded in the plastic cover at the interface." Specific bedding and backfill materials around the pipe tend to accumulate and concentrate additional loading on the pipeline. In PGD-prone areas, the use of controlled low-strength material (CLSM), which may also be known as controlled density fill in some regions, should be considered carefully. Seismic design standards for the Willamette Water Supply Program only allow for the use of CLSM in select cases in areas where PGD is expected (Britch et al. 2020). CLSM around the pipe may result in accumulation of greater loading on the pipe and/or concentrate loading on the pipe where it cracks, which is not a desirable condition.

As one designs a portion of a water system, it is important to recognize that elements of the system may behave differently than other portions of the system, which can result in unexpected outcomes. These may be in the form of discontinuities between elements of the system or where portions of the system perform differently and act as anchors for the system in the ground. Two examples are fittings and casings where the pipe may be grouted in place. With fittings, for example, long straight runs of pipe are expected to behave differently than pipe sections joined together from different directions as with a tee or a cross. With casings used in a trenchless crossing, a pipe that is located within a casing that is grouted in place will have a much higher mass than the rest of the system and accumulate loads differently than the pipe sections away from the casing. This may serve to create a condition where the pipe is not allowed to move as much as ground conditions require.

A final design consideration relates to creating redundant supplies to critical customers or working with them specifically to make sure adequate emergency response plans are developed. This will help mitigate any shortfalls in supply. Understanding a customer's needs and critical demand requirements is in itself a step towards achieving system resilience.

13.9.4 Installation

Construction specifications for the installation of special pipe fittings should be consistent with the recommendations from the manufacturer to achieve the desired system performance. With fittings that allow axial movement, the common joint installation position is "at the mid-point, which allows for both joint expansion and contraction" (Gaston and Ratcliffe 2016). Thrust restraint will need to be accounted for at bends and fittings, but away from those locations, a neutral setting on joints will preserve greater system flexibility.

In some instances, construction practices related to joint assembly by contractors have involved full insertion of the spigot into the bell, primarily the result of using excavators or other equipment to install the pipe and pushing the joint home. This results in an installation where much of the joint flexibility has been lost and the overall system resilience reduced. Such practices should be avoided. All installations should be performed in accordance with AWWA standards and the specific written recommendations of the manufacturers. Construction specifications that are consistent with these recommendations should be utilized and appropriate field inspections performed to confirm compliance.

A strategy to use thrust blocks to position the joints to a certain initial setting may be employed. Where joints that allow movement are utilized, thrust blocks can be used to set movable restrained joints at a desired position.

This is approach is not a standard practice in the industry. The designer should conduct an appropriate geotechnical investigation to assess ground conditions and expected movement; consult with the pipe manufacturers to understand limitations associated with the pipe, joints, and fittings; and size thrust blocks for the intended service and desired performance objectives. Furthermore, it may be desirable to make sure thrust blocks are not attached to the pipe in such a way that the weight of the thrust block will not create an additional hazard for the pipe.

One way to do this would be to preset the joints in a neutral position to allow expansion and contraction of the joint. Designers can also use thrust blocks to preset a waterline across a fault with each of the joints set to accommodate the expected movement. In the event of a large earthquake and PGD, the thrust blocks may fall away from the line if designed to do so and allow the line to move and articulate with the deformed ground. Using thrust blocks allows for expansion in areas that would be fully engaged with normal joint restraint.

13.9.5 Critical Repair Parts

Understanding the nature of the expected seismic event and how repair efforts will be affected is a critical part of the planning and design process. Following a seismic event, transportation infrastructure may be damaged, and access to the locations in need of repair may be limited. To minimize recovery time, an agency should consider which critical parts it should maintain in its inventory. As appropriate, those parts may be kept at multiple key stockpile locations that allow access from different strategic locations.

REFERENCES

American Lifelines Alliance/Federal Emergency Management Agency/National Institute of Building Sciences. 2005. *Seismic Guidelines for Water Pipelines*. Washington: ALA.

American Society of Civil Engineers (ASCE). 2018. "Manual of Practice on Seismic Design of Buried Water/Wastewater Pipelines" (Working draft), UESI Task Committee on Seismic Design of Buried Water/Wastewater Pipelines.

Britch, M., M. Havekost, and B. Robless. 2020. *Seismic Guidelines and Minimum Design Requirements.* Willamette Water Supply Program. Version 1.1.

Cascadia Region Earthquake Workgroup (CREW). 2013. *Cascadia Subduction Zone Earthquakes: A Magnitude 9.0 Earthquake Scenario (Update).* Oregon Department of Geology and Mineral Industries Open-File Report 0-13-22. Seattle: CREW.

Davis, C.A. 2008. "Assessing Geotechnical Earthquake Hazards for Water Lifeline Systems with Uniform Confidence." *Geotechnical Earthquake Engineering and Soil Dynamics IV*, doi.org/10.1061/40975(318)19.

Department of Homeland Security, Infrastructure Sector Assessment (DHS). 2015. *Infrastructure System Overview: Water systems* Office of Cyber and Infrastructure Analysis. www.cisa.gov/topics/critical-infrastructure-security-and-resilience/critical-infrastructure-sectors/water-and-wastewater-sector (accessed October 4, 2024).

Elhmadi, K. and M.J. O'Rourke. 1989. *Seismic Wave Propagation Effects on Straight Jointed Buried Pipelines.* Technical Report NCEER-89-0022. Buffalo, NY: NCEER.

Federal Emergency Management Agency (FEMA). 2014. Cascadia Subduction Zone Earthquakes: A Magnitude 9.0 Earthquake Scenario. Emergency Management Institute Community-specific integrated emergency management course, Emmitsburg, MD.

Gaston, M.D. and R.S. Ratcliffe. 2016. "Earthquake Resistant Ductile Iron Pipe, Valve, and Hydrant System." *Pipelines*. Reston, VA: ASCE: 1061–1071.

International Organization for Standards. 2020. ISO 16134. Earthquake-resistant and Subsidence-resistant Design of Ductile Iron Pipelines. Geneva: ISO.

Jaiswal, K. S., M.D. Petersen, K. Rukstales, and W.S. Leith. 2015. "Earthquake Shaking Hazard Estimates and Exposure Changes in the Conterminous United States." *Earthquake Spectra,* 31:S1:S201–S220.

Japan Water Works Association. 1997. *Seismic Design and Construction Guidelines for Water Supply Facilities.* Tokyo: JWWA.

Karamanos, S. A., G. C. Sarvanis, B.D. Keil, and R.J. Card. 2017. "Analysis and Design of Buried Steel Water Pipelines in Seismic Areas." *Journal of Pipeline System Engineering Practice,* 8:4:doi.org/10.1061/(ASCE)PS.1949-1204.00002.

Meis, R.D., E.M. Maragakis, and R. Siddharthan. 2003. *Behavior of Underground Piping Due to Static and Dynamic Loading.* Multidisciplinary Center for Earthquake Engineering Research. Buffalo, NY: MCEER.

The National Fire Protection Association (NFPA). 2016. Largest fire losses in the United States. Retrieved October 09, 2017, from www.nfpa.org/news-and-research/fire-statistics-and-reports/fire-statistics/fires-in-the-us/large-property-loss/largest-fire-losses-in-the-united-states

Oregon Seismic Safety Policy Advisory Commission (OSSPAC). 2013 *The Oregon Resilience Plan: Reducing Risk and Improving Recovery for the Next Cascadia Earthquake and Tsunami.* Salem, OR: OSSPAC. www.oregon.gov/OMD/OEM/osspac/docs/Oregon_Resilience_Plan_Final.pdf (accessed October 4, 2024).

O'Rourke, M. J., and X. Liu. 2012. *Seismic Design of Buried and Offshore Pipelines* (Monograph MCEER-12-MN04). Buffalo, NY: MCEER.

O'Rourke, T.D., J.K. Jung, and C. Argyrou C. 2016. "Underground Pipeline Response to Earthquake-Induced Ground Deformation." *Soil Dynamics and Earthquake Engineering.* www.sciencedirect.com/science/article/abs/pii/S026772611630241X (accessed June 9, 2025).

O'Rourke, T.D. and D. Honegger. 2017. WWSP - Seismic Design Consultation Meeting #2 [Interview by M. Britch, M. Havekost, and B. Robless]

Petersen, M. D., A. M. Shumway, P.M. Powers, C.S. Mueller, M.P. Moschetti, A.D. Frankel, S. Rezaeian, D.E. McNamara, N. Luco, O.S. Boyd, K.S. Rukstales, K.S Jaiswal, E.M. Thompson, S.M. Hoover, B.S. Clayton, E. H. Field, and Y. Zeng. 2020. "The 2018 Update of the US National Seismic Hazard Model: Overview of Model and Implications." *Earthquake Spectra,* 36:1:5–41. doi.org/10.1177/8755293019878199

The White House Office of the Press Secretary. 2013. *Presidential Policy Directive/PPD-21.* February12,2013.https://obamawhitehouse.archives.gov/the-press-office/2013/02/12/presidential-policy-directive-critical-infrastructure-security-and-resil (accessed October 4, 2024).

Ripley, A. 2009. *The Unthinkable Who Survives When Disaster Strikes—and Why.* New York: Three Rivers Press.

Singhal, A. and J. Benavides, J. 1983. "Axial and Bending Behavior of Pipeline Joints." *Journal AWWA*. 75:11:572–578.

Singhal, A. 1984. "Behavior of Jointed Ductile Iron Pipelines." *Journal of Transportation Engineering*, 110:2:235–250.

Tualatin Valley Water District (TVWD). 2015. *Water System Master Plan* (Rep.). Prepared by Carollo Engineers.

US Department of Commerce, National Institute of Standards and Technology (NIST), 2020. *Community Resilience Planning Guide for Buildings and Infrastructure Systems* https://nvlpubs.nist.gov/nistpubs/SpecialPublications/NIST.SP.1190GB-16.pdf (accessed June 9, 2025).

US Geological Survey (USGS) Earthquake Hazards Program. 2019. 2018 United States (Lower 48) Seismic Hazard Long-term Model. USGS. www.usgs.gov/programs/earthquake-hazards/science/2018-united-states-lower-48-seismic-hazard-long-term-model (accessed October 3, 2024).

US Geological Survey (USGS). Modified Mercalli Intensity Scale. USGS. https://www.usgs.gov/media/images/modified-mercalli-intensity-scale. (accessed October 3, 2024).

Wham, B.P. and T.D. O'Rourke. 2015. "Jointed Pipeline Response to Large Ground Deformation." *Journal of Pipeline Systems Engineering and Practice*, 7:1:doi.org/10.1061/(ASCE)PS.1949-1204.0000207.

Wham, B.P., C. Pariya-Ekkasut, C. Argyrou, A. Lederman, T. D. O'Rourke, and H. E. Stewart. 2017. "Experimental Characterization of Hazard-Resilient Ductile Iron Pipe Soil/Structure Interaction under Axial Displacement." Proc. ASCE Congress on Technical Advancement, Duluth, Minn.

Wham, B.P., B.A. Berger, C. Pariya-Ekkasut, T.D. O'Rourke, H.E. Stewart, T.K. Bond, and C. Argyrou. 2018. "Achieving Resilient Water Networks–Experimental Performance Evaluation." Proc. 11th National Conference on Earthquake Engineering, Los Angeles, Calif.

Chapter 14

Guidelines for Purchase of Pipe and Fittings

14.1 GENERAL

14.1.1 Standards

Ductile-iron pipe and accessories should be purchased in accordance with specifications that refer to AWWA standards. The material may be purchased directly by an owner or through a distributor or a contractor. Specifications need not repeat requirements included in AWWA standards if those standards are properly referenced.

14.1.2 Design

The owner or the owner's representative has the obligation to ensure that the material, designed in accordance with AWWA standards, is adequate for the service intended. This includes but is not limited to consideration of working pressures and surges, external loads, soil and bedding conditions, installation practices, conveyed liquid, and thrust forces (see Chapter 4).

14.1.3 Design Submittals

If the material is specified to be provided with pressure classes and joints shown in AWWA standards, design submittals are not necessary. If the design deviates from ANSI/AWWA C150/A21.50, Thickness Design of Ductile-Iron Pipe, submittals may be required by the owner or the owner's representative.

14.1.4 Layout Drawings

Ductile-iron pipe can usually be adapted to fit most underground situations with standard fittings because of the deflections available in the joints and the ability to cut the pipe in the field. Therefore, layout drawings are normally not required for underground installations. Laying schedules may be required for installations involving large numbers of restrained joints or multiple classes of pipe.

Layout drawings are normally required for interior or aboveground piping that must be fabricated to exact lengths. In such cases, the owner should provide information including dimensions and joint types of connecting equipment, structure dimensions, type of wall or floor penetrations desired, and any other details necessary to complete drawings. In the case of wall piping specified to be provided with collars, the owner should specify whether the collars are to be used as seep rings only or also to resist thrust. The owner should check all drawings carefully and agree on the bill of materials to be supplied.

14.1.5 Specials

Specials, fittings, joints, or linings not shown in AWWA standards may be provided in accordance with the manufacturer's standards, if approved by the owner. Drawings or submittals may be required if such specials are not shown in the manufacturer's catalog or literature.

14.1.6 Markings

Materials should be marked in accordance with applicable AWWA standards. In addition, special markings may be required, such as mark numbers, paint marks, or lettering to identify different classes, special linings, or joints.

14.1.7 Certification

Certification may be required to show that material provided is in accordance with the contract specifications. In most cases, a general certification before shipping material and covering all of the material to be provided by a supplier is adequate.

If the owner desires to have outside inspection during manufacturing or fabrication of the materials, the specification or purchase documents need to define the requirement. The cost of outside inspection is the responsibility of the owner/purchaser. When outside inspection is included in the purchase documents, the manufacturer will provide the owner's inspector appropriate access to those parts of the manufacturer's plant necessary to ensure compliance with the specifications. The manufacturer will also make available necessary gauges that might be required by the inspector and provide assistance as necessary for handling the material. Material may be supplied from inventory if it meets specifications.

14.1.8 Shipment and Delivery

Dates of shipment and rate of shipment are to be agreed on by the vendor and purchaser at the time of purchase. The supplier will be able to assist in filing a claim on the customer's behalf for shortage of or damaged material at the agreed point of delivery only if such is recorded on the delivery receipt or similar document by the carrier's agent (Figure 14-1).

Regarding shipments, freight on board shipping point, freight collect or freight prepaid, customer pick-up, title passes at the shipping point, and responsibility for delivery and condition of shipment all rest with the purchaser and/or the freight carrier.

14.1.9 Sales Order Contract

If a sales order contract is provided by the supplier in confirmation of a purchase agreement, the purchaser shall promptly bring to the vendor's attention any errors or changes

Source: Ductile Iron Pipe Research Association

Figure 14-1 A full truck load of ductile-iron pipe

before manufacturing and shipping material. The purchaser shall check with the vendor before canceling any ordered material to determine whether manufacture has begun and whether cancellation charges may be incurred.

14.2 DUCTILE-IRON PIPE

Ductile-iron pipe is designed in accordance with ANSI/AWWA C150/A21.50 and manufactured in accordance with ANSI/AWWA C151/A21.51, Ductile-Iron Pipe, Centrifugally Cast, or ANSI/AWWA C115/A21.15, Flanged Ductile-Iron Pipe With Ductile-Iron or Gray-Iron Threaded Flanges, unless otherwise specified. Except as modified by the provisions above, the purchase agreement for ductile-iron pipe should include the following:

- Applicable standard(s)
- Sizes
- Quantity
- Length, per pipe (flanged pipe only)
- Joint type
- Nominal thickness or class
- Outside coating
- Inside lining

- Standard accessories
- Special accessories, if required
- Special testing requirements, if necessary
- Threaded-on flanges, if required
  - Type of material (ductile iron or gray iron)
  - Type of flange (solid or hollow-back)

14.3 DUCTILE-IRON AND GRAY-IRON FITTINGS AND SPECIALS

Fittings will be provided in accordance with ANSI/AWWA C153/A21.53, Ductile-Iron Compact Fittings, or ANSI/AWWA C110/A21.10, Ductile-Iron and Gray-Iron Fittings, unless otherwise specified. Fittings not included in these standards can be provided in accordance with manufacturer's standards. Except as modified by the provisions above, the purchase agreement for fittings and specials should include the following:

- Applicable standard(s)
- Size(s)
- Quantity
- Joint types and end combinations (Note: All bell ends are preferred.)
- Pressure class
- Type of iron, ductile or gray
- Outside coating
- Inside lining
- Special bolt hole alignment or orientation, if required
- Standard accessories
- Special accessories, if required
- Special testing requirements, if necessary
- Method of calculating payment, e.g., per each or per published body weight excluding accessories

14.4 POLYETHYLENE ENCASEMENT

When required, polyethylene encasement is provided in accordance with ANSI/AWWA C105/A21.5, unless otherwise specified. Except as modified by the provisions above, the purchase agreement for polyethylene encasement should include the following:

- Size(s)
- Quantity
- Type of material (8-mil linear low density or 4-mil high density, cross laminated)
- Color of polyethylene material, if desired
- Installation method (a, b, or c), if one method is preferred
- Encasement style (collapsed-folded or sheet), if preferred
- Method of securing wrap (tape or straps), if preferred

Appendix **A**

Illustrations of Proprietary Joints for Ductile-Iron Pipe and Fittings

This appendix contains drawings of a number of proprietary joints used on or with ductile-iron pipe and ductile-iron and gray-iron fittings. These drawings are included to help the user identify these joints. Trademarks or descriptive names are shown under each illustration. Please note that whereas trade names are proprietary, actual joint designs may not be proprietary or exclusive unless patented.

The inclusion of any joint illustration in this appendix is not to be construed as a recommendation of this joint by AWWA or AWWA Committee A21, nor is the failure to include any joint illustration meant to imply that such joint is unsuitable or not equal to those shown.

Although an attempt was made to make this list of illustrations current at the time it was compiled, some of the illustrated joints may already be obsolete, and other joints may be on the market and not shown. For more information about a specific joint, contact the manufacturer.

Note: The mechanical joint and flanged joints shown in Figures 5-1 and 5-2, respectively, are not included in this appendix because they are standardized and identified in AWWA standards.

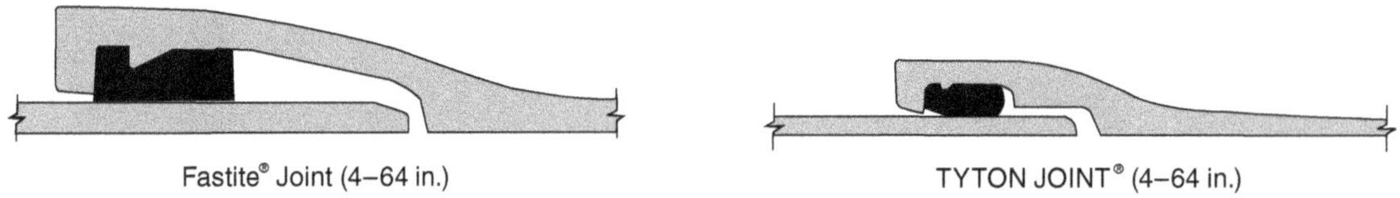

Source: Ductile Iron Pipe Research Association

Figure A-1 Push-on joints

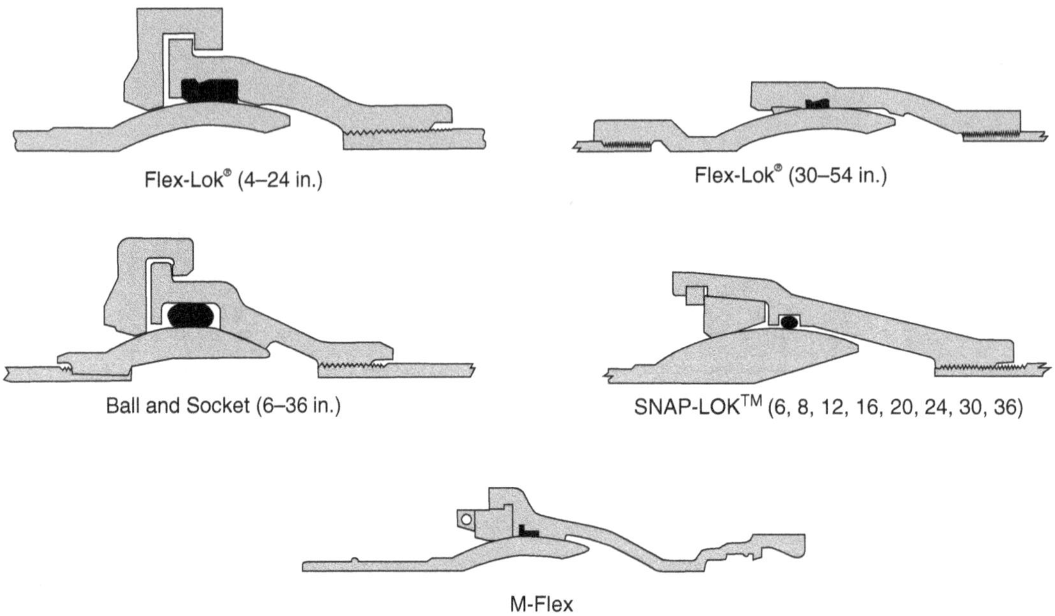

Source: Ductile Iron Pipe Research Association

Figure A-2 Ball-and-socket joints

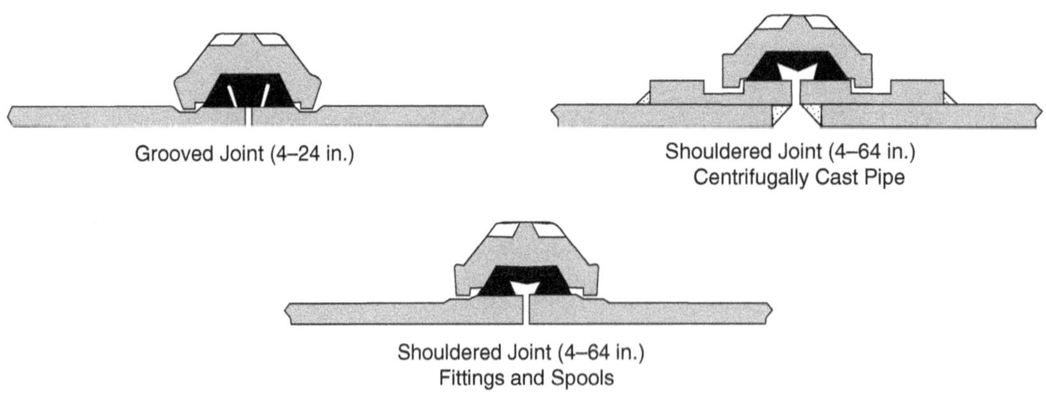

Source: Ductile Iron Pipe Research Association

Figure A-3 Grooved-and-shouldered joints

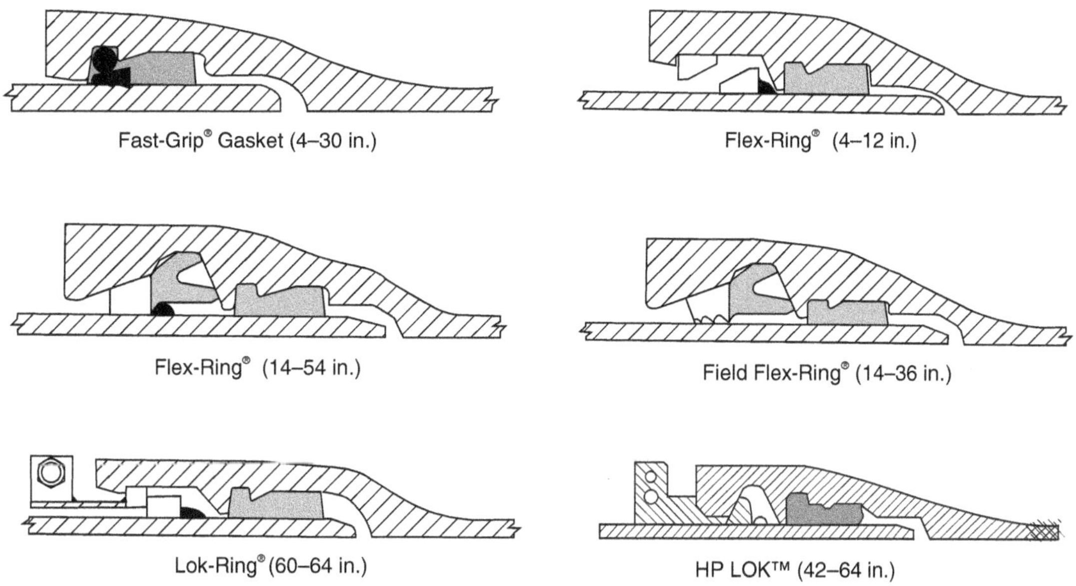

Source: Ductile Iron Pipe Research Association

Figure A-4a Restrained joints—push-on type

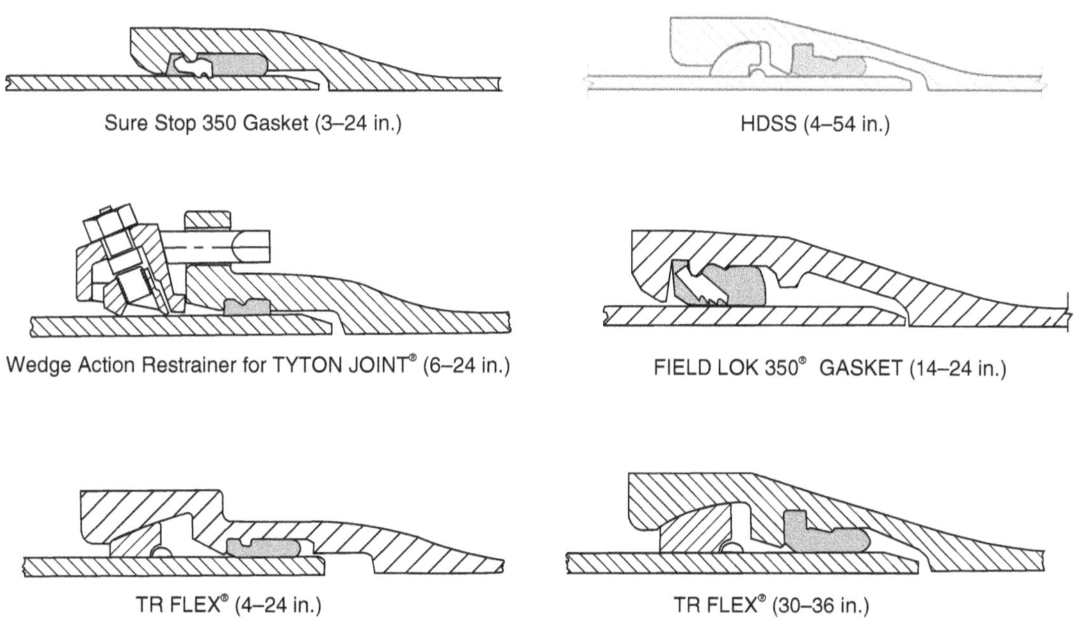

Source: Ductile Iron Pipe Research Association

Figure A-4b Restrained joints—push-on type

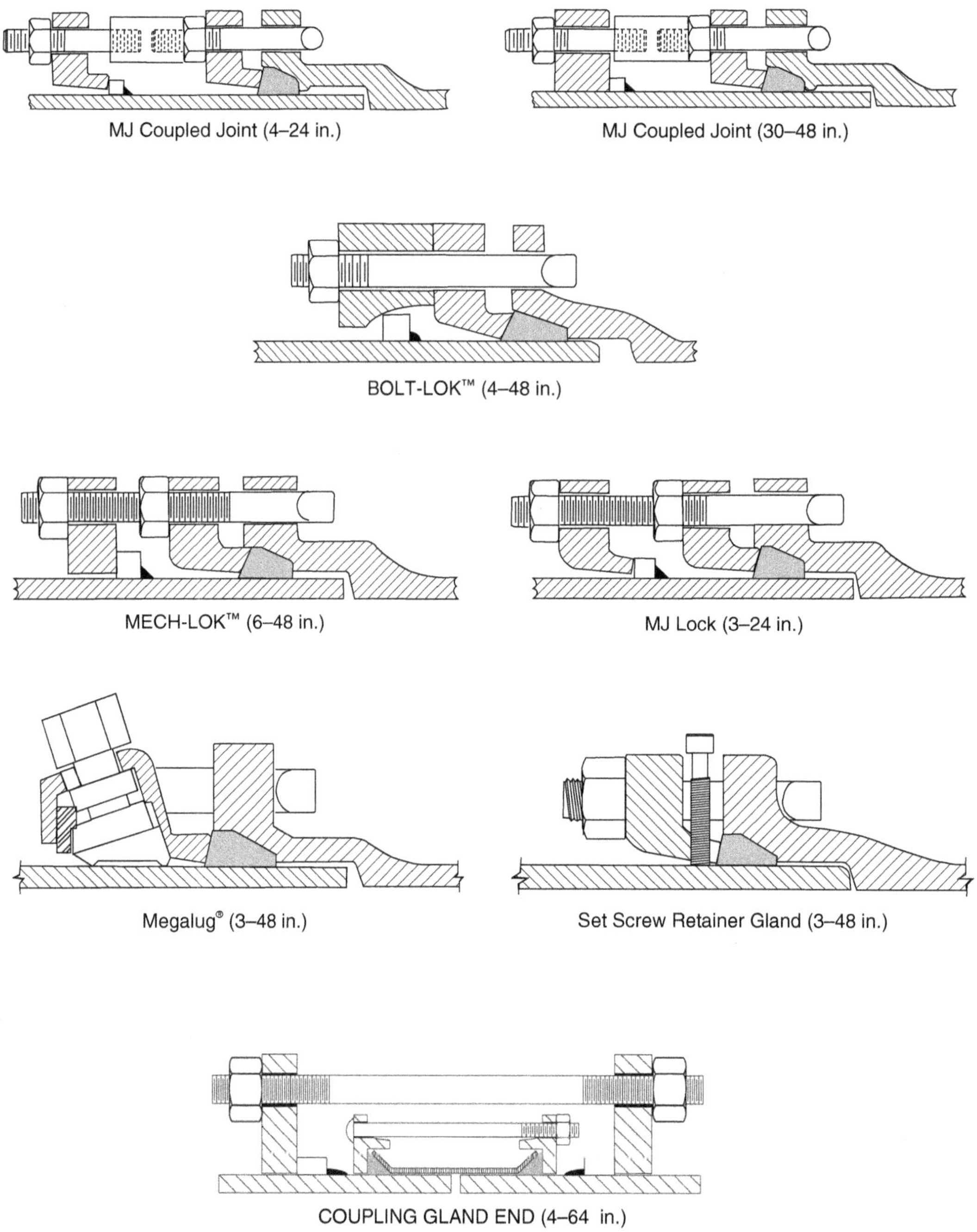

Source: Ductile Iron Pipe Research Association

Figure A-5 Restrained joints—mechanical joint and other joint type

EQ Coupling

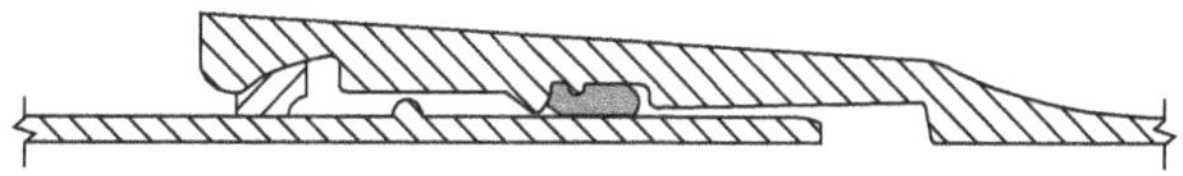

TR-XTREME® (6, 8, 12, 16, 20, 24 in.)

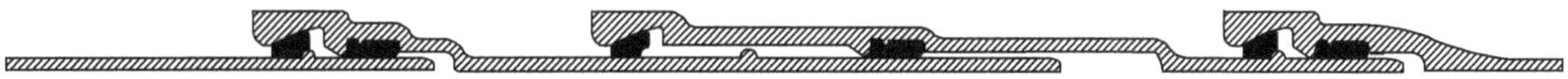

Seismic Flex Coupling™ (4–36 in.)

Source: Ductile Iron Pipe Research Association

Figure A-6 Seismic Joints

This page intentionally blank.

Index

Note: *f* indicates figure; *t* indicates table

E

F

O

P

R

S

T

U

V

W

AWWA Manuals

M1, *Principles of Water Rates, Fees, and Charges*, #30001
M2, *Instrumentation and Control*, #30002
M3, *Safety Management for Water Utilities*, #30003
M4, *Water Fluoridation Principles and Practices*, #30004
M5, *Water Utility Management*, #30005
M6, *Water Meters—Selection, Installation, Testing, and Maintenance*, #30006
M7, *Problem Organisms in Water: Identification and Treatment*, #30007
M9, *Concrete Pressure Pipe*, #30009
M11, *Steel Pipe—A Guide for Design and Installation*, #30011
M12, *Simplified Procedures for Water Examination*, #30012
M14, *Backflow Prevention and Cross-Connection Control: Recommended Practices*, #30014
M17, *Fire Hydrants: Installation, Field Testing, and Maintenance*, #30017
M19, *Emergency Planning for Water and Wastewater Utilities*, #30019
M20, *Water Chlorination/Chloramination Practices and Principles*, #30020
M21, *Groundwater*, #30021
M22, *Sizing Water Service Lines and Meters*, #30022
M23, *PVC Pipe—Design and Installation*, #30023
M24, *Planning for the Distribution of Reclaimed Water*, #30024
M25, *Flexible-Membrane Covers and Linings for Potable-Water Reservoirs*, #30025
M27, *External Corrosion Control for Infrastructure Sustainability*, #30027
M28, *Rehabilitation of Water Mains*, #30028
M29, *Water Utility Capital Financing*, #30029
M30, *Precoat Filtration*, #30030
M31, *Distribution System Requirements for Fire Protection*, #30031
M32, *Computer Modeling of Water Distribution Systems*, #30032
M33, *Flowmeters in Water Supply*, #30033
M36, *Water Audits and Loss Control Programs*, #30036
M37, *Operational Control of Coagulation and Filtration Processes*, #30037
M38, *Electrodialysis and Electrodialysis Reversal*, #30038
M41, *Ductile-Iron Pipe and Fittings*, #30041
M42, *Steel Water-Storage Tanks*, #30042
M44, *Distribution Valves: Selection, Installation, Field Testing, and Maintenance*, #30044
M41, *Fiberglass Pipe Design*, #30045
M46, *Reverse Osmosis and Nanofiltration*, #30046
M47, *Capital Project Delivery*, #30047
M48, *Waterborne Pathogens*, #30048
M49, *Quarter-Turn Valves: Head Loss, Torque, and Cavitation Analysis*, #30049
M50, *Water Resources Planning*, #30050
M51, *Air Valves: Air-Release, Air/Vacuum and Combination*, #30051
M52, *Water Conservation Programs—A Planning Manual*, #30052
M53, *Microfiltration and Ultrafiltration Membranes for Drinking Water*, #30053
M54, *Developing Rates for Small Systems*, #30054
M55, *PE Pipe—Design and Installation*, #30055
M56, *Nitrification Prevention and Control in Drinking Water*, #30056
M57, *Algae: Source to Treatment*, #30057
M58, *Internal Corrosion Control in Water Distribution Systems*, #30058
M60, *Drought Preparedness and Response*, #30060
M61, *Desalination of Seawater*, #30061
M62, *Membrane Applications for Water Reuse*, #30062
M63, *Aquifer Storage and Recovery*, #30063
M65, *On-Site Generation of Hypochlorite*, #30065

M66, *Cylinder and Vane Actuators and Controls—Design and Installation,* #30066

M68, *Water Quality in Distribution Systems,* #30068

M69, *Inland Desalination and Concentrate Management,* #30069

M77, *Condition Assessment of Water Mains,* #30077

M80, *Biological Drinking Water Treatment* #30080

www.ingramcontent.com/pod-product-compliance
Lightning Source LLC
Chambersburg PA
CBHW082115210126
38574CB00022B/258
* 9 7 8 1 6 4 7 1 7 2 0 1 5 *